Introduction to the
Fractional Calculus
of Variations

Introduction to the
Fractional Calculus
of Variations

Agnieszka B. Malinowska
Bialystok University of Technology, Poland

Delfim F. M. Torres
University of Aveiro, Portugal

Imperial College Press

ICP

Published by

Imperial College Press
57 Shelton Street
Covent Garden
London WC2H 9HE

Distributed by

World Scientific Publishing Co. Pte. Ltd.
5 Toh Tuck Link, Singapore 596224
USA office: 27 Warren Street, Suite 401-402, Hackensack, NJ 07601
UK office: 57 Shelton Street, Covent Garden, London WC2H 9HE

British Library Cataloguing-in-Publication Data
A catalogue record for this book is available from the British Library.

INTRODUCTION TO THE FRACTIONAL CALCULUS OF VARIATIONS

ISBN 978-1-84816-966-1

Printed in Singapore by B & Jo Enterprise Pte Ltd

Preface

The *calculus of variations* deals with the extremization of functionals, that is, minimization or maximization of functions whose domain is some set of functions. It is an old branch of optimization theory that has many applications in physics, geometry, engineering, dynamics, control theory, biology, and economics. Apart from a few examples known since ancient times, such as Queen Dido's isoperimetric problem reported in *The Aeneid* by Virgil, in the late 1st century BC, the problem of finding optimal curves and surfaces has been posed first by physicists such as Galileo (1564–1642), Huygens (1629–1695), and Newton (1643–1727). Their contemporary mathematicians, starting with Leibniz (1646–1716) and the Bernoulli brothers, Jacob (1654–1705) and Johann (1667–1748), and followed by Euler (1707–1783) and Lagrange (1736–1813), invented the calculus of variations in order to solve those problems. This field of mathematics is still, in the XXI century, an extremely active area of research.

The *Fractional Calculus of Variations* (FCV), the subject of this book, is fifteen years old. It unifies the calculus of variations and the fractional calculus; by inserting fractional derivatives into the variational functionals. This occurs naturally in many problems of physics, mechanics, and engineering, in order to provide more accurate models of physical phenomena. The physical reasons for the appearance of fractional equations are, in general, long-range dissipation and non-conservatism. There are many books about the calculus of variations, many others about fractional calculus and its applications, but none about the FCV. This was a strong motivation for us to write the book that the reader has now in her/his hands. The literature on the FCV is vast, with several interesting and useful results, but they are spread through many specialized and technical journals. Sometimes it is difficult to find them, and a unified mathematical perspective to the subject is missing. These facts provide the main goal of the present book: to give, for the first time in the literature, a unified mathematical treatment to the fractional calculus of variations, making the subject accessible to a wider community. The book was written having in mind advanced undergraduate, graduate students and researchers, in mathematics, physics, operations research, and applied sciences. It gives our personal view on the subject, the fruit of our own research and experience within the mathematical research group of the University of Aveiro. We do not try to provide a complete review of the subject, but rather a coherent and self-contained introduction to the FCV. The interested reader can find representative references in the Bibliography, where the results of this book first appeared, and many others from the references therein.

Fractional Calculus (FC) is the branch of mathematics that deals with derivatives and integrals of a non-integer (real or complex) order. Like the calculus of variations, its origin goes back more than three centuries, when, in 1695, L'Hopital (1661–1704) made some remarks to Leibniz about the mathematical meaning of a fractional derivative of order $1/2$. After that, several famous mathematicians contributed to the development of FC, among them Euler, Laplace (1749–1827), Fourier (1768–1830), Abel (1802–1829), Liouville (1809–1882), Riemann (1826–1866), and Weyl (1885–1955). FC is nowadays an important mathematical discipline, with different notions of fractional-order derivatives available. The most popular ones are those of Riemann–Liouville and Caputo. Particularly since the last decade of the XX century, numerous applications and physical manifestations of fractional calculus have been found in various seemingly diverse and widespread fields of science and engineering as turbulence and fluid dynamics, chaotic dynamics, solid state physics, chemistry, stochastic dynamical systems, plasma physics and controlled thermonuclear fusion, kinetic theory, quantization, field theory, control theory, signal and image processing, dynamics of earthquakes, biological systems, material sciences, micro structures, rheological properties of rocks, scaling phenomena, astrophysics, and economics. Fractional differentiation and fractional integration are now recognized as vital mathematical tools to model behavior and to understand complex systems, classical or quantum, conservative or non-conservative, with or without constraints. Several recent books on the subject have been written, illustrating the usefulness of the theory in applications. Some researchers refer to FC as the calculus of the XXI century.

The fractional operators are non-local, therefore they are suitable for constructing models possessing memory effect. The first application of FC belongs to Niels Henrik Abel and goes back to 1823. Abel applied the FC to the solution of an integral equation which arises in the formulation of the *tautochrone problem*. This problem, sometimes also called the *isochrone problem*, is that of finding the shape of a frictionless wire lying in a vertical plane such that the time of a bead placed on the wire slides to the lowest point of the wire in the *same time* regardless of where the bead is placed. The cycloid is the isochrone as well as the *brachistochrone* curve: it gives the *shortest time* of slide and is usually associated with the birth of the *calculus of variations*. One can then claim that the calculus of variations and fractional calculus are connected from the very beginning! However, a theory of the fractional calculus of variations started only in 1996 with the work of Riewe, in order to better describe non-conservative systems in mechanics (Riewe, 1996, 1997). Riewe formulated the problem of the calculus of variations with fractional derivatives, and obtained the respective Euler–Lagrange equation, combining both conservative and non-conservative cases.

The inclusion of non-conservatism in the theory of the calculus of variations is extremely important from the point of view of applications. To illustrate this, let us consider the notion of *conservation law* or *constant of motion*. When a closed system is characterized by a quantity which remains unchangeable in the course of time, no matter what kind of processes take place in the system, such quantity defines a conservation law. Some fundamental conservation laws include the conservation of energy, impulse, momentum impulse, motion of the centre of gravity, electrical charge, and others. All physical laws are described in terms of differential equations (equations of motion). The conservation laws represent the first integrals of the equations of motion and are important for three reasons. Firstly, the task of solving the equations of motion explicitly is not always possible and knowledge of the first integrals may considerably simplify that task. Secondly, often there is no further necessity to solve the equation of motion, as the useful information is contained in the conservation laws. Thirdly, the conservation laws have a deep physical meaning and can be measured directly. Typical application of the constants of motion in the calculus of variations is to reduce the number of degrees of freedom, thus reducing the problems to a lower dimension, facilitating the integration of the differential equations given by the necessary optimality conditions (i.e., the Euler–Lagrange equations). One of the most important conservation laws is the integral of energy, discovered by Leonhard Euler in 1744: when a Lagrangian $L(t, q, \dot{q})$ does not depend explicitly on the independent variable

t, then

$$-L(q(t), \dot{q}(t)) + \partial_2 L(q(t), \dot{q}(t)) \cdot \dot{q}(t) \equiv \text{constant} \qquad \text{(CE)}$$

holds along all the solutions q of the Euler–Lagrange equations (along the extremals of the autonomous variational problem), where $\partial_2 L(\cdot, \cdot)$ denote the partial derivative of Lagrangian $L(\cdot, \cdot)$ with respect to its second argument. The conservation law (CE) is known in the calculus of variations as the 2nd Erdmann necessary condition; in concrete applications, it gains different interpretations: conservation of energy in mechanics; income-wealth law in economics; first law of thermodynamics; etc. Emmy Noether (1882–1935) was the first who proved, in 1918, that conservation laws like (CE) are a particular case of an universal principle connecting the notion of symmetry with the one of conservation law. All differential equations of physics (i.e., the equations of motion of physical systems) have a variational structure. In other words, the equations of motion of a physical system are the Euler–Lagrange equations of a certain variational problem. It turns out that the conservation laws are the result of the invariance of the functional to be extremized with respect to a continuous group of transformations, given by some symmetry principle. The more general expression of the interrelation symmetry/variational structure/conservation, is given by Noether's theorem. Noether's theorem asserts that the conservation laws for a system of differential equations, which correspond to the Euler–Lagrange equations of a certain variational problem, come from the invariance of the variational functional with respect to a one-parameter continuous group of transformations. The group of symmetry transformations requested by Noether's theorem depend on the physical properties of the system. One can see that the autonomous Lagrangian $L(q, \dot{q})$ is invariant under time-translations (time-homogeneity symmetry), and (CE) follows from Noether's theorem: in mechanics the total energy of a conservative system always remain constant in time, "it cannot be created or destroyed, but only transferred from one form into another". If instead of time-invariance, the physical system is invariant with respect to spatial translation, then one gets conservation of linear momentum; while invariance with respect to rotation gives conservation of angular momentum.

In the last decades, Noether's principle has been formulated in various contexts (Torres, 2002, 2004c). What is important to remark here is that constants of motion appear naturally in closed systems, and that in practical terms such systems do not exist: forces that do not store energy, so-called non-conservative or dissipative forces, are always present in real systems. Friction is an example of a non-conservative force. More generally, any friction-type force, like air resistance, is a non-conservative force. Non-conservative forces remove energy from the systems and, as a consequence, the constant of motion (CE) is broken. This explains, for instance, why the innumerable "perpetual motion machines" that have been proposed in the past, by the most creative and ingenious minds, fail. In the presence of external non-conservative forces, Noether's theorem and respective constants of motion cease to be valid. However, it is still possible to obtain a Noether-type theorem which covers both conservative (closed system) and non-conservative cases. Roughly speaking, one can prove that Noether's conservation laws are still valid if a new term, involving the non-conservative forces, is added to the standard constants of motion. The seminal work (Frederico and Torres, 2007a) makes use of the notion of Euler–Lagrange fractional extremal introduced by Riewe, to prove a Noether-type theorem that combines conservative and non-conservative cases.

Fractional versions of Noether's theorem have a central role in our book. Conservative physical systems imply frictionless motion and are a simplification of the real dynamical world. Almost all systems contain internal damping and are subject to external forces. For non-conservative dynamical systems, i.e., in the presence of non-conservative forces (forces that do not store energy and which are not equivalent to the gradient of a potential), the conservation laws are broken so that the standard Lagrangian or Hamiltonian formalism is no longer valid for describing the behavior of the system. Methodologically, Newtonian dissipative dynamical systems are a complement to conservative systems, because not only energy, but also other physical quantities such as linear or angular momentums, are not conserved. As we have already mentioned, in this case the classical Noether's theorem ceases to be valid. However, it is still possible to obtain the validity of Noether's principle using the fractional calculus of variations. Therefore, the new fractional variational calculus provide a more realistic approach to physics, permitting the consideration of non-conservative systems in a natural way.

In recent years numerous works on the fractional calculus of variations, fractional optimal control and its applications, have been written. Some of them deal with Riemann–Liouville fractional derivatives, others with Caputo, Riesz derivatives, and others. Depending on the type of variational functional being considered, different fractional Euler–Lagrange type equations are obtained. Other variational formulations consider functionals with a Lagrangian containing not only a fractional derivative but also a fractional integral and sometimes also the classical derivative. We believe that there is no one approach to fractional calculus that is better than all the others. All fractional derivatives have some advantages and disadvantages. This means that for a given classical Lagrangian we have at our disposal several different methods to obtain the fractional Euler–Lagrange equations and corresponding Hamiltonians. The equations of motion depend on the fractional derivatives used, and the existence of several options can be used to treat different physical systems with different specifications and characteristics. The most popular of such problems are considered and studied in our book.

We prove Euler–Lagrange necessary conditions and corresponding Noether theorems for several types of fractional variational problems of the calculus of variations, with and without constraints, using both Lagrangian and Hamiltonian formalisms. Sufficient optimality conditions are also obtained under appropriate convexity assumptions, and Leitmann's direct method is discussed in the framework of the fractional variational calculus. The text is organized as follows. In Chapter 1 we present a short introduction to the classical calculus of variations. The book is then divided into four further chapters: Chapter 2 develops the FCV via Riemann–Liouville operators, Chapter 3 uses Caputo derivatives, other fractional approaches are given in Chapter 4, while Chapter 5 considers some applications to classical and quantum mechanical problems. The reader with a background on the calculus of variations can skip Chapter 1 and begin immediately with the fractional part (Chapters 2, 3 and 4). The book was written in such a way that it can be read from the beginning to the end, or chapter by chapter. This is particularly useful, in our opinion, since different definitions for fractional derivatives and integrals are used depending on the purpose under study, and a reader interested, let us say, in the fractional approach via Caputo operators, can go immediately to Chapter 3.

This short book gives a gentle but solid introduction to the FCV. The audience is primarily advanced undergraduate and graduate students of mathematics. However, the book also provides an opportunity for an introduction to the fractional variational calculus even for experienced researchers. Our aim is to introduce the theory of the fractional calculus of variations in a book suitable for self-study, and to give the reader the state of the art of a very active and promising area of research. We will be extremely happy if the present book motivates and encourages some readers to follow a research activity in the area, and take part in the exploration of this exciting subject. We would like to mention that although the literature on FCV is already vast, in fact the theory is still in its childhood, and that much remains to be done. We finish this preface by pointing out two possible topics for further research: multiple integral fractional calculus of variations, and fractional optimal control. (i) The fractional variational theory involving multiple integrals was initiated in the paper (El-Nabulsi and Torres, 2008), where some important consequences of such a theory in mechanical problems involving dissipative systems with infinitely many degrees of freedom are given, but a formal theory for that is missing. Generalizing the results presented in this book to multiple fractional variational integrals gives interesting and challenging open questions. The recent results proved (see Almeida, Malinowska and Torres, 2010; Malinowska, 2012c; Odzijewicz and Torres, 2011) may be useful to that objective. (ii) The calculus of variations is now part of a more vast discipline, called optimal control. To the best of the authors' knowledge, there is no general proof of a fractional version of the Pontryagin Maximum Principle, the central result of optimal control theory. To extend some of the results of the book to the more general context of fractional optimal control also provides interesting open questions for research.

This book would not have been possible without the help, support, and encouragement of many people. We are particularly grateful to all our co-authors that have worked with us for years on the development of the fractional calculus of variations. It was a pleasure for us to share ideas and insights with them. Our deep thanks to Ricardo Almeida, Nuno Bastos, Rui Ferreira, Gastão Frederico, Dorota Mozyrska, Tatiana Odzijewicz, Shakoor Pooseh, and Moulay Rchid Sidi Ammi: this book is also their book. To Dr. Xin-She Yang, Editor-in-Chief of Int. J. of Mathematical Modelling and Numerical Optimisation, our immense gratitude for important advice and contacts, and for being always available to help. We cannot forget to thank Dr. Kellye Curtis, Commissioning Editor of Imperial College Press, World Scientific Publishing, London, UK, who kindly and patiently guided us on how to introduce our book ideas to the editorial board committee and on how to write a publication proposal for consideration. Imperial College Press / World Scientific Publishing has proved to be a proactive and supportive publisher, and we are grateful to all the editors who worked with us during the process, particularly to Jacqueline Downs for carefully reading all the manuscript and improving our English. Last but not least, we would like to thank the Department of Mathematics of the University of Aveiro for the excellent working conditions, and to Białystok University of Technology and The Center for Research and Development in Mathematics and Applications (CIDMA) of the University of Aveiro for financial support throughout the years.

Any comments or suggestions related to the material here contained are more than welcome, and may be submitted by post or by electronic mail to the authors.

Please contact the authors at:

Agnieszka B. Malinowska <a.malinowska@pb.edu.pl>
Department of Mathematics, Faculty of Computer Science
Białystok University of Technology
15-351 Białystok, Poland
http://www.researcherid.com/rid/G-5162-2010
http://arxiv.org/a/malinowska_a_1.html

Delfim F. M. Torres <delfim@ua.pt>
Center for Research and Development in Mathematics and Applications
Department of Mathematics, University of Aveiro
3810-193 Aveiro, Portugal
http://researcherid.com/rid/A-7682-2008
http://arxiv.org/a/torres_d_1.html

Agnieszka B. Malinowska and Delfim F. M. Torres
March 2012, Aveiro, Portugal

Contents

Chapter 1

The Classical Calculus of Variations

For the convenience of the reader, we begin with some well-known definitions and facts from the classical calculus of variations. With the exception of Section 1.5, results are given without proofs. For proofs and detailed discussions, we refer the reader to one of the many books on the subject (e.g., Giaquinta and Hildebrandt, 1996; Troutman, 1996; van Brunt, 2004). Here we follow Chachuat, 2007, which gives all the necessary background for our purposes.

1.1   Problem Statement

We are concerned with the problem of finding minima (or maxima) of a functional $\mathcal{J} : \mathcal{D} \to \mathbb{R}$, where $\mathcal{D}$ is a subset of a (normed) linear space $\mathbf{D}$ of real-valued (or real-vector-valued) functions. The formulation of a problem of the calculus of variations requires two steps: the specification of a performance criterion, and the statement of physical constraints that should be satisfied. The performance criterion $\mathcal{J}$, also called cost functional (or objective), must be specified for evaluating quantitatively the performance of the system under study. The typical form of the cost is

$$\mathcal{J}(y) = \int_a^b L(t, y(t), y'(t)) \, dt,$$

where $t \in [a, b]$ is the independent variable, usually called time; $y(t) \in \mathbb{R}^N$, $N \geq 1$, is a real vector variable, the functions $y(t)$, $a \leq t \leq b$, are generally called trajectories or curves; $y'(t) \in \mathbb{R}^N$ stands for the derivative of $y(t)$ with respect to time t; and $L : [a, b] \times \mathbb{R}^{2N} \to \mathbb{R}$ is a real-valued function, called the Lagrangian.

1

Enforcing constraints in the optimization problem reduces the set of candidate functions and leads to the following definition.

Definition 1.1. *A trajectory $y \in \mathbf{D}$ is said to be an admissible trajectory (or admissible function), provided it satisfies all the constraints of the problem along the interval $[a, b]$. The set of admissible trajectories is denoted by $\mathcal{D}$.*

A great variety of boundary conditions is of interest. The simplest one is to enforce both end-points fixed, e.g., $y(a) = y_a$ and $y(b) = y_b$, $y_a, y_b \in \mathbb{R}^N$. Alternatively, we may require that the trajectory $y \in \mathbf{D}$ joins a fixed point (a, y_a) to a specified curve $f(t)$, $a \leq t \leq T$. In this case, not only the optimal trajectory y shall be determined, but also the optimal value of b. Besides boundary constraints, another type of constraints is often required,

$$\mathcal{G}^j(y) = \int_a^b G^j(t, y(t), y'(t))dt = l_j, \quad j = 1, \ldots, r, \quad r \geq 1,$$

where $G^j : [a, b] \times \mathbb{R}^{2N} \to \mathbb{R}$, $j = 1, \ldots, r$. These constraints are often referred to as isoperimetric constraints. Similar constraints with $\leq$ sign can be considered. More generally, constraints of the form

$$G^j(t, y(t), y'(t))dt = 0, \quad j = 1, \ldots, r, \quad r \geq 1,$$

are called constraints of Lagrange form.

Having defined an objective functional $\mathcal{J}$ and constraints, one must then decide about the class of functions with respect to which the optimization shall be performed. The traditional choice in the calculus of variations is to consider the class of continuously differentiable functions, e.g., $C^1([a, b])$. We endow $C^1([a, b])$ with a norm. The most natural choice for a norm on $C^1([a, b])$ is

$$\|y\|_{1,\infty} := \max_{a \leq t \leq b} \|y(t)\| + \max_{a \leq t \leq b} \|y'(t)\|,$$

where $\| \cdot \|$ stands for the Euclidean norm in $\mathbb{R}^N$. The class of functions $C^1([a, b])$ endowed with $\| \cdot \|_{1,\infty}$ is a Banach space.

We now define what is meant by a minimum of $\mathcal{J}$ on $\mathcal{D}$.

Definition 1.2. *A trajectory $\bar{y} \in \mathcal{D}$ is said to be a local minimizer (resp. local maximizer) for $\mathcal{J}$ on $\mathcal{D}$, if there exists $\delta > 0$ such that $\mathcal{J}(\bar{y}) \leq \mathcal{J}(y)$ (resp. $\mathcal{J}(\bar{y}) \geq \mathcal{J}(y)$) for all $y \in \mathcal{D}$ with $\|y - \bar{y}\|_{1,\infty} < \delta$.*

The concept of variation of a functional is central to the solution of problems of the calculus of variations.

Definition 1.3. *The first variation of $\mathcal{J}$ at $y \in \mathbf{D}$ in the direction $y \in \mathbf{D}$ is defined as*

$$\delta\mathcal{J}(y; h) := \lim_{\varepsilon \to 0} \frac{\mathcal{J}(y + \varepsilon h) - \mathcal{J}(y)}{\varepsilon} = \frac{\partial}{\partial \varepsilon}\mathcal{J}(y + \varepsilon h)\bigg|_{\varepsilon=0},$$

provided the limit exists.

Definition 1.4. *A direction $h \in \mathbf{D}$, $h \neq 0$, is said to be an admissible variation for $\mathcal{J}$ at $y \in \mathcal{D}$ if*

(i) $\delta\mathcal{J}(y; h)$ *exists; and*
(ii) $y + \varepsilon h \in \mathcal{D}$ *for all sufficiently small ε.*

The following well-known result offers a necessary optimality condition for the problems of the calculus of variations, based on the concept of variation.

Theorem 1.1. *Let $\mathcal{J}$ be a functional defined on $\mathcal{D}$. Suppose that y is a local minimizer (or local maximizer) for $\mathcal{J}$ on $\mathcal{D}$. Then, $\delta\mathcal{J}(y; h) = 0$ for each admissible variation h at y.*

1.2 The Euler–Lagrange Equations

In this section, we present a first-order necessary optimality condition for a problem which is known as the elementary (or basic or fundamental) problem of the calculus of variations.

The next lemma is an essential result upon which the calculus of variations depends. It is called the fundamental lemma of the calculus of variations, sometimes also called the DuBois–Reymond lemma.

Lemma 1.1 (The fundamental lemma of the calculus of variations). *If $g(t)$ is a continuous function of t for $a \leq t \leq b$, and if*

$$\int_a^b g(t)h(t)\, dt = 0$$

for all functions $h(t)$ that are continuous for $a \leq t \leq b$ and are zero at $t = a$ and $t = b$, then $g(t) = 0$ for all $a \leq t \leq b$.

We denote by $\partial_i K$, $i = 1, \ldots, M$ ($M \in \mathbb{N}$), the partial derivative of a function $K : \mathbb{R}^M \to \mathbb{R}$ with respect to its ith argument. The following theorem provides a necessary optimality condition for the elementary problem of the calculus of variations.

Theorem 1.2 (The Euler–Lagrange equations). *Consider the problem of minimizing (or maximizing) the functional*

$$\mathcal{J}(y) = \int_a^b L(t, y(t), y'(t)) \, dt$$

on $\mathcal{D} = \{y \in \mathbf{D} : y(a) = y_a, \, y(b) = y_b\}$, where $L : [a,b] \times \mathbb{R}^{2N} \to \mathbb{R}$ is a twice continuously differentiable function. Suppose that y gives a (local) minimum (or maximum) to $\mathcal{J}$ on $\mathcal{D}$. Then,

$$\partial_i L(t, y(t), y'(t)) = \frac{d}{dt}\partial_{N+i} L(t, y(t), y'(t)), \quad i = 2, \ldots N+1, \qquad (1.1)$$

for all $t \in [a,b]$.

Definition 1.5. *A function y that satisfies the system of Euler–Lagrange equations (1.1) on $[a,b]$ is called an extremal for the functional $\mathcal{J}$.*

If one of the boundary conditions $y(a) = y_a$ or $y(b) = y_b$ is not present in the problem (it is possible that all of them are not present), then in order to find the extremizers we must add another necessary condition, usually called the natural boundary condition (or transversality condition).

Theorem 1.3 (Natural boundary conditions). *If y is a local minimizer (or maximizer) to the functional*

$$\mathcal{J}(y) = \int_a^b L(t, y(t), y'(t)) \, dt,$$

then y satisfies the Euler–Lagrange equations (1.1). Moreover,
(i) if $y(a) = y_a$ is free, then the natural boundary conditions

$$\partial_{N+i} L(a, y(a), y'(a)) = 0, \quad i = 2, \ldots N+1, \qquad (1.2)$$

hold;
(ii) if $y(b)$ is free, then the natural boundary conditions

$$\partial_{N+i} L(b, y(b), y'(b)) = 0, \quad i = 2, \ldots N+1, \qquad (1.3)$$

hold.

1.3 Problems with Isoperimetric Constraints

An isoperimetric problem of the calculus of variations is a problem wherein one or more constraints involve the integral of a given function over part or all of the integration horizon $[a, b]$. Such isoperimetric constraints arise frequently in geometry problems, such as the determination of the curve (resp. surface) enclosing the largest surface (resp. volume) subject to a fixed perimeter (resp. area). The following theorems provide a characterization of the extremals for isoperimetric problems, based on the method of Lagrange multipliers.

Theorem 1.4. *Consider the problem of minimizing (or maximizing) the functional*

$$\mathcal{J}(y) = \int_a^b L(t, y(t), y'(t)) \, dt$$

on $\mathcal{D}$ given by those $y \in \mathbf{D}$ such that $y(a) = y_a$, $y(b) = y_b$, and

$$\mathcal{G}(y) = \int_a^b G(t, y(t), y'(t)) dt = l, \tag{1.4}$$

where $L, G : [a, b] \times \mathbb{R}^{2N} \to \mathbb{R}$ are twice continuously differentiable functions. Suppose that y gives a (local) minimum (or maximum) to this problem. Assume that $\delta\mathcal{G}(y; h)$ does not vanish for all $h \in \mathbf{D}$. Then there exists a constant $\lambda \in \mathbb{R}$ such that y is a solution of the Euler–Lagrange equations

$$\partial_i F(t, y(t), y'(t), \lambda) = \frac{d}{dt} \partial_{N+i} F(t, y(t), y'(t), \lambda), \quad i = 2, \dots N + 1,$$

where $F(t, y, y', \lambda) = L(t, y, y') - \lambda G(t, y, y')$.

Remark 1.1. The equality (1.4) is called an isoperimetric constraint. Observe that $\delta\mathcal{G}(y; h)$ does not vanish for all $h \in \mathbf{D}$ if, and only if, y is not an extremal for $\mathcal{G}$.

Theorem 1.4 can be generalized to r conditions of integral type (to r isoperimetric constraints).

Theorem 1.5. *Consider the problem of minimizing (or maximizing) the functional*

$$\mathcal{J}(y) = \int_a^b L(t, y(t), y'(t)) \, dt$$

on $\mathcal{D}$ given by those $y \in \mathbf{D}$ such that $y(a) = y_a$, $y(b) = y_b$, and

$$\mathcal{G}^j(y) = \int_a^b G^j(t, y(t), y'(t))dt = l_j, \quad j = 1, \ldots, r,$$

where $L, G^j : [a, b] \times \mathbb{R}^{2N} \to \mathbb{R}$, $j = 1, \ldots, r$, are twice continuously differentiable functions. Suppose that y gives a (local) minimum (or maximum) to this problem. Assume that there are functions $h^1, \ldots, h^r \in \mathbf{D}$ such that the matrix

$$A = (a_{kl}), \quad a_{kl} := \delta\mathcal{G}^k(y; h^l), \tag{1.5}$$

has maximal rank r. Then there exist constants $\lambda_1, \ldots, \lambda_r \in \mathbb{R}$ such that y is a solution of the Euler–Lagrange equations

$$\partial_i F(t, y(t), y'(t), \lambda) = \frac{d}{dt}\partial_{N+i} F(t, y(t), y'(t), \lambda), \quad i = 2, \ldots N+1,$$

where
$$F(t, y, y', \lambda) = F(t, y, y', \lambda_1, \ldots, \lambda_r) = L(t, y, y') - \sum_{j=1}^r \lambda_j G^j(t, y, y').$$

1.4　Sufficient Optimality Conditions via Joint Convexity

In this section we present a sufficient condition for an extremal to be a global extremizer (minimizer or maximizer).

Definition 1.6. *Given a function $f \in C^1([a, b] \times \mathbb{R}^{2N}; \mathbb{R})$, we say that $f(\underline{x}, y, v)$ is jointly convex (resp. jointly concave) in (y, v), if*

$$f(x, y + y^0, v + v^0) - f(x, y, v) \geq (\leq) \sum_{i=2}^{N+1} \partial_i f(x, y, v)y^0_{i-1}$$

$$+ \sum_{i=2}^{N+1} \partial_{N+i} f(x, y, v)v^0_{i-1}$$

for all $(x, y, v), (x, y + y^0, v + v^0) \in [a, b] \times \mathbb{R}^{2N}$.

Theorem 1.6. *Let $L(\underline{x}, y, v)$ be jointly convex (resp. jointly concave) in (y, v). If y satisfies the system of N Euler–Lagrange equations (1.1) with $y(a) = y_a$ and $y(b) = y_b$, then y is a global minimizer (resp. global maximizer) to*

$$\mathcal{J}(y) = \int_a^b L(t, y(t), y'(t))\, dt$$

on $\mathcal{D} = \{y \in \mathbf{D} : y(a) = y_a, y(b) = y_b\}$.

1.5   Noether's Theorem

We now review one of the most beautiful results of the calculus of variations: the classical theorem of Emmy Noether. This result explains all the conservation laws in mechanics (e.g., conservation of momentum or conservation of energy). There exist several ways to prove this result. In this section we recall one of those proofs. The proof is done in two steps: we begin by proving Noether's theorem without transformation of the time (without transformation of the independent variable); then, using a technique of time-reparameterization, we obtain Noether's theorem in its general form. This technique is not so popular while proving Noether's theorem (Torres, 2004c), but it turns out to be, as we shall see, very useful in the context of the fractional calculus of variations.

We begin by formulating the fundamental problem of the calculus of variations, using now the usual notation of physics:

$$I[q(\cdot)] = \int_a^b L\left(t, q(t), \dot{q}(t)\right) dt \longrightarrow \min \qquad (1.6)$$

subject to the boundary conditions $q(a) = q_a$ and $q(b) = q_b$, and where $\dot{q} = \frac{dq}{dt}$. The Lagrangian $L : [a, b] \times \mathbb{R}^n \times \mathbb{R}^n \to \mathbb{R}$ is assumed here to be a C^2-function with respect to all its arguments.

Definition 1.7 (Invariance without transforming the time).
Functional (1.6) is said to be invariant under an ε-parameter group of infinitesimal transformations

$$\bar{q}(t) = q(t) + \varepsilon\xi(t, q) + o(\varepsilon) \qquad (1.7)$$

if

$$\int_{t_a}^{t_b} L\left(t, q(t), \dot{q}(t)\right) dt = \int_{t_a}^{t_b} L\left(t, \bar{q}(t), \dot{\bar{q}}(t)\right) dt \qquad (1.8)$$

for any subinterval $[t_a, t_b] \subseteq [a, b]$.

We denote by $\partial_i L$ the partial derivative of L with respect to its ith argument, $i = 1, 2, 3$.

Theorem 1.7 (Necessary condition of invariance). *If (1.6) is invariant under transformations (1.7), then*

$$\partial_2 L\left(t, q, \dot{q}\right) \cdot \xi + \partial_3 L\left(t, q, \dot{q}\right) \cdot \dot{\xi} = 0. \qquad (1.9)$$

Proof. Having in mind that condition (1.8) is valid for any subinterval $[t_a, t_b] \subseteq [a, b]$, we can get rid off the integral signs in (1.8): equation (1.8) is equivalent to

$$L\left(t, q, \dot{q}\right) = L(t, q + \varepsilon \xi + o(\varepsilon), \dot{q} + \varepsilon \dot{\xi} + o(\varepsilon)) \, . \tag{1.10}$$

Differentiating both sides of equation (1.10) with respect to ε, then substituting $\varepsilon = 0$, we obtain equality (1.9). $\square$

Definition 1.8 (Conserved quantity). *A quantity $C(t, q(t), \dot{q}(t))$ is said to be conserved if $\frac{d}{dt} C(t, q(t), \dot{q}(t)) = 0$ along all the solutions of the Euler–Lagrange equations*

$$\frac{d}{dt} \partial_3 L\left(t, q, \dot{q}\right) = \partial_2 L\left(t, q, \dot{q}\right) \, . \tag{1.11}$$

Theorem 1.8 (Noether's theorem without transforming time). *If functional (1.6) is invariant under the one-parameter group of transformations (1.7), then*

$$C(t, q, \dot{q}) = \partial_3 L\left(t, q, \dot{q}\right) \cdot \xi(t, q) \tag{1.12}$$

is conserved.

Proof. Using the Euler–Lagrange equations (1.11) and the necessary condition of invariance (1.9), we obtain:

$$\frac{d}{dt} \left(\partial_3 L\left(t, q, \dot{q}\right) \cdot \xi(t, q) \right)$$

$$= \frac{d}{dt} \partial_3 L\left(t, q, \dot{q}\right) \cdot \xi(t, q) + \partial_3 L\left(t, q, \dot{q}\right) \cdot \dot{\xi}(t, q)$$

$$= \partial_2 L\left(t, q, \dot{q}\right) \cdot \xi(t, q) + \partial_3 L\left(t, q, \dot{q}\right) \cdot \dot{\xi}(t, q)$$

$$= 0 \, . \qquad\qquad \square$$

Remark 1.2. In classical mechanics, $\partial_3 L\left(t, q, \dot{q}\right)$ is interpreted as the generalized momentum.

Definition 1.9 (Invariance of (1.6)). *Functional (1.6) is said to be invariant under the one-parameter group of infinitesimal transformations*

$$\begin{cases} \bar{t} = t + \varepsilon \tau(t, q) + o(\varepsilon) \, , \\ \bar{q}(t) = q(t) + \varepsilon \xi(t, q) + o(\varepsilon) \, , \end{cases} \tag{1.13}$$

if

$$\int_{t_a}^{t_b} L\left(t, q(t), \dot{q}(t)\right) dt = \int_{\bar{t}(t_a)}^{\bar{t}(t_b)} L\left(\bar{t}, \bar{q}(\bar{t}), \dot{\bar{q}}(\bar{t})\right) d\bar{t}$$

for any subinterval $[t_a, t_b] \subseteq [a, b]$.

Theorem 1.9 (Noether's theorem). *If functional* (1.6) *is invariant, in the sense of Definition 1.9, then*

$$C(t, q, \dot{q}) = \partial_3 L\,(t, q, \dot{q}) \cdot \xi(t, q) + (L(t, q, \dot{q}) - \partial_3 L\,(t, q, \dot{q}) \cdot \dot{q})\, \tau(t, q) \quad (1.14)$$

is conserved.

Proof. Every non-autonomous problem (1.6) is equivalent to an autonomous one, considering t as a dependent variable. For that we consider a Lipschitzian one-to-one transformation $[a, b] \ni t \longmapsto \sigma \in [\sigma_a, \sigma_b]$ such that

$$I\,[q(\cdot)] = \int_a^b L\,(t, q(t), \dot{q}(t))\, dt = \int_{\sigma_a}^{\sigma_b} L\left(t(\sigma), q(t(\sigma)), \frac{\frac{dq(t(\sigma))}{d\sigma}}{\frac{dt(\sigma)}{d\sigma}}\right) \frac{dt(\sigma)}{d\sigma} d\sigma$$

$$= \int_{\sigma_a}^{\sigma_b} L\left(t(\sigma), q(t(\sigma)), \frac{q'_\sigma}{t'_\sigma}\right) t'_\sigma\, d\sigma \doteq \int_{\sigma_a}^{\sigma_b} \bar{L}\left(t(\sigma), q(t(\sigma)), t'_\sigma, q'_\sigma\right) d\sigma$$

$$\doteq \bar{I}\,[t(\cdot), q(t(\cdot))]\ ,$$

where $t(\sigma_a) = a$, $t(\sigma_b) = b$, $t'_\sigma = \frac{dt(\sigma)}{d\sigma}$, and $q'_\sigma = \frac{dq(t(\sigma))}{d\sigma}$. If functional $I[q(\cdot)]$ is invariant in the sense of Definition 1.9, then functional $\bar{I}[t(\cdot), q(t(\cdot))]$ is invariant in the sense of Definition 1.7. Applying Theorem 1.8, we obtain that

$$C\left(t, q, t'_\sigma, q'_\sigma\right) = \partial_4 \bar{L} \cdot \xi + \partial_3 \bar{L} \tau \quad (1.15)$$

is a conserved quantity. Since

$$\partial_4 \bar{L} = \partial_3 L\,(t, q, \dot{q})\ ,$$

$$\partial_3 \bar{L} = -\partial_3 L\,(t, q, \dot{q}) \cdot \frac{q'_\sigma}{t'_\sigma} + L(t, q, \dot{q}) \quad (1.16)$$

$$= L(t, q, \dot{q}) - \partial_3 L\,(t, q, \dot{q}) \cdot \dot{q}\ ,$$

substituting (1.16) into (1.15), we arrive at (1.14). $\qquad\qquad\square$

Chapter 2

Fractional Calculus of Variations via Riemann–Liouville Operators

In this chapter we are concerned with problems of the calculus of variations with Riemann–Liouville fractional derivatives and integrals. The Riemann–Liouville operators are described in Section 2.1. Then, in Section 2.2, we prove fractional Euler–Lagrange equations for the basic problem of the calculus of variations, and in Section 2.3 we consider isoperimetric problems. A brief discussion of sufficient conditions of optimality is given in Section 2.4. A Noether-like theorem for variational problems with fractional derivatives is presented in Section 2.5, while in Section 2.6 we consider more general multidimensional fractional variational problems. In Section 2.7 some fundamental results of fractional linear control systems with Riemann–Liouville derivatives are proved. Section 2.8 deals with fractional optimal control problems. Finally, it is shown in Section 2.9 that the direct method of Leitmann can also be applied to fractional problems of the calculus of variations as well as fractional optimal control problems. The results presented in this chapter appeared in Almeida, Ferreira and Torres, 2012; Almeida and Torres, 2009a,b, 2010; Frederico and Torres, 2007a, 2008a; Malinowska, 2012c; Mozyrska and Torres, 2011; Odzijewicz and Torres, 2010.

2.1 Riemann–Liouville Fractional Integrals and Derivatives

In this section we review the necessary definitions and facts about Riemann–Liouville fractional operators. For more on the subject we refer the reader to the books by Kilbas, Srivastava and Trujillo, 2006; Podlubny, 1999; Samko, Kilbas and Marichev, 1993.

Definition 2.1. *Let $\varphi \in L_1\left([t_0, t_1], \mathbb{R}\right)$. The integrals*

$$_{t_0}I_t^\alpha \varphi(t) := \frac{1}{\Gamma(\alpha)} \int_{t_0}^t \varphi(\tau)(t - \tau)^{\alpha - 1} d\tau, \quad t > t_0,$$

and

$$_tI_{t_1}^\alpha \varphi(t) := \frac{1}{\Gamma(\alpha)} \int_t^{t_1} \varphi(\tau)(\tau - t)^{\alpha - 1} d\tau, \quad t < t_1,$$

where $\alpha > 0$ and $\Gamma(\cdot)$ is the Gamma function, i.e.,

$$\Gamma(z) = \int_0^\infty t^{z-1} e^{-t} dt, \quad z > 0,$$

are called, respectively, the left and the right fractional integrals of order α. Additionally, we define $_{t_0}I_t^0 = {_tI_{t_1}^0} := I$ (identity operator).

The proof of the following results can be found, e.g., in Kilbas, Srivastava and Trujillo, 2006.

Proposition 2.1 (Integration by parts for fractional integrals).
Let $\alpha > 0$ and $1/p + 1/q \leq 1 + \alpha$, $p \geq 1$, $q \geq 1$, with $p \neq 1$ and $q \neq 1$ in the case $1/p + 1/q = 1 + \alpha$. For $\varphi \in L_p\left([t_0, t_1], \mathbb{R}\right)$ and $\psi \in L_q\left([t_0, t_1], \mathbb{R}\right)$ the following equality holds:

$$\int_{t_0}^{t_1} \varphi(\tau) {_{t_0}I_t^\alpha} \psi(\tau) d\tau = \int_{t_0}^{t_1} \psi(\tau) {_tI_{t_1}^\alpha} \varphi(\tau) d\tau. \tag{2.1}$$

Remark 2.1. *When $t_0 = 0$ we write $_0I_t^\alpha = I^\alpha$ and then $I^\alpha f(t) = (f * \varphi_\alpha)(t)$, where $\varphi_\alpha(t) = \frac{t^{\alpha-1}}{\Gamma(\alpha)}$ for $t > 0$, $\varphi_\alpha(t) = 0$ for $t \leq 0$, and $\varphi_\alpha \to \delta(t)$ as $\alpha \to 0$, with δ the delta Dirac pseudo function.*

Proposition 2.2. *If $f \in L_p(a, b) \, (1 \leq p \leq \infty)$, $\alpha > 0$ and $\beta > 0$, then*

$$_aI_t^\alpha {_aI_t^\beta} f(t) = {_aI_t^{\alpha+\beta}} f(t)$$

almost everywhere on $[a, b]$.

Definition 2.2. *Let φ be defined on the interval $[t_0, t_1]$. The left Riemann–Liouville derivative of order α and the lower limit t_0 are defined through the following:*

$$_{t_0}D_t^\alpha \varphi(t) := \frac{1}{\Gamma(n - \alpha)} \left(\frac{d}{dt}\right)^n \int_{t_0}^t \varphi(\tau)(t - \tau)^{n - \alpha - 1} d\tau,$$

where n is a natural number satisfying $n = [\alpha] + 1$ with $[\alpha]$ denoting the integer part of α. Similarly, the right Riemann–Liouville derivative of order α and the upper limit t_1 are defined by

$$_tD_{t_1}^\alpha \varphi(t) := \frac{1}{\Gamma(n - \alpha)} \left(-\frac{d}{dt}\right)^n \int_t^{t_1} \varphi(\tau)(\tau - t)^{n - \alpha - 1} d\tau.$$

For $\alpha = 0$, we set $_tD_{t_1}^\alpha = {_{t_0}D_t^\alpha} := I$, the identity operator.

Remark 2.2. If α is an integer, then one obtains the standard derivatives, that is,

$$_aD_t^\alpha f(t) = \left(\frac{d}{dt}\right)^\alpha f(t), \quad _tD_b^\alpha f(t) = \left(-\frac{d}{dt}\right)^\alpha f(t).$$

Proposition 2.3. *Let f and g be two functions defined on $[a, b]$ such that $_aD_t^\alpha f$ and $_aD_t^\alpha g$ exist almost everywhere. Moreover, let $c_1, c_2 \in \mathbb{R}$. Then, $_aD_t^\alpha (c_1 f + c_2 g)$ exists almost everywhere, and $_aD_t^\alpha (c_1 f + c_2 g) = c_1{}_aD_t^\alpha f + c_2{}_aD_t^\alpha g$.*

Similar results hold for right fractional derivatives.

Remark 2.3. In general, the fractional derivative of a constant is not equal to zero.

Remark 2.4. The fractional derivative of order $\alpha > 0$ of function $(t - a)^\upsilon$, $\upsilon > -1$, is given by

$$_aD_t^\alpha (t - a)^\upsilon = \frac{\Gamma(\upsilon + 1)}{\Gamma(-\alpha + \upsilon + 1)}(t - a)^{\upsilon - \alpha}.$$

Remark 2.5. In the literature, when one reads "Riemann–Liouville fractional derivative", one usually means the "left Riemann–Liouville fractional derivative". In physics, if t denotes the time-variable, the right Riemann–Liouville fractional derivative of $f(t)$ is interpreted as a future state of the process $f(t)$. For this reason, the right-derivative is usually neglected in applications, when the present state of the process does not depend on the results of the future development. From a mathematical point of view, both derivatives appear naturally in the fractional calculus of variations.

Remark 2.6. Let $\alpha \in (0, 1)$. For functions φ given in the interval $[t_0, t_1]$, each of the following expressions

$$_{t_0}D_t^\alpha \varphi(t) = \frac{1}{\Gamma(1 - \alpha)}\frac{d}{dt}\int_{t_0}^t \varphi(\tau)(t - \tau)^{-\alpha}d\tau = \frac{d}{dt}\left(_{t_0}I_t^{1-\alpha}\varphi(t)\right),$$

$$_tD_{t_1}^\alpha \varphi(t) = -\frac{1}{\Gamma(1 - \alpha)}\frac{d}{dt}\int_t^{t_1} \varphi(\tau)(\tau - t)^{-\alpha}d\tau = -\frac{d}{dt}\left(_tI_{t_1}^{1-\alpha}\varphi(t)\right)$$

are called a Riemann–Liouville fractional derivative of order α, left and right respectively.

The following properties hold (see Theorem 2.4 of Kilbas, Srivastava and Trujillo, 2006).

Proposition 2.4. *If $\alpha > 0$, then $_{t_0}D_t^\alpha{}_{t_0}I_t^\alpha\varphi(t) = \varphi(t)$ for any $\varphi \in L_1(t_0,t_1)$, while $_{t_0}I_t^\alpha{}_{t_0}D_t^\alpha\varphi(t) = \varphi(t)$ is satisfied for $\varphi \in {}_{t_0}I_t^\alpha(L_1(t_0,t_1))$ with*

$$_{t_0}I_t^\alpha(L_1(t_0,t_1)) = \{\varphi(t) : \varphi(t) = {}_{t_0}I_t^\alpha\psi(t),\ \psi \in L_1(t_0,t_1)\}.$$

However,

$$_{t_0}I_t^\alpha{}_{t_0}D_t^\alpha\varphi(t) = \varphi(t) - \sum_{k=0}^{n-1} \frac{(t-t_0)^{\alpha-k-1}}{\Gamma(\alpha-k)} \left({}_{t_0}I_t^{n-\alpha}\varphi(t)\right)_{t=t_0}.$$

In particular, for $\alpha \in (0,1]$ we have

$$_{t_0}I_t^\alpha{}_{t_0}D_t^\alpha\varphi(t) = \varphi(t) - \frac{(t-t_0)^{\alpha-1}}{\Gamma(\alpha)} \left({}_{t_0}I_t^{1-\alpha}\varphi(t)\right)_{t=t_0}.$$

Similar results hold for right fractional derivatives. There exist also a formula for the composition of Riemann–Liouville derivatives (cf. Podlubny, 1999).

The next proposition is particularly useful for our purposes.

Proposition 2.5. *The formula*

$$\int_{t_0}^{t_1} f(t)_{t_0}D_t^\alpha g(t)dt = \int_{t_0}^{t_1} g(t)_t D_{t_1}^\alpha f(t)dt, \quad 0 < \alpha < 1, \qquad (2.2)$$

is valid under the assumption that $f(t) \in {}_tI_{t_1}^\alpha(L_p)$ and $g(t) \in {}_{t_0}I_t^\alpha(L_q)$ with $1/p + 1/q \leq 1 + \alpha$.

Proof. The equality (2.2) follows from (2.1) if we denote $_tD_{t_1}^\alpha f(t) = \varphi(t)$, $_{t_0}D_t^\alpha g(t) = \psi(t)$, and take into account that $_{t_0}I_t^\alpha{}_{t_0}D_t^\alpha f(t) = f(t)$ is valid for $f(t) \in {}_{t_0}I_t^\alpha(L_1)$. (cf. Samko, Kilbas and Marichev, 1993). $\qquad\square$

Another important result is the formula for the Laplace transformation of the derivative of a function φ. Like the Laplace transformation of an integer order derivative, it is easy to show that the Laplace transformation of a fractional order Riemann–Liouville derivative is given by

$$\mathcal{L}\left[{}_0D_t^\alpha\varphi(t)\right](s) = s^\alpha\mathcal{L}\left[\varphi(t)\right](s) - \sum_{k=0}^{n-1} c_k s^k, \quad n-1 < \alpha \leq n,$$

where $c_k = \left({}_0I_t^{k+1-\alpha}\varphi(t)\right)_{t=0}$. For $\alpha \in (0,1]$, we have $\mathcal{L}\left[{}_0D_t^\alpha\varphi(t)\right](s) = s^\alpha\mathcal{L}\left[\varphi(t)\right](s) - a$.

By $d_t^\alpha \varphi$ we denote the derivative without initialization or just the Liouville derivative defined by the inverse Laplace transform:

$$_0 d_t^\alpha \varphi(t) := \mathcal{L}^{-1}\left[s^\alpha F(s)\right](t),$$

where $F(s) = \mathcal{L}\left[\varphi(t)\right](s)$. For a fixed function φ and $\alpha > 0$, by initialization function $\psi = \psi(t)$ we mean a function that satisfies the following:

$$_0 D_t^\alpha \varphi(t) = {}_0 d_t^\alpha(\varphi)(t) + \psi(t).$$

For problems starting at 0 we have

$$\mathcal{L}\left[{}_0 D_t^\alpha \varphi(t)\right](s) = s^\alpha F(s) + \mathcal{L}\left[\psi(t)\right](s).$$

For a function $x : [0, T] \to \mathbb{R}^n$, we use similar notation as in the classical case:

$$_0 D_t^\alpha x(t) = {}_0 D_t^\alpha \begin{pmatrix} x_1(t) \\ x_2(t) \\ \vdots \\ x_n(t) \end{pmatrix} = \begin{pmatrix} {}_0 D_t^\alpha x_1(t) \\ {}_0 D_t^\alpha x_2(t) \\ \vdots \\ {}_0 D_t^\alpha x_n(t) \end{pmatrix}.$$

Such a situation, when for each component we use the same fractional order α of differentiation, is called in the literature the fractional-order derivative with commensurate order. The notion of commensurate or non-commensurate order is used in connection with Riemann–Liouville or other types of fractional derivatives, as are the ones we will consider later.

We mention now two important functions and their extensions to matrices.

Definition 2.3. *The two-parameter Mittag–Leffler type function is defined by the series expansion*

$$E_{\alpha,\beta}(z) := \sum_{k=0}^{\infty} \frac{z^k}{\Gamma(k\alpha + \beta)}, \quad \alpha > 0, \quad \beta > 0.$$

Let $A \in \mathbb{R}^{n \times n}$. By

$$E_{\alpha,\beta}(At^\alpha) = \sum_{k=0}^{\infty} A^k \frac{t^{k\alpha}}{\Gamma(k\alpha + \beta)}$$

it is denoted the extension of the two-parameter Mittag–Leffler function to matrices.

For $\beta = 1$ we obtain the Mittag–Leffler function of one parameter:

$$E_\alpha(z) = \sum_{k=0}^\infty \frac{z^k}{\Gamma(k\alpha + 1)}.$$

In the matrical case, $E_\alpha(At^\alpha) = \sum_{k=0}^\infty A^k \frac{t^{k\alpha}}{\Gamma(k\alpha+1)}$.

Definition 2.4. *Let $A \in \mathbb{R}^{n \times n}$. Then,*

$$e_\alpha^{At} = t^{\alpha-1} \sum_{k=0}^\infty A^k \frac{t^{k\alpha}}{\Gamma[(k+1)\alpha]} = \sum_{k=0}^\infty A^k \frac{t^{(k+1)\alpha-1}}{\Gamma[(k+1)\alpha]} = t^{\alpha-1} E_{\alpha,\alpha}(At^\alpha)$$

denote the α-exponential matrix function.

For $\alpha = 1$ both functions given in Definitions 2.3 and 2.4 are equal and coincide with the classical exponential matrix: $E_\alpha(At^\alpha) = e_\alpha^{At} = \exp(At)$.

Proposition 2.6. *Let $\alpha > 0$. Then,*

(a) $_0D_t^\alpha e_\alpha^{At} = Ae_\alpha^{At}$;
(b) $_tD_T^\alpha \varphi(t) = A\varphi(t)$, *where* $\varphi(t) = S(T - t) = e_\alpha^{A(T-t)}$;
(c) $I + \int_0^t Ae_\alpha^{A\tau}d\tau = E_\alpha(At^\alpha)$;
(d) $\frac{d}{dt}E_\alpha(At^\alpha) = Ae_\alpha^{A\tau}$, $t > 0$.

Proof. See proofs of Propositions 3.4 and 3.5 in Chapter 3. $\square$

Note that, item b) of Proposition 2.6 is also true for

$$S(t) = t^{\alpha+\beta-1} E_{\alpha,\alpha+\beta}(At^\alpha).$$

It is easy to explain the following properties of fractional integrals of the α-exponential function. If $t_0 = 0$ we write $I^\alpha = {}_0I_t^\alpha$.

(a) Let $k \in \mathbb{N} \cup \{0\}$. Then, $I^{1-\alpha} \frac{t^{(k+1)\alpha-1}}{\Gamma[(k+1)\alpha]} = \frac{t^{k\alpha}}{\Gamma(k\alpha+1)}$;
(b) $I^{1-\alpha}e_\alpha^{At} = E_\alpha(At)$;
(c) $I^\beta e_\alpha^{At} = t^{\alpha+\beta-1}E_{\alpha,\alpha+\beta}(At^\alpha)$;
(d) $I^\beta e^{At} = I^\beta e_{\alpha=1}^{At} = t^\beta E_{1,\beta+1}(At)$;
(e) $\left(I^\beta e_\alpha^{At}\right)_{t=0} = \mathbf{0}$, $\beta > 1 - \alpha$;
(f) $\left(I^{1-\alpha}e_\alpha^{At}a\right)_{t=0} = a$, $a \in \mathbb{R}^n$.

Let $S(t) = t^{\alpha+\beta-1}E_{\alpha,\alpha+\beta}(At^\alpha)$. The next result follows from (2.2) and item (b) of Proposition 2.6.

Lemma 2.1. *Let $0 < \alpha < 1$ and $\mu(t) \in {}_0I_t^\alpha(L_q)$, where $1/p + 1/q \le 1 + \alpha$ for p such that all components of $S(T - t)$ belong to $_tI_T^\alpha(L_p)$. Then,*

$$\int_0^T S(T - \tau)D_0^\alpha\mu(\tau)d\tau = \int_0^T AS(T - \tau)\mu(\tau)d\tau.$$

2.2   Fundamental Problem of the Calculus of Variations

Depending on the type of functional being considered, different fractional Euler–Lagrange type equations are obtained. Here we propose two kinds of functionals. In Subsection 2.2.1 we consider dependence of the integrands on the independent variable t, unknown function y, and $y' + k\,_aD_t^\alpha y$ with k a real parameter. As a consequence, one gets a proper extension of the classical calculus of variations, in the sense that the classical theory is recovered with the particular situation $k = 0$. Subsection 2.2.2 is devoted to problems with a Lagrangian containing not only a Riemann–Liouville fractional derivative (RLFD) but also a Riemann–Liouville fractional integral (RLFI).

2.2.1   *Problems with Classical and Fractional Derivatives*

Let $0 < \alpha < 1$. Consider the following problem: find a function $y \in C^1[a, b]$ for which the functional

$$\mathcal{J}(y) = \int_a^b F\left(t, y(t), y'(t) + k\,_aD_t^\alpha y(t)\right) dt \qquad (2.3)$$

subject to given boundary conditions

$$y(a) = y_a, \quad y(b) = y_b, \qquad (2.4)$$

has an extremum. We assume that k is a fixed real number, $F \in C^2([a, b] \times \mathbb{R}^2; \mathbb{R})$, and $\partial_3 F$ (the partial derivative of $F(\cdot, \cdot, \cdot)$ with respect to its third argument) has a continuous right Riemann–Liouville fractional derivative of order α.

Definition 2.5. *A function $y \in C^1([a, b])$ that satisfies the given boundary conditions* (2.4) *is said to be an admissible trajectory for problem* (2.3)–(2.4).

For simplicity of notation we introduce the operator $[\cdot]_k^\alpha$ defined by

$$[y]_k^\alpha(t) = \left(t, y(t), y'(t) + k\,_aD_t^\alpha y(t)\right).$$

With this notation we can write (2.3) simply as

$$\mathcal{J}(y) = \int_a^b F[y]_k^\alpha(t)dt.$$

Theorem 2.1 (Fractional Euler–Lagrange equation). *If y is an extremizer (minimizer or maximizer) to problem (2.3)–(2.4), then y satisfies the Euler–Lagrange equation*

$$\partial_2 F[y]_k^\alpha(t) - \frac{d}{dt}\partial_3 F[y]_k^\alpha(t) + k\,_tD_b^\alpha \partial_3 F[y]_k^\alpha(t) = 0 \qquad (2.5)$$

for all $t \in [a, b]$.

Proof. Suppose that y is a solution of (2.3)–(2.4). Note that admissible functions $\hat{y}$ can be written in the form $\hat{y}(t) = y(t) + \epsilon\eta(t)$, where $\eta \in C^1[a, b]$, $\eta(a) = \eta(b) = 0$, and $\epsilon \in \mathbb{R}$. Let $J(\epsilon) = \int_a^b F(t, y(t) + \epsilon\eta(t), \frac{d}{dt}(y(t) + \epsilon\eta(t)) + k_aD_t^\alpha(y(t) + \epsilon\eta(t)))dt$. Since $_aD_t^\alpha$ is a linear operator, we know that

$$_aD_t^\alpha(y(t) + \epsilon\eta(t)) =_a D_t^\alpha y(t) + \epsilon_aD_t^\alpha\eta(t).$$

On the other hand,

$$\left.\frac{dJ}{d\epsilon}\right|_{\epsilon=0} = \int_a^b \frac{d}{d\epsilon}F[\hat{y}]_k^\alpha(t)dt\bigg|_{\epsilon=0}$$

$$= \int_a^b \left(\partial_2 F[y]_k^\alpha(t)\cdot\eta(t) + \partial_3 F[y]_k^\alpha(t)\frac{d\eta(t)}{dt}\right. \qquad (2.6)$$

$$\left. + k\partial_3 F[y]_k^\alpha(t)_aD_t^\alpha\eta(t)\right)dt.$$

Using integration by parts we get

$$\int_a^b \partial_3 F\frac{d\eta}{dt}dt = \partial_3 F\eta\big|_a^b - \int_a^b \left(\eta\frac{d}{dt}\partial_3 F\right)dt \qquad (2.7)$$

and

$$\int_a^b \partial_3 F_aD_t^\alpha\eta dt = \int_a^b \eta\,_tD_b^\alpha\partial_3 F dt. \qquad (2.8)$$

Substituting (2.7) and (2.8) into (2.6), and having in mind that $\eta(a) = \eta(b) = 0$, it follows that

$$\left.\frac{dJ}{d\epsilon}\right|_{\epsilon=0} = \int_a^b \eta(t)\left(\partial_2 F[y]_k^\alpha(t) - \frac{d}{dt}\partial_3 F[y]_k^\alpha(t) + k\,_tD_b^\alpha\partial_3 F[y]_k^\alpha(t)\right)dt.$$

A necessary optimality condition is given by $\left.\frac{dJ}{d\epsilon}\right|_{\epsilon=0} = 0$. Hence,

$$\int_a^b \eta(t)\left(\partial_2 F[y]_k^\alpha(t) - \frac{d}{dt}\partial_3 F[y]_k^\alpha(t) + k\,_tD_b^\alpha\partial_3 F[y]_k^\alpha(t)\right)dt = 0. \qquad (2.9)$$

We obtain (2.5) applying the fundamental lemma of the calculus of variations to (2.9). □

Remark 2.7. Note that for $k = 0$ our necessary optimality condition (2.5) reduces to the classical Euler–Lagrange equation (1.1).

2.2.2 Problems with Fractional Derivatives and Integrals

Let us consider the following problem:

$$\mathcal{J}(y) = \int_a^b L(x, {_aI_x^{1-\alpha}}y(x), {_aD_x^\beta}y(x))\,dx \longrightarrow \min, \qquad (2.10)$$

where $L(\cdot, \cdot, \cdot) \in C^1([a, b] \times \mathbb{R}^2; \mathbb{R})$, $x \to \partial_2 L(x, {_aI_x^{1-\alpha}}y(x), {_aD_x^\beta}y(x))$ has continuous right RLFI of order $1 - \alpha$, and $x \to \partial_3 L(x, {_aI_x^{1-\alpha}}y(x), {_aD_x^\beta}y(x))$ has continuous right RLFD of order β, with α and β real numbers in the interval $(0, 1)$.

Remark 2.8. We are assuming that the admissible trajectory y are such that ${_aI_x^{1-\alpha}}y(x)$ and ${_aD_x^\beta}y(x)$ exist on the closed interval $[a, b]$. We also note that as α and β goes to 1 our fractional functional $\mathcal{J}$ tends to the classical functional $\int_a^b L(x, y(x), y'(x))\,dx$ of the calculus of variations.

Remark 2.9. We consider functionals $\mathcal{J}$ containing the left RLFI and the left RLFD only. These comprise the important cases in applications. The results are easily generalized for functionals containing also the right RLFI and/or right RLFD.

Theorem 2.2 (Fractional Euler–Lagrange equation). *Let $y(\cdot)$ be a local minimizer to problem* (2.10). *Then, $y(\cdot)$ satisfies the fractional Euler–Lagrange equation*

$$_xI_b^{1-\alpha}\partial_2 L(x, {_aI_x^{1-\alpha}}y(x), {_aD_x^\beta}y(x)) + {_xD_b^\beta}\partial_3 L(x, {_aI_x^{1-\alpha}}y(x), {_aD_x^\beta}y(x)) = 0 \tag{2.11}$$

for all $x \in [a, b]$.

Proof. Since y is an extremizer for $\mathcal{J}$, by a well-known result of the calculus of variations the first variation of $\mathcal{J}(\cdot)$ is zero at y, i.e.,

$$0 = \delta\mathcal{J}(\eta, y) = \int_a^b ({_aI_x^{1-\alpha}}\eta\,\partial_2 L + {_aD_x^\beta}\eta\,\partial_3 L)\,dx\,. \tag{2.12}$$

Integrating by parts,

$$\int_a^b {_aI_x^{1-\alpha}}\eta\,\partial_2 L\,dx = \int_a^b \eta\,{_xI_b^{1-\alpha}}\partial_2 L\,dx \tag{2.13}$$

and

$$\int_a^b {_aD_x^\beta}\eta\,\partial_3 L\,dx = \int_a^b \eta\,{_xD_b^\beta}\partial_3 L\,dx. \tag{2.14}$$

Substituting (2.13) and (2.14) into equation (2.12), we find that $\int_a^b (\,_xI_b^{1-\alpha}\partial_2 L + \,_xD_b^\beta \partial_3 L)\eta\,dx = 0$ for each η. Since η is an arbitrary function, by the fundamental lemma of the calculus of variations we deduce that $\,_xI_b^{1-\alpha}\partial_2 L + \,_xD_b^\beta \partial_3 L = 0$. $\hfill\square$

Remark 2.10. As α and β go to 1, the Euler–Lagrange equation (2.11) becomes the classical Euler–Lagrange equation $\partial_2 L - d/dx\partial_3 L = 0$.

Note that, in the case α goes to 1, we obtain the following result.

Theorem 2.3. *If y an extremum of functional $\mathcal{J}$, where*

$$\mathcal{J}(y) = \int_a^b L(x, y(x), \,_aD_x^\beta y(x))\,dx \longrightarrow \min,$$

then y satisfies the following Euler–Lagrange equation:

$$\partial_2 L + \,_xD_b^\beta \partial_3 L = 0. \tag{2.15}$$

A function that is a solution to the fractional differential equation (2.11) will be called an extremal of $\mathcal{J}$. Extremals also play an important role in the solution of the fractional isoperimetric problem. We note that equation (2.11) contains right RLFI and right RLFD, which are not present in the formulation of problem (2.10).

Extension to variational problems of non-commensurate order

We now consider problems of the calculus of variations with Riemann–Liouville derivatives and integrals of non-commensurate order, i.e., we consider functionals containing RLFI and RLFD of different fractional orders. Let

$$\mathcal{J}(y) = \int_a^b L(x, \,_aI_x^{1-\alpha_1}y(x), \ldots, \,_aI_x^{1-\alpha_n}y(x), \,_aD_x^{\beta_1}y(x), \ldots, \,_aD_x^{\beta_m}y(x))\,dx, \tag{2.16}$$

where n and m are two positive integers and $\alpha_i, \beta_j \in (0,1)$, $i = 1, \ldots, n$ and $j = 1, \ldots, m$. Following the proof of Theorem 2.2, we deduce the following result.

Theorem 2.4. *If $y(\cdot)$ is a local minimizer of (2.16), then $y(\cdot)$ satisfies the Euler–Lagrange equation*

$$\sum_{i=1}^n \,_xI_b^{1-\alpha_i}\partial_{i+1}L + \sum_{j=1}^m \,_xD_b^{\beta_j}\partial_{j+n+1}L = 0$$

for all $x \in [a, b]$.

Extension to several dependent variables

We now study the case of multiple unknown functions $y_1, \ldots, y_n$.

Theorem 2.5. *Let $\mathcal{J}$ be the functional given by the expression*

$$\mathcal{J}(y) = \int_a^b L(x, {_aI_x^{1-\alpha}}y_1(x), \ldots, {_aI_x^{1-\alpha}}y_n(x), {_aD_x^{\beta}}y_1(x), \ldots, {_aD_x^{\beta}}y_n(x))\, dx.$$

If $y = (y_1(\cdot), \ldots, y_n(\cdot))$ is a local minimizer of $\mathcal{J}$, then it satisfies, for all $x \in [a, b]$, the following system of n fractional differential equations:

$${_xI_b^{1-\alpha}}\partial_{k+1}L + {_xD_b^{\beta}}\partial_{n+k+1}L = 0, \quad k = 1, \ldots, n.$$

Proof. Denote by y and η the vectors $(y_1, \ldots, y_n)$ and $(\eta_1, \ldots, \eta_n)$, respectively. For a parameter ϵ, we consider a new function

$$J(\epsilon) = \mathcal{J}(y + \epsilon\eta). \tag{2.17}$$

Since $y_1(\cdot), \ldots, y_n(\cdot)$ is an extremizer of $\mathcal{J}$, $J'(0) = 0$. Differentiating equation (2.17) with respect to ϵ, at $\epsilon = 0$, we obtain

$$\int_a^b \left[{_aI_x^{1-\alpha}}\eta_1\, \partial_2 L + \cdots + {_aI_x^{1-\alpha}}\eta_n\, \partial_{n+1} L + {_aD_x^{\beta}}\eta_1\, \partial_{n+2} L \right.$$

$$\left. + \cdots + {_aD_x^{\beta}}\eta_n\, \partial_{2n+1}L \right]\, dx = 0.$$

Integrating by parts leads to

$$\int_a^b \left[{_xI_b^{1-\alpha}}\partial_2 L + {_xD_b^{\beta}}\partial_{n+2}L \right] \eta_1$$

$$+ \cdots + \left[{_xI_b^{1-\alpha}}\partial_{n+1}L + {_xD_b^{\beta}}\partial_{2n+1}L \right] \eta_n\, dx = 0.$$

Considerer a variation $\eta = (\eta_1, 0, \ldots, 0)$, η_1 arbitrary; then by the fundamental lemma of the calculus of variations we obtain ${_xI_b^{1-\alpha}}\partial_2 L + {_xD_b^{\beta}}\partial_{n+2}L = 0$. Selecting appropriate variations η, one deduce the remaining formulas. □

2.3   Fractional Isoperimetric Problems

This section is devoted to the study of isoperimetric problems of the calculus of variations with Riemann–Liouville fractional operators. Both situations, when the lower bound of the variational integrals coincide and do not coincide with the lower bound of the fractional derivatives, are considered. We also provide results for problems with fractional and classical derivatives, as well as problems with fractional derivatives and integrals.

2.3.1 *Problems with Fractional Derivatives*

Let us fix $\alpha, \beta \in (0, 1)$. Consider functionals $\mathcal{J}$ of the form

$$\mathcal{J}(y) = \int_a^b L(x, y, \, _aD_x^\alpha y, \, _xD_b^\beta y)dx \tag{2.18}$$

defined on the set of admissible functions y that have continuous left fractional derivatives of order α and continuous right fractional derivatives of order β in $[a, b]$, and where $(x, y, u, v) \to L(x, y, u, v)$ is a function with continuous first and second partial derivatives with respect to all its arguments such that $\frac{\partial L}{\partial u}(x, y, \, _aD_x^\alpha y, \, _xD_b^\beta y)$ has a continuous right fractional derivative of order α for all $x \in [a, b]$ and $\frac{\partial L}{\partial v}(x, y, \, _aD_x^\alpha y, \, _xD_b^\beta y)$ has a continuous left fractional derivative of order β in $[a, b]$.

Remark 2.11. The left Riemann–Liouville fractional derivative is infinite at $x = a$ if $y(a) \neq 0$. If $y(b) \neq 0$, then the right Riemann–Liouville fractional derivative is also not finite at $x = b$. For this reason, by considering that the admissible functions y have continuous left fractional derivatives, then necessarily $y(a) = 0$; by considering that the admissible functions y have continuous right fractional derivatives, then necessarily $y(b) = 0$. This fact seems to have been neglected in some of the initial work on the calculus of variations with Riemann–Liouville fractional derivatives. Alternatively, we can consider the general case of boundary conditions, say $y(a) = y_a$ and $y(b) = y_b$, and study functionals of type

$$\mathcal{J}(y) = \int_a^b L(x, y(x), \, _aD_x^\alpha(y(x) - y_a), \, _xD_b^\beta(y(x) - y_b))dx.$$

This needs, however, a modified fractional calculus (Almeida, Malinowska and Torres, 2010; Malinowska, Sidi Ammi and Torres, 2010).

We introduce the fractional isoperimetric problem as follows: find the functions y that satisfy boundary conditions

$$y(a) = y_a, \quad y(b) = y_b \tag{2.19}$$

($y_a = 0$ if left Riemann–Liouville fractional derivatives are present in (2.18); $y_b = 0$ if right Riemann–Liouville fractional derivatives are present in (2.18) — cf. Remark 2.11), the integral constraint

$$\mathcal{I}(y) = \int_a^b g(x, y, \, _aD_x^\alpha y, \, _xD_b^\beta y)dx = l, \tag{2.20}$$

and give a minimum or a maximum to (2.18). We assume that l is a specified real constant, functions y have continuous left and right fractional derivatives (if present in (2.18)), and $(x, y, u, v) \to g(x, y, u, v)$ is a function with continuous first and second partial derivatives with respect to all its arguments such that $\frac{\partial g}{\partial u}(x, y, {}_aD_x^\alpha y, {}_xD_b^\beta y)$ has continuous right fractional derivative of order α for all $x \in [a, b]$ and $\frac{\partial g}{\partial v}(x, y, {}_aD_x^\alpha y, {}_xD_b^\beta y)$ has continuous left fractional derivative of order β in $[a, b]$. Theorem 2.3 is the basis of the following definition (cf. Theorem 2.10 with $A = a$ and $B = b$).

Definition 2.6. *An admissible function y is an* extremal *for $\mathcal{I}$ in* (2.20) *if it satisfies the equation*

$$\frac{\partial g}{\partial y} + {}_xD_b^\alpha \frac{\partial g}{\partial u} + {}_aD_x^\beta \frac{\partial g}{\partial v} = 0$$

for all $x \in [a, b]$.

The following theorem gives a necessary condition for y to be a solution of the fractional isoperimetric problem defined by (2.18)–(2.19)–(2.20) under the assumption that y is not an extremal for $\mathcal{I}$.

Theorem 2.6. *Suppose that $\mathcal{J}$ given by* (2.18) *has a local minimum or a local maximum at y, subject to the boundary conditions* (2.19) *and the isoperimetric constraint* (2.20). *Further, suppose that y is not an extremal for the functional $\mathcal{I}$. Then there exists a constant λ such that y satisfies the fractional differential equation*

$$\frac{\partial F}{\partial y} + {}_xD_b^\alpha \frac{\partial F}{\partial u} + {}_aD_x^\beta \frac{\partial F}{\partial v} = 0 \tag{2.21}$$

with $F = L - \lambda g$.

Proof. Consider neighboring functions of the form

$$\hat{y} = y + \epsilon_1 \eta_1 + \epsilon_2 \eta_2, \tag{2.22}$$

where for each $i \in \{1, 2\}$ ϵ_i is a sufficiently small parameter, η_i have continuous left and right fractional derivatives, and $\eta_i(a) = \eta_i(b) = 0$.

First we will show that (2.22) has a subset of admissible functions for the fractional isoperimetric problem. Consider the quantity

$$\mathcal{I}(\hat{y}) = \int_a^b g(x, y + \epsilon_1 \eta_1 + \epsilon_2 \eta_2, {}_aD_x^\alpha y + \epsilon_1 {}_aD_x^\alpha \eta_1 + \epsilon_2 {}_aD_x^\alpha \eta_2,$$

$$ {}_xD_b^\beta y + \epsilon_1 {}_xD_b^\beta \eta_1 + \epsilon_2 {}_xD_b^\beta \eta_2) dx.$$

Then we can regard $\mathcal{I}(\hat{y})$ as a function of ϵ_1 and ϵ_2. Define $\hat{I}(\epsilon_1, \epsilon_2) = \mathcal{I}(\hat{y}) - l$. Thus,

$$\hat{I}(0, 0) = 0. \tag{2.23}$$

On the other hand, we have

$$\left.\frac{\partial \hat{I}}{\partial \epsilon_2}\right|_{(0,0)} = \int_a^b \left[\frac{\partial g}{\partial y}\eta_2 + \frac{\partial g}{\partial u}{_aD_x^\alpha}\eta_2 + \frac{\partial g}{\partial v}{_xD_b^\beta}\eta_2\right] dx$$

$$= \int_a^b \left[\frac{\partial g}{\partial y} + {_xD_b^\alpha}\frac{\partial g}{\partial u} + {_aD_x^\beta}\frac{\partial g}{\partial v}\right]\eta_2 dx, \qquad (2.24)$$

where (2.24) follows from the integration by parts formula (2.2). Since y is not an extremal for $\mathcal{I}$, by the fundamental lemma of the calculus of variations, there exists a function η_2 such that

$$\left.\frac{\partial \hat{I}}{\partial \epsilon_2}\right|_{(0,0)} \neq 0. \qquad (2.25)$$

Using (2.23) and (2.25), the implicit function theorem asserts that there exists a function $\epsilon_2(\cdot)$, defined in a neighborhood of zero, such that $\hat{I}(\epsilon_1, \epsilon_2(\epsilon_1)) = 0$. We are now in a position to derive the necessary condition (2.21). Consider the real function $\hat{J}(\epsilon_1, \epsilon_2) = \mathcal{J}(\hat{y})$. By hypothesis, $\hat{J}$ has minimum (or maximum) at $(0,0)$ subject to the constraint $\hat{I}(0,0) = 0$, and we have proved that $\nabla \hat{I}(0,0) \neq \mathbf{0}$. We can appeal to the Lagrange multiplier rule (see, e.g., p. 77 of van Brunt, 2004) to assert the existence of a number λ such that $\nabla(\hat{J}(0,0) - \lambda \hat{I}(0,0)) = \mathbf{0}$. Repeating the calculations as before,

$$\left.\frac{\partial \hat{J}}{\partial \epsilon_1}\right|_{(0,0)} = \int_a^b \left[\frac{\partial L}{\partial y} + {_xD_b^\alpha}\frac{\partial L}{\partial u} + {_aD_x^\beta}\frac{\partial L}{\partial v}\right]\eta_1(x)dx$$

and

$$\left.\frac{\partial \hat{I}}{\partial \epsilon_1}\right|_{(0,0)} = \int_a^b \left[\frac{\partial g}{\partial y} + {_xD_b^\alpha}\frac{\partial g}{\partial u} + {_aD_x^\beta}\frac{\partial g}{\partial v}\right]\eta_1(x)dx.$$

Therefore, one has

$$\int_a^b \left[\frac{\partial L}{\partial y} + {_xD_b^\alpha}\frac{\partial L}{\partial u} + {_aD_x^\beta}\frac{\partial L}{\partial v} - \lambda\left(\frac{\partial g}{\partial y} + {_xD_b^\alpha}\frac{\partial g}{\partial u} + {_aD_x^\beta}\frac{\partial g}{\partial v}\right)\right]\eta_1(x)dx$$
$$= 0. \qquad (2.26)$$

Since (2.26) holds for any function η_1, we obtain (2.21):

$$\frac{\partial L}{\partial y} + {_xD_b^\alpha}\frac{\partial L}{\partial u} + {_aD_x^\beta}\frac{\partial L}{\partial v} - \lambda\left(\frac{\partial g}{\partial y} + {_xD_b^\alpha}\frac{\partial g}{\partial u} + {_aD_x^\beta}\frac{\partial g}{\partial v}\right) = 0.$$

$\square$

Example 2.1. Let α be a given number in the interval $(0,1)$. We consider the following fractional isoperimetric problem:

$$\int_0^1 (x^4 + (_0D_x^\alpha y)^2)dx \longrightarrow \min$$

$$\int_0^1 x^2 {_0D_x^\alpha y}\,dx = \frac{1}{5} \tag{2.27}$$

$$y(0) = 0\,, \quad y(1) = \frac{2}{2\alpha + 3\alpha^2 + \alpha^3}\,.$$

The augmented Lagrangian is

$$F(x, y, {_0D_x^\alpha y}, {_xD_1^\beta y}) = x^4 + (_0D_x^\alpha y)^2 - \lambda\, x^2 {_0D_x^\alpha y}$$

and it is a simple exercise to see that

$$y(x) = \frac{1}{\Gamma(\alpha)} \int_0^x \frac{t^2}{(x - t)^{1-\alpha}}dt = \frac{1}{\Gamma(\alpha)} \frac{2x^{\alpha+2}}{2\alpha + 3\alpha^2 + \alpha^3} \tag{2.28}$$

(i) is not an extremal for the isoperimetric functional,
(ii) satisfy $_0D_x^\alpha y = x^2$,
(iii) (2.21) holds for $\lambda = 2$, i.e., $_xD_1^\alpha(2\,_0D_x^\alpha y - 2x^2) = 0$. We remark that

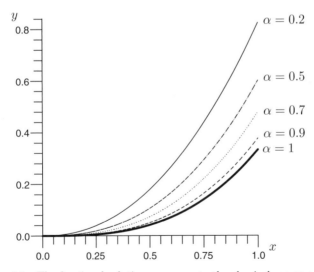

Fig. 2.1   The fractional solution converges to the classical one as $\alpha \to 1$.

for $\alpha = 1$ (2.28) gives $y(x) = x^3/3$, which coincides with the solution of the associated classical variational problem (Fig. 2.1). Indeed, for $\alpha \to 1$

our fractional problem (2.27) tends to the classical isoperimetric problem of minimizing the functional $\int_0^1 (x^4 + (y')^2) dx$ subject to the isoperimetric constraint $\int_0^1 x^2 y' \, dx = \frac{1}{5}$ and the boundary conditions $y(0) = 0$ and $y(1) = 1/3$. Then, $F = x^4 + (y')^2 - \lambda x^2 y'$ and the classical Euler–Lagrange equation is

$$\frac{\partial F}{\partial y} - \frac{d}{dx}\left(\frac{\partial F}{\partial y'}\right) = 0 \Leftrightarrow -2y'' + 2\lambda x = 0. \qquad (2.29)$$

The solution of (2.29) subject to $y(0) = 0$, $y(1) = 1/3$, and $\int_0^1 x^2 y' \, dx = \frac{1}{5}$ is $\lambda = 2$ and $y = x^3/3$.

Introducing a multiplier λ_0 associated with the cost functional (2.18), we can easily include in Theorem 2.6 the situation when the solution of the fractional isoperimetric problem, defined by (2.18)–(2.19)–(2.20), is an extremal for the fractional isoperimetric functional. This is done in Theorem 2.7.

Theorem 2.7. *If y is a local minimizer or a local maximizer of (2.18) subject to the boundary conditions (2.19) and the isoperimetric constraint (2.20), then there exist two constants, λ_0 and λ, not both zero, such that*

$$\frac{\partial K}{\partial y} + {}_x D_b^\alpha \frac{\partial K}{\partial u} + {}_a D_x^\beta \frac{\partial K}{\partial v} = 0 \qquad (2.30)$$

with $K = \lambda_0 L - \lambda g$.

Proof. Using the same notation as in the proof of Theorem 2.6, we have that $(0,0)$ is an extremal of $\hat{J}$ subject to the constraint $\hat{I} = 0$. Then, by the abnormal Lagrange multiplier rule (see, e.g., p. 82 of van Brunt, 2004), there exist two reals, λ_0 and λ, not both zero, such that $\nabla(\lambda_0 \hat{J}(0,0) - \lambda \hat{I}(0,0)) = \mathbf{0}$. Therefore,

$$\lambda_0 \left.\frac{\partial \hat{J}}{\partial \epsilon_1}\right|_{(0,0)} - \lambda \left.\frac{\partial \hat{I}}{\partial \epsilon_1}\right|_{(0,0)} = 0.$$

Applying the same reasoning as in the proof of Theorem 2.6, we end up with (2.30). □

2.3.2   Problems with Classical and Fractional Derivatives

As before, let $0 < \alpha < 1$. We now consider the problem of extremizing a functional

$$\mathcal{J}(y) = \int_a^b F\left(t, y(t), y'(t) + k \, {}_a D_t^\alpha y(t)\right) dt \qquad (2.31)$$

in the class $y \in C^1[a, b]$ when subject to given boundary conditions

$$y(a) = y_a, \quad y(b) = y_b, \tag{2.32}$$

and an isoperimetric constraint

$$\mathcal{I}(y) = \int_a^b G(t, y(t), y'(t) + k\, {}_aD_t^\alpha y(t)) dt = \xi. \tag{2.33}$$

We assume that k and ξ are fixed real numbers, $F, G \in C^2([a, b] \times \mathbb{R}^2; \mathbb{R})$, and $\partial_3 F$ and $\partial_3 G$ have continuous right Riemann–Liouville fractional derivatives of order α.

The next theorem gives a necessary optimality condition for the fractional isoperimetric problem (2.31)–(2.33).

Theorem 2.8. *Let y be an extremizer to the functional* (2.31) *subject to the boundary conditions* (2.32) *and the isoperimetric constraint* (2.33). *If y is not an extremal for $\mathcal{I}$, then there exists a constant λ such that*

$$\partial_2 H[y]_k^\alpha(t) - \frac{d}{dt}\partial_3 H[y]_k^\alpha(t) + k\, {}_tD_b^\alpha \partial_3 H[y]_k^\alpha(t) = 0 \tag{2.34}$$

for all $t \in [a, b]$, where $H(t, y, v, \lambda) = F(t, y, v) - \lambda G(t, y, v)$.

Proof. Is similar to that of Theorem 2.6 and can be found in Odzijewicz and Torres, 2010. □

Example 2.2. Let $\alpha \in (0, 1)$ and $k, \xi \in \mathbb{R}$. Consider the following fractional isoperimetric problem:

$$\mathcal{J}(y) = \int_0^1 \left(y' + k\, {}_0D_t^\alpha y\right)^2 dt \longrightarrow \min$$

$$\mathcal{I}(y) = \int_0^1 \left(y' + k\, {}_0D_t^\alpha y\right) dt = \xi \tag{2.35}$$

$$y(0) = 0, \ y(1) = \int_0^1 E_{1-\alpha, 1}\left(-k\left(1 - \tau\right)^{1-\alpha}\right) \xi d\tau.$$

In this case the augmented Lagrangian H of Theorem 2.8 is given by

$$H(t, y, v) = v^2 - \lambda v.$$

One can easily check that

$$y(t) = \int_0^t E_{1-\alpha, 1}\left(-k\left(t - \tau\right)^{1-\alpha}\right) \xi d\tau \tag{2.36}$$

(i) is not an extremal for $\mathcal{I}$;

(ii) satisfies $y' + k\,{}_0D_t^\alpha y = \xi$.

Moreover, (2.36) satisfies (2.34) for $\lambda = 2\xi$, i.e.,

$$-\frac{d}{dt}\left(2\left(y' + k\,{}_0D_t^\alpha y\right) - 2\xi\right) + k\,{}_tD_1^\alpha\left(2\left(y' + k\,{}_0D_t^\alpha y\right) - 2\xi\right) = 0.$$

We conclude that (2.36) is the extremal for problem (2.35).

Choose $k = 0$. In this case the isoperimetric constraint is trivially satisfied, (2.35) is reduced to the classical problem of the calculus of variations

$$\mathcal{J}(y) = \int_0^1 (y'(t))^2 dt \longrightarrow \min$$

$$y(0) = 0\,, \quad y(1) = \xi, \tag{2.37}$$

and our general extremal (2.36) simplifies to the well-known minimizer $y(t) = \xi t$ of (2.37).

When $\alpha \to 1$ the isoperimetric constraint is redundant with the boundary conditions, and the fractional problem (2.35) simplifies to the classical variational problem

$$\mathcal{J}(y) = (k+1)^2 \int_0^1 y'(t)^2 dt \longrightarrow \min$$

$$y(0) = 0\,, \quad y(1) = \frac{\xi}{k+1}. \tag{2.38}$$

Our fractional extremal (2.36) gives $y(t) = \frac{\xi}{k+1}t$, which is exactly the minimizer of (2.38).

Choose $k = \xi = 1$. When $\alpha \to 0$ one gets from (2.35) the classical isoperimetric problem

$$\mathcal{J}(y) = \int_0^1 (y'(t) + y(t))^2 dt \longrightarrow \min$$

$$\mathcal{I}(y) = \int_0^1 y(t)dt = \frac{1}{e}$$

$$y(0) = 0\,, \quad y(1) = 1 - \frac{1}{e}. \tag{2.39}$$

Our extremal (2.36) is then reduced to the classical extremal $y(t) = 1 - e^{-t}$ of (2.39).

Choose $k = 1$ and $\alpha = \frac{1}{2}$. Then (2.35) gives the following fractional isoperimetric problem:

$$\mathcal{J}(y) = \int_0^1 \left(y' + {_0}D_t^{\frac{1}{2}} y \right)^2 dt \longrightarrow \min$$

$$\mathcal{I}(y) = \int_0^1 \left(y' + {_0}D_t^{\frac{1}{2}} y \right) dt = \xi \qquad (2.40)$$

$$y(0) = 0, \quad y(1) = -\xi \left(1 - \operatorname{erfc}(1) + \frac{2}{\sqrt{\pi}} \right),$$

where erfc is the complementary error function. The extremal (2.36) for the particular fractional problem (2.40) is

$$y(t) = -\xi \left(1 - e^t \operatorname{erfc}(\sqrt{t}) + \frac{2\sqrt{t}}{\sqrt{\pi}} \right).$$

2.3.3 Problems with Fractional Derivatives and Integrals

We consider now the problem of minimizing the functional $\mathcal{J}$ given by (2.10) subject to an integral constraint

$$\mathcal{I}(y) = \int_a^b g(x, {_a}I_x^{1-\alpha} y(x), {_a}D_x^{\beta} y(x)) \, dx = l, \qquad (2.41)$$

where l is a prescribed value.

Theorem 2.9. *Consider the problem of minimizing the functional $\mathcal{J}$ as in (2.10) on the set of functions y satisfying condition $\mathcal{I}(y) = l$ given by (2.41). Let y be a local minimizer to the problem. Then, there exist two constants λ_0 and λ, not both zero, such that y satisfies the Euler–Lagrange equation ${_x}I_b^{1-\alpha} \partial_2 K + {_x}D_b^{\beta} \partial_3 K = 0$ for all $x \in [a, b]$, where $K = \lambda_0 L + \lambda g$.*

Proof. Is similar to that of Theorem 2.6 and can be found in Almeida and Torres, 2009b. □

Remark 2.12. If y is not an extremal for $\mathcal{I}$, then one can choose $\lambda_0 = 1$ in Theorem 2.9: there exists a constant λ such that y satisfies ${_x}I_b^{1-\alpha} \partial_2 F + {_x}D_b^{\beta} \partial_3 F = 0$ for all $x \in [a, b]$, where $F = L + \lambda g$.

2.3.4 Some Extensions

We consider now a fractional functional

$$\mathcal{L}(y) = \int_A^B L(x, y, {_a}D_x^{\alpha} y) \, dx \qquad (2.42)$$

with $[A, B] \subset [a, b]$, i.e., with the lower bound of the integral not coinciding with the lower bound of the fractional derivative. We will prove an Euler–Lagrange equation for functionals containing both left and right Riemann–Liouville fractional derivatives, i.e., for fractional functionals of the form

$$\mathcal{J}(y) = \int_A^B L(x, y, {}_aD_x^\alpha y, {}_xD_b^\beta y)dx, \tag{2.43}$$

where the integrand L satisfies the same conditions as before. Let y be a local extremum of $\mathcal{J}$ such that $y(a) = y_a$ and $y(b) = y_b$, and let $\hat{y} = y + \epsilon\eta$ with $\eta(a) = \eta(b) = 0$. Consider the function $\hat{J}(\epsilon) = \mathcal{J}(y + \epsilon\eta)$. Since $\hat{J}(\epsilon)$ has a local extremum at $\epsilon = 0$, then

$$0 = \int_A^B \left[\frac{\partial L}{\partial y} \cdot \eta + \frac{\partial L}{\partial u} \cdot {}_aD_x^\alpha\eta + \frac{\partial L}{\partial v} \cdot {}_xD_b^\beta\eta \right] dx$$

$$= \int_A^B \frac{\partial L}{\partial y} \cdot \eta dx + \left[\int_a^B \frac{\partial L}{\partial u} \cdot {}_aD_x^\alpha\eta dx - \int_a^A \frac{\partial L}{\partial u} \cdot {}_aD_x^\alpha\eta dx \right]$$

$$+ \left[\int_A^b \frac{\partial L}{\partial v} \cdot {}_xD_b^\beta\eta dx - \int_B^b \frac{\partial L}{\partial v} \cdot {}_xD_b^\beta\eta dx \right]$$

$$= \int_A^B \frac{\partial L}{\partial y} \cdot \eta dx + \left[\int_a^B \eta \cdot {}_xD_B^\alpha \frac{\partial L}{\partial u} dx - \int_a^A \eta \cdot {}_xD_A^\alpha \frac{\partial L}{\partial u} dx \right]$$

$$+ \left[\int_A^b \eta \cdot {}_AD_x^\beta \frac{\partial L}{\partial v} dx - \int_B^b \eta \cdot {}_BD_x^\beta \frac{\partial L}{\partial v} dx \right].$$

Continuing in a similar way,

$$0 = \int_A^B \frac{\partial L}{\partial y} \cdot \eta dx$$

$$+ \left[\int_a^A \eta \cdot {}_xD_B^\alpha \frac{\partial L}{\partial u} dx + \int_A^B \eta \cdot {}_xD_B^\alpha \frac{\partial L}{\partial u} dx - \int_a^A \eta \cdot {}_xD_A^\alpha \frac{\partial L}{\partial u} dx \right]$$

$$+ \left[\int_A^B \eta \cdot {}_AD_x^\beta \frac{\partial L}{\partial v} dx + \int_B^b \eta \cdot {}_AD_x^\beta \frac{\partial L}{\partial v} dx - \int_B^b \eta \cdot {}_BD_x^\beta \frac{\partial L}{\partial v} dx \right]$$

$$= \int_a^A \left[{}_xD_B^\alpha \frac{\partial L}{\partial u} - {}_xD_A^\alpha \frac{\partial L}{\partial u} \right] \eta dx + \int_A^B \left[\frac{\partial L}{\partial y} + {}_xD_B^\alpha \frac{\partial L}{\partial u} + {}_AD_x^\beta \frac{\partial L}{\partial v} \right] \eta dx$$

$$+ \int_B^b \left[{}_AD_x^\beta \frac{\partial L}{\partial v} - {}_BD_x^\beta \frac{\partial L}{\partial v} \right] \eta dx.$$

Let $\eta_1 : [a, A] \to \mathbb{R}$ be any function satisfying $\eta_1(a) = 0$, and η be given by

$$\eta(x) = \begin{cases} \eta_1(x) & \text{if } x \in [a, A], \\ 0 & \text{elsewhere.} \end{cases}$$

Therefore,

$$0 = \int_a^A \left[{}_xD_B^\alpha \frac{\partial L}{\partial u} - {}_xD_A^\alpha \frac{\partial L}{\partial u} \right] \eta_1 dx.$$

By the arbitrariness of η_1 and the fundamental lemma of the calculus of variations,

$${}_xD_B^\alpha \frac{\partial L}{\partial u} - {}_xD_A^\alpha \frac{\partial L}{\partial u} = 0 \text{ for all } x \in (a, A).$$

Analogously, we have

$$\frac{\partial L}{\partial y} + {}_xD_B^\alpha \frac{\partial L}{\partial u} + {}_AD_x^\beta \frac{\partial L}{\partial v} = 0 \text{ for all } x \in [A, B],$$

and

$${}_AD_x^\beta \frac{\partial L}{\partial v} - {}_BD_x^\beta \frac{\partial L}{\partial v} = 0 \text{ for all } x \in (B, b).$$

We have just proved the following.

Theorem 2.10. *Let y be a local extremizer of (2.43). Then, y satisfies the following equations:*

$$\begin{cases} \dfrac{\partial L}{\partial y} + {}_xD_B^\alpha \dfrac{\partial L}{\partial u} + {}_AD_x^\beta \dfrac{\partial L}{\partial v} = 0 & \text{for all } x \in [A, B], \\[2mm] {}_xD_B^\alpha \dfrac{\partial L}{\partial u} - {}_xD_A^\alpha \dfrac{\partial L}{\partial u} = 0 & \text{for all } x \in (a, A), \\[2mm] {}_AD_x^\beta \dfrac{\partial L}{\partial v} - {}_BD_x^\beta \dfrac{\partial L}{\partial v} = 0 & \text{for all } x \in (B, b). \end{cases}$$

We will study now the fractional isoperimetric problem for functionals of type (2.43) subject to the isoperimetric constraint

$$\mathcal{I}(y) = \int_A^B g(x, y, {}_aD_x^\alpha y, {}_xD_b^\beta y)dx = l. \qquad (2.44)$$

Definition 2.7. *We say that y is an extremal for functional $\mathcal{I}$ given in (2.44) if*

$$\frac{\partial g}{\partial y} + {}_xD_B^\alpha \frac{\partial g}{\partial u} + {}_AD_x^\beta \frac{\partial g}{\partial v} = 0 \quad \text{for all } x \in [A, B].$$

Theorem 2.11. *Let y give a local minimum or a local maximum to the fractional functional (2.43) subject to the constraint (2.44). If y is not an extremal for $\mathcal{I}$, then there exists a constant λ such that*

$$\begin{cases} \dfrac{\partial F}{\partial y} + {}_xD_B^\alpha \dfrac{\partial F}{\partial u} + {}_AD_x^\beta \dfrac{\partial F}{\partial v} = 0 & \text{for all } x \in [A, B] \\[2mm] {}_xD_B^\alpha \dfrac{\partial F}{\partial u} - {}_xD_A^\alpha \dfrac{\partial F}{\partial u} = 0 & \text{for all } x \in (a, A) \\[2mm] {}_AD_x^\beta \dfrac{\partial F}{\partial v} - {}_BD_x^\beta \dfrac{\partial F}{\partial v} = 0 & \text{for all } x \in (B, b) \end{cases} \qquad (2.45)$$

with $F = L - \lambda g$.

Proof.   Consider a variation $(\epsilon_1, \epsilon_2) \mapsto \hat{y} = y + \epsilon_1\eta_1 + \epsilon_2\eta_2$ where $\eta_1(a) = \eta_1(b) = \eta_2(a) = \eta_2(b) = 0$. Let

$$\hat{I}(\epsilon_1, \epsilon_2) = \int_A^B g(x, \hat{y}, \,_aD_x^\alpha\hat{y}, \,_xD_b^\beta\hat{y})dx - l.$$

Then, $\hat{I}(0,0) = 0$ and

$$\left.\frac{\partial\hat{I}}{\partial\epsilon_2}\right|_{(0,0)} = \int_A^B \left[\frac{\partial g}{\partial y}\eta_2 + \frac{\partial g}{\partial u}\,_aD_x^\alpha\eta_2 + \frac{\partial g}{\partial v}\,_xD_b^\beta\eta_2\right] dx$$

$$= \int_a^A \left[\,_xD_B^\alpha\frac{\partial g}{\partial u} - \,_xD_A^\alpha\frac{\partial g}{\partial u}\right]\eta_2 dx$$

$$+ \int_A^B \left[\frac{\partial g}{\partial y} + \,_xD_B^\alpha\frac{\partial g}{\partial u} + \,_AD_x^\beta\frac{\partial g}{\partial v}\right]\eta_2 dx$$

$$+ \int_B^b \left[\,_AD_x^\beta\frac{\partial g}{\partial v} - \,_BD_x^\beta\frac{\partial g}{\partial v}\right]\eta_2 dx.$$

Since y is not an extremal for $\mathcal{I}$, there exists a function η_2 such that $\left.\frac{\partial\hat{I}}{\partial\epsilon_2}\right|_{(0,0)} \neq 0$. By the implicit function theorem, there exists a subset of curves $\{y + \epsilon_1\eta_1 + \epsilon_2\eta_2 \mid (\epsilon_1, \epsilon_2) \in \mathbb{R}^2\}$ admissible for the fractional isoperimetric problem. Let $\hat{J}(\epsilon_1, \epsilon_2) = \mathcal{J}(\hat{y})$. Then, there exists a real λ such that $\nabla(\hat{J}(0,0) - \lambda\hat{I}(0,0)) = \mathbf{0}$. Because

$$\left.\frac{\partial\hat{J}}{\partial\epsilon_1}\right|_{(0,0)} = \int_a^A \left[\,_xD_B^\alpha\frac{\partial L}{\partial u} - \,_xD_A^\alpha\frac{\partial L}{\partial u}\right]\eta_1 dx$$

$$+ \int_A^B \left[\frac{\partial L}{\partial y} + \,_xD_B^\alpha\frac{\partial L}{\partial u} + \,_AD_x^\beta\frac{\partial L}{\partial v}\right]\eta_1 dx$$

$$+ \int_B^b \left[\,_AD_x^\beta\frac{\partial L}{\partial v} - \,_BD_x^\beta\frac{\partial L}{\partial v}\right]\eta_1 dx,$$

$$\left.\frac{\partial\hat{I}}{\partial\epsilon_1}\right|_{(0,0)} = \int_a^A \left[\,_xD_B^\alpha\frac{\partial g}{\partial u} - \,_xD_A^\alpha\frac{\partial g}{\partial u}\right]\eta_1 dx$$

$$+ \int_A^B \left[\frac{\partial g}{\partial y} + \,_xD_B^\alpha\frac{\partial g}{\partial u} + \,_AD_x^\beta\frac{\partial g}{\partial v}\right]\eta_1 dx$$

$$+ \int_B^b \left[\,_AD_x^\beta\frac{\partial g}{\partial v} - \,_BD_x^\beta\frac{\partial g}{\partial v}\right]\eta_1 dx,$$

and

$$\left.\frac{\partial\hat{J}}{\partial\epsilon_1}\right|_{(0,0)} - \lambda\left.\frac{\partial\hat{I}}{\partial\epsilon_1}\right|_{(0,0)} = 0,$$

(2.45) follows from the fact that η_1 is an arbitrary function. $\square$

Similarly as before, we can include in Theorem 2.11 the situation when the solution y is an extremal for $\mathcal{I}$ (abnormal extremizer). For that we introduce a new multiplier, λ_0, that will be zero when the solution y is an extremal for $\mathcal{I}$, and one otherwise.

Theorem 2.12. *If y is a local minimizer or a local maximizer of* (2.43) *subject to the isoperimetric constraint* (2.44)*, then there exist two constants λ_0 and λ, not both zero, such that*

$$\begin{cases} \dfrac{\partial K}{\partial y} + {}_xD_B^\alpha \dfrac{\partial K}{\partial u} + {}_AD_x^\beta \dfrac{\partial K}{\partial v} = 0 & \text{for all } x \in [A,B] \\[2ex] {}_xD_B^\alpha \dfrac{\partial K}{\partial u} - {}_xD_A^\alpha \dfrac{\partial K}{\partial u} = 0 & \text{for all } x \in (a,A) \\[2ex] {}_AD_x^\beta \dfrac{\partial K}{\partial v} - {}_BD_x^\beta \dfrac{\partial K}{\partial v} = 0 & \text{for all } x \in (B,b) \end{cases}$$

with $K = \lambda_0 L - \lambda g$.

Dependence on a parameter

Consider the following fractional problem of the calculus of variations: to extremize the functional

$$\Psi(y) = \int_0^1 \left[\frac{x^\alpha}{\Gamma(\alpha+1)} ({}_0D_x^\alpha y)^2 - 2\overline{y}\, {}_0D_x^\alpha y \right]^2 dx$$

when subject to the boundary conditions

$$y(0) = 0, \quad y(1) = 1.$$

Here, $\overline{y} := x^\alpha, x \in [0,1]$. The fractional Euler–Lagrange associated to this problem is

$${}_xD_1^\alpha \left(2 \left[\frac{x^\alpha}{\Gamma(\alpha+1)} ({}_0D_x^\alpha y)^2 - 2\overline{y}\, {}_0D_x^\alpha y \right] \cdot \left[\frac{2x^\alpha}{\Gamma(\alpha+1)} {}_0D_x^\alpha y - 2\overline{y} \right] \right) = 0.$$

$$(2.46)$$

Replacing y by $\overline{y}$, and since ${}_0D_x^\alpha \overline{y} = \Gamma(\alpha+1)$, we conclude that $\overline{y}$ is a solution of (2.46).

Consider now the following problem: what is the order of the derivative α, such that $\Psi(\overline{y})$ attains a maximum or a minimum? In other words, find the extremizers for $\psi(\alpha) = \Psi(\overline{y})$ (Fig. 2.2). Direct computations show that

$$\psi(\alpha) = \int_0^1 [x^\alpha \Gamma(\alpha+1)]^2 dx.$$

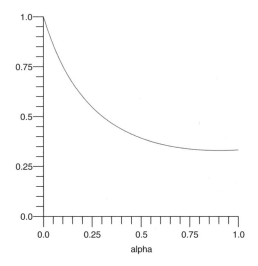

Fig. 2.2 Graph of $\Psi(\overline{y})$ for $\alpha \in [0, 1]$.

Evaluating its derivative,

$$\psi'(\alpha) = \int_0^1 \frac{d}{d\alpha} \left[x^\alpha \Gamma(\alpha + 1)\right]^2 dx$$

$$= \int_0^1 2x^\alpha \Gamma(\alpha + 1) \left[x^\alpha \ln x \, \Gamma(\alpha + 1) + x^\alpha \int_0^\infty t^\alpha \ln t \, e^{-t} dt\right] dx.$$

We have that $\alpha \approx 0.901$ is a solution of the equation $\psi'(\alpha) = 0$, and such value is precisely where $\Psi(\overline{y})$ attains a minimum.

More generally, consider the functional

$$\Phi(y, \alpha) = \int_a^b L(x, y(x), \, _aD_x^\alpha y(x))dx. \tag{2.47}$$

Functional (2.47) contains the left Riemann–Liouville derivative only, but we can consider functionals containing right Riemann–Liouville derivatives or both in a similar way. Let h be a curve such that $h(a) = h(b) = 0$, δ be a real number, and (y, α) be an extremal for Φ. Then,

$$\Phi(y + h, \alpha + \delta) - \Phi(y, \alpha)$$

$$= \int_a^b \frac{\partial L}{\partial y} \cdot h + \frac{\partial L}{\partial u} \cdot {}_aD_x^{\alpha+\delta} h + \frac{\partial L}{\partial u} \cdot ({}_aD_x^{\alpha+\delta} y - {}_aD_x^\alpha y)dx + O|(h, \delta)|^2.$$

For $\delta = 0$, using the fractional integration by parts formula and the fundamental lemma of the calculus of variations, we obtain the known fractional Euler–Lagrange equation:

$$\frac{\partial L}{\partial y}(x, y(x),\ _aD_x^\alpha y(x)) + {}_xD_b^\alpha \frac{\partial L}{\partial u}(x, y(x),\ _aD_x^\alpha y(x)) = 0.$$

For $h = 0$, we obtain the relation

$$\int_a^b \frac{\partial L}{\partial u}(x, y(x),\ _aD_x^\alpha y(x))\phi'(\alpha)dx = 0,$$

where $\phi(\alpha) = {}_aD_x^\alpha y(x)$. In summary, we have:

Theorem 2.13. *If (y, α) is an extremal of Φ given by (2.47), satisfying the boundary conditions $y(a) = 0$ and $y(b) = y_b$, then y satisfies the system*

$$\begin{cases} \dfrac{\partial L}{\partial y}(x, y(x),\ _aD_x^\alpha y(x)) + {}_xD_b^\alpha \dfrac{\partial L}{\partial u}(x, y(x),\ _aD_x^\alpha y(x)) = 0 \\ \displaystyle\int_a^b \dfrac{\partial L}{\partial u}(x, y(x),\ _aD_x^\alpha y(x))\phi'(\alpha)dx = 0 \end{cases} \tag{2.48}$$

where $\phi(\alpha) = {}_aD_x^\alpha y(x)$.

In the previous example, the solution obtained satisfies system (2.48) since

$$\frac{\partial L}{\partial u}(x, \overline{y}(x),\ _aD_x^\alpha \overline{y}(x)) = 0.$$

2.4 Sufficient Conditions for Optimality

In this section we prove sufficient conditions that ensure the existence of minimizers. As in the classical calculus of variations, some conditions of convexity are in order.

Theorem 2.14. *Let $L(\underline{x}, u, v)$ be a jointly convex function in (u, v) and let y_0 be a curve satisfying the fractional Euler–Lagrange equation (2.11). Then, y_0 minimizes (2.10).*

Proof. The following holds:

$$
\begin{aligned}
\mathcal{J}&(y_0 + \eta) - \mathcal{J}(y_0) \\
&= \int_a^b \left[L(x, {}_aI_x^{1-\alpha}y_0(x) + {}_aI_x^{1-\alpha}\eta(x), {}_aD_x^{\beta}y_0(x) + {}_aD_x^{\beta}\eta(x)) \right. \\
&\qquad\qquad \left. - L(x, {}_aI_x^{1-\alpha}y_0(x), \, {}_aD_x^{\beta}y_0(x)) \right] dx \\
&\geq \int_a^b \left[\partial_2 L(x, {}_aI_x^{1-\alpha}y_0(x), \, {}_aD_x^{\beta}y_0(x)) \, {}_aI_x^{1-\alpha}\eta \right. \\
&\qquad\qquad \left. + \partial_3 L(x, {}_aI_x^{1-\alpha}y_0(x), \, {}_aD_x^{\beta}y_0(x)) \, {}_aD_x^{\beta}\eta \right] dx \\
&= \int_a^b \left[{}_xI_b^{1-\alpha}\partial_2 L + {}_xD_b^{\beta}\partial_3 L \right]_{(x, {}_aI_x^{1-\alpha}y_0(x), \, {}_aD_x^{\beta}y_0(x))} \eta \, dx = 0.
\end{aligned}
$$

Thus, $\mathcal{J}(y_0 + \eta) \geq \mathcal{J}(y_0)$. $\qquad\qquad\qquad\qquad\qquad\qquad\square$

We now present a sufficient condition for convex Lagrangians on the third variable only. First we recall the notion of exact field.

Definition 2.8. *Let $D \subseteq \mathbb{R}^2$ and let $\Phi : D \to \mathbb{R}$ be a function of class C^1. We say that Φ is an exact field for L covering D if there exists a function $S \in C^1(D, \mathbb{R})$ such that*

$$
\begin{aligned}
\partial_1 S(x, y) &= L(x, y, \Phi(x, y)) - \partial_3 L(x, y, \Phi(x, y))\Phi(x, y), \\
\partial_2 S(x, y) &= \partial_3 L(x, y, \Phi(x, y)).
\end{aligned}
$$

Remark 2.13. This definition is motivated by the classical Euler–Lagrange equation. Indeed, every solution $y_0 \in C^2[a, b]$ of the differential equation $y' = \Phi(x, y(x))$ satisfies the (classical) Euler–Lagrange equation $\partial_2 L - \frac{d}{dx}\partial_3 L = 0$.

Theorem 2.15. *Let $L(\underline{x}, \underline{u}, v)$ be a convex function in $[a, b] \times \mathbb{R}^2$, Φ an exact field for L covering $[a, b] \times \mathbb{R} \subseteq D$, and y_0 a solution of the fractional equation*

$$
{}_aD_x^{\alpha}y(x) = \Phi(x, {}_aI_x^{1-\alpha}y(x)). \tag{2.49}
$$

Then, y_0 is a minimizer for $\mathcal{J}(y) = \int_a^b L(x, {}_aI_x^{1-\alpha}y(x), \, {}_aD_x^{\alpha}y(x)) \, dx$ subject to the constraint

$$
\left\{ y : [a, b] \to \mathbb{R} \mid {}_aI_a^{1-\alpha}y(a) = {}_aI_a^{1-\alpha}y_0(a), \, {}_aI_b^{1-\alpha}y(b) = {}_aI_b^{1-\alpha}y_0(b) \right\}. \tag{2.50}
$$

Proof. Let $E(x, y, z, w) = L(x, y, w) - L(x, y, z) - \partial_3 L(x, y, z)(w - z)$. First observe that

$$
\begin{aligned}
\frac{d}{dx}S(x, {}_aI_x^{1-\alpha}y(x)) &= \partial_1 S(x, {}_aI_x^{1-\alpha}y(x)) + \partial_2 S(x, {}_aI_x^{1-\alpha}y(x))\frac{d}{dx}{}_aI_x^{1-\alpha}y(x) \\
&= \partial_1 S(x, {}_aI_x^{1-\alpha}y(x)) + \partial_2 S(x, {}_aI_x^{1-\alpha}y(x)){}_aD_x^{\alpha}y(x).
\end{aligned}
$$

Since $E \geq 0$, it follows that

$$
\begin{aligned}
\mathcal{J}(y) &= \int_a^b \Big[E(x, {}_aI_x^{1-\alpha}y, \Phi(x, {}_aI_x^{1-\alpha}y), {}_aD_x^\alpha y) \\
&\quad + L(x, {}_aI_x^{1-\alpha}y, \Phi(x, {}_aI_x^{1-\alpha}y)) \\
&\quad + \partial_3 L(x, {}_aI_x^{1-\alpha}y, \Phi(x, {}_aI_x^{1-\alpha}y))({}_aD_x^\alpha y - \Phi(x, {}_aI_x^{1-\alpha}y)) \Big]\, dx \\
&\geq \int_a^b \Big[L(x, {}_aI_x^{1-\alpha}y, \Phi(x, {}_aI_x^{1-\alpha}y)) \\
&\quad + \partial_3 L(x, {}_aI_x^{1-\alpha}y, \Phi(x, {}_aI_x^{1-\alpha}y))({}_aD_x^\alpha y - \Phi(x, {}_aI_x^{1-\alpha}y)) \Big]\, dx \\
&= \int_a^b \Big[\partial_1 S(x, {}_aI_x^{1-\alpha}y) + \partial_2 S(x, {}_aI_x^{1-\alpha}y)\, {}_aD_x^\alpha y \Big]\, dx \\
&= \int_a^b \frac{d}{dx} S(x, {}_aI_x^{1-\alpha}y)\, dx \\
&= S(b, {}_aI_b^{1-\alpha}y(b)) - S(a, {}_aI_a^{1-\alpha}y(a)).
\end{aligned}
$$

Because y_0 is a solution of (2.49),

$$
E(x, {}_aI_x^{1-\alpha}y_0, \Phi(x, {}_aI_x^{1-\alpha}y_0), {}_aD_x^\alpha y_0) = 0.
$$

With similar calculations as before, one has $\mathcal{J}(y_0) = S(b, {}_aI_b^{1-\alpha}y_0(b)) - S(a, {}_aI_a^{1-\alpha}y_0(a))$. We just proved that $\mathcal{J}(y_0) \leq \mathcal{J}(y)$ when subject to the constraint (2.50). $\qquad\square$

2.5 Fractional Noether's Theorem

In this section, we generalize Noether's theorem for problems having fractional derivatives. Here we use the notion of Euler–Lagrange fractional extremal to prove a Noether-type theorem. For that we propose a generalization of the classical concept of conservation law, introducing an appropriate fractional operator.

The fundamental functional of the fractional calculus of variations is defined as follows:

$$
I[q(\cdot)] = \int_a^b L\left(t, q(t), {}_aD_t^\alpha q(t), {}_tD_b^\beta q(t)\right) dt \longrightarrow \min, \qquad (2.51)
$$

where the Lagrangian $L : [a, b] \times \mathbb{R}^n \times \mathbb{R}^n \times \mathbb{R}^n \to \mathbb{R}$ is a C^2 function with respect to all its arguments, and $0 < \alpha, \beta \leq 1$.

Remark 2.14. In the case $(\alpha, \beta) \to (1, 1)$, problem (2.51) is reduced to problem (1.6):

$$
I[q(\cdot)] = \int_a^b \mathcal{L}\left(t, q(t), \dot{q}(t)\right) dt \longrightarrow \min
$$

with

$$\mathcal{L}(t, q, \dot{q}) = L(t, q, \dot{q}, -\dot{q}). \tag{2.52}$$

Theorem 2.16. *If q is a minimizer to problem* (2.51), *then it satisfies the fractional Euler–Lagrange equations:*

$$\partial_2 L\left(t, q, {}_aD_t^\alpha q, {}_tD_b^\beta q\right) + {}_tD_b^\alpha \partial_3 L\left(t, q, {}_aD_t^\alpha q, {}_tD_b^\beta q\right)$$
$$+ {}_aD_t^\beta \partial_4 L\left(t, q, {}_aD_t^\alpha q, {}_tD_b^\beta q\right) = 0. \tag{2.53}$$

Definition 2.9 (cf. Definition 1.7). *We say that functional* (2.51) *is invariant under the transformation* (1.7) *if, and only if,*

$$\int_{t_a}^{t_b} L\left(t, q(t), {}_aD_t^\alpha q(t), {}_tD_b^\beta q(t)\right) dt$$
$$= \int_{t_a}^{t_b} L\left(t, \bar{q}(t), {}_aD_t^\alpha \bar{q}(t), {}_tD_b^\beta \bar{q}(t)\right) dt \tag{2.54}$$

for any subinterval $[t_a, t_b] \subseteq [a, b]$.

The next theorem establishes a necessary condition of invariance, of extreme importance for our objectives.

Theorem 2.17 (cf. Theorem 1.7). *If functional* (2.51) *is invariant under transformations* (1.7), *then*

$$\partial_2 L\left(t, q, {}_aD_t^\alpha q, {}_tD_b^\beta q\right) \cdot \xi(t, q) + \partial_3 L\left(t, q, {}_aD_t^\alpha q, {}_tD_b^\beta q\right) \cdot {}_aD_t^\alpha \xi(t, q)$$
$$+ \partial_4 L\left(t, q, {}_aD_t^\alpha q, {}_tD_b^\beta q\right) \cdot {}_tD_b^\beta \xi(t, q) = 0. \tag{2.55}$$

Remark 2.15. In the particular case $(\alpha, \beta) \to (1, 1)$, we obtain from (2.55) the necessary condition (1.9) applied to $\mathcal{L}$ given by (2.52).

Proof.   Having in mind that condition (2.54) is valid for any subinterval $[t_a, t_b] \subseteq [a, b]$, we can get rid of the integral signs in (2.54). Differentiating this condition with respect to ε, substituting $\varepsilon = 0$, and using the definitions and properties of the Riemann–Liouville fractional derivatives, we arrive at

$$0 = \partial_2 L\left(t, q, {}_aD_t^\alpha q, {}_tD_b^\beta q\right) \cdot \xi(t,q)$$

$$+ \partial_3 L\left(t, q, {}_aD_t^\alpha q, {}_tD_b^\beta q\right) \cdot \frac{d}{d\varepsilon}\left[\frac{1}{\Gamma(n-\alpha)}\left(\frac{d}{dt}\right)^n \int_a^t (t-\theta)^{n-\alpha-1} q(\theta)d\theta\right.$$

$$\left. + \frac{\varepsilon}{\Gamma(n-\alpha)}\left(\frac{d}{dt}\right)^n \int_a^t (t-\theta)^{n-\alpha-1}\xi(\theta,q)d\theta\right]_{\varepsilon=0}$$

$$+ \partial_4 L\left(t, q, {}_aD_t^\alpha q, {}_tD_b^\beta q\right) \cdot \frac{d}{d\varepsilon}\left[\frac{1}{\Gamma(n-\beta)}\left(-\frac{d}{dt}\right)^n \int_t^b (\theta-t)^{n-\beta-1} q(\theta)d\theta\right.$$

$$\left. + \frac{\varepsilon}{\Gamma(n-\beta)}\left(-\frac{d}{dt}\right)^n \int_t^b (\theta-t)^{n-\beta-1}\xi(\theta,q)d\theta\right]_{\varepsilon=0}. \quad (2.56)$$

Expression (2.56) is equivalent to (2.55). $\qquad\square$

The following definition is useful in order to introduce an appropriate concept of fractional conserved quantity.

Definition 2.10. *Given two functions f and g of class C^1 in the interval $[a,b]$, we introduce the following notation:*

$$D_t^\gamma (f,g) = -g\, {}_tD_b^\gamma f + f\, {}_aD_t^\gamma g,$$

where $t \in [a,b]$ and $\gamma \in \mathbb{R}_0^+$.

Remark 2.16. If $\gamma = 1$, the operator D_t^γ is reduced to

$$D_t^1 (f,g) = -g\, {}_tD_b^1 f + f\, {}_aD_t^1 g = \dot{f}g + f\dot{g} = \frac{d}{dt}(fg).$$

In particular, $D_t^1 (f,g) = D_t^1 (g,f)$.

Remark 2.17. In the fractional case ($\gamma \neq 1$), functions f and g do not commute: in general $D_t^\gamma (f,g) \neq D_t^\gamma (g,f)$.

Remark 2.18. The linearity of the operators ${}_aD_t^\gamma$ and ${}_tD_b^\gamma$ imply the linearity of the operator D_t^γ.

Definition 2.11 (Fractional-conserved quantity – cf. Definition 1.8). *We say that $C_f\left(t, q, {}_aD_t^\alpha q, {}_tD_b^\beta q\right)$ is fractional-conserved if, and only if, it is possible to write C_f in the form*

$$C_f (t, q, d_l, d_r) = \sum_{i=1}^m C_i^1 (t, q, d_l, d_r) \cdot C_i^2 (t, q, d_l, d_r) \quad (2.57)$$

for some $m \in \mathbb{N}$ *and some functions* C_i^1 *and* C_i^2, $i = 1, \ldots, m$, *where each pair* C_i^1 *and* C_i^2, $i = 1, \ldots, m$, *satisfy*

$$D_t^{\gamma_i} \left(C_i^{j_i^1} \left(t, q, {}_aD_t^\alpha q, {}_tD_b^\beta q \right), C_i^{j_i^2} \left(t, q, {}_aD_t^\alpha q, {}_tD_b^\beta q \right) \right) = 0 \qquad (2.58)$$

with $\gamma_i \in \{\alpha, \beta\}$, $j_i^1 = 1$ *and* $j_i^2 = 2$ *or* $j_i^1 = 2$ *and* $j_i^2 = 1$, *along all the fractional Euler–Lagrange extremals (i.e., along all the solutions of the fractional Euler–Lagrange equations* (2.53)*).*

Remark 2.19. Noether conserved quantities (1.14) are a sum of products, like the structure (2.57) we are assuming in Definition 2.11. For $\alpha = \beta = 1$ (2.58) is equivalent to the standard definition

$$\frac{d}{dt} \left[C_f \left(t, q(t), \dot{q}(t), -\dot{q}(t) \right) \right] = 0.$$

Example 2.3. Let C_f be a fractional-conserved quantity written in the form (2.57) with $m = 1$, that is, let $C_f = C_1^1 \cdot C_1^2$ be a fractional-conserved quantity for some given functions C_1^1 and C_1^2. Condition (2.58) of Definition 2.11 means one of four things: $D_t^\alpha \left(C_1^1, C_1^2 \right) = 0$, or $D_t^\alpha \left(C_1^2, C_1^1 \right) = 0$, or $D_t^\beta \left(C_1^1, C_1^2 \right) = 0$, or $D_t^\beta \left(C_1^2, C_1^1 \right) = 0$.

Remark 2.20. Given a fractional-conserved quantity C_f, the definition of C_i^1 and C_i^2, $i = 1, \ldots, m$, is never unique. In particular, one can always choose C_i^1 to be C_i^2 and C_i^2 to be C_i^1. Definition 2.11 is imune to the arbitrariness in defining the C_i^j, $i = 1, \ldots, m$, $j = 1, 2$.

Remark 2.21. Due to the simple fact that the same function can be written in several different but equivalent ways, to a given fractional-conserved quantity C_f there corresponds an integer value, m, in (2.57) which is, in general, also not unique (see Example 2.4).

Example 2.4. Let f, g and h be functions satisfying $D_t^\alpha (g, f) = 0$, $D_t^\alpha (f, g) \neq 0$, $D_t^\alpha (h, f) = 0$, $D_t^\alpha (f, h) \neq 0$ along all the fractional Euler–Lagrange extremals of a given fractional variational problem. One can provide different proofs to the fact that $C = f(g + h)$ is a fractional-conserved quantity:

(i) C is fractional-conserved because we can write C in the form (2.57) with $m = 2$, $C_1^1 = g$, $C_1^2 = f$, $C_2^1 = h$, and $C_2^2 = f$, satisfying (2.58) $(D_t^\alpha (C_1^1, C_1^2) = 0$ and $D_t^\alpha (C_2^1, C_2^2) = 0)$;

(ii) C is fractional-conserved because we can write C in the form (2.57) with $m = 1$, $C_1^1 = g + h$, and $C_1^2 = f$, satisfying (2.58) $(D_t^\alpha (C_1^1, C_1^2) = D_t^\alpha (g + h, f) = D_t^\alpha (g, f) + D_t^\alpha (h, f) = 0)$.

Theorem 2.18 (cf. Theorem 1.8). *If the functional* (2.51) *is invariant under the transformations* (1.7), *in the sense of Definition 2.9, then*

$$C_f\left(t, q, {}_aD_t^\alpha q, {}_tD_b^\beta q\right)$$

$$= \left[\partial_3 L\left(t, q, {}_aD_t^\alpha q, {}_tD_b^\beta q\right) - \partial_4 L\left(t, q, {}_aD_t^\alpha q, {}_tD_b^\beta q\right)\right] \cdot \xi(t, q) \quad (2.59)$$

is fractional-conserved.

Remark 2.22. In the particular case $\alpha = \beta = 1$, we obtain from (2.59) the conserved quantity (1.12) applied to $\mathcal{L}$ given by (2.52).

Proof. We use the fractional Euler–Lagrange equations

$$\partial_2 L\left(t, q, {}_aD_t^\alpha q, {}_tD_b^\beta q\right)$$

$$= -{}_tD_b^\alpha \partial_3 L\left(t, q, {}_aD_t^\alpha q, {}_tD_b^\beta q\right) - {}_aD_t^\beta \partial_4 L\left(t, q, {}_aD_t^\alpha q, {}_tD_b^\beta q\right)$$

in (2.55), obtaining

$$-{}_tD_b^\alpha \partial_3 L\left(t, q, {}_aD_t^\alpha q, {}_tD_b^\beta q\right) \cdot \xi(t, q) + \partial_3 L\left(t, q, {}_aD_t^\alpha q, {}_tD_b^\beta q\right) \cdot {}_aD_t^\alpha \xi(t, q)$$

$$-{}_aD_t^\beta \partial_4 L\left(t, q, {}_aD_t^\alpha q, {}_tD_b^\beta q\right) \cdot \xi(t, q) + \partial_4 L\left(t, q, {}_aD_t^\alpha q, {}_tD_b^\beta q\right) \cdot {}_tD_b^\beta \xi(t, q)$$

$$= D_t^\alpha\left(\partial_3 L\left(t, q, {}_aD_t^\alpha q, {}_tD_b^\beta q\right), \xi(t, q)\right)$$

$$- D_t^\beta\left(\xi(t, q), \partial_4 L\left(t, q, {}_aD_t^\alpha q, {}_tD_b^\beta q\right)\right) = 0.$$

The proof is complete.   □

Definition 2.12 (Invariance of (2.51) – cf. Definition 1.9).
Functional (2.51) *is said to be invariant under the one-parameter group of infinitesimal transformations* (1.13) *if, and only if,*

$$\int_{t_a}^{t_b} L\left(t, q(t), {}_aD_t^\alpha q(t), {}_tD_b^\beta q(t)\right) dt$$

$$= \int_{\bar{t}(t_a)}^{\bar{t}(t_b)} L\left(\bar{t}, \bar{q}(\bar{t}), {}_aD_{\bar{t}}^\alpha \bar{q}(\bar{t}), {}_{\bar{t}}D_b^\beta \bar{q}(\bar{t})\right) d\bar{t}$$

for any subinterval $[t_a, t_b] \subseteq [a, b]$.

    The next theorem provides the extension of Noether's theorem for fractional problems of the calculus of variations.

Theorem 2.19 (Fractional Noether's theorem). *If functional* (2.51) *is invariant, in the sense of Definition 2.12, then*

$$
C_f\left(t, q, {}_aD_t^\alpha q, {}_tD_b^\beta q\right)
$$
$$
= \left[\partial_3 L\left(t, q, {}_aD_t^\alpha q, {}_tD_b^\beta q\right) - \partial_4 L\left(t, q, {}_aD_t^\alpha q, {}_tD_b^\beta q\right)\right] \cdot \xi(t, q)
$$
$$
+ \left[L\left(t, q, {}_aD_t^\alpha q, {}_tD_b^\beta q\right) - \alpha\partial_3 L\left(t, q, {}_aD_t^\alpha q, {}_tD_b^\beta q\right) \cdot {}_aD_t^\alpha q\right.
$$
$$
\left. -\beta\partial_4 L\left(t, q, {}_aD_t^\alpha q, {}_tD_b^\beta q\right) \cdot {}_tD_b^\beta q\right] \tau(t, q) \quad (2.60)
$$

is fractional-conserved (Definition 2.11).

Remark 2.23. If $\alpha = \beta = 1$, the fractional conserved quantity (2.60) gives (1.14) applied to $\mathcal{L}$ given by (2.52).

Proof.   Our proof is an extension of the method used in the proof of Theorem 1.9. For that we reparameterize the time (the independent variable t) by the Lipschitzian transformation

$$
[a, b] \ni t \longmapsto \sigma f(\lambda) \in [\sigma_a, \sigma_b]
$$

that satisfies $t_\sigma^{'} = f(\lambda) = 1$ if $\lambda = 0$. Functional (2.51) is reduced, in this way, to an autonomous functional:

$$
\bar{I}[t(\cdot), q(t(\cdot))]
$$
$$
= \int_{\sigma_a}^{\sigma_b} L\left(t(\sigma), q(t(\sigma)), {}_{\sigma_a}D_{t(\sigma)}^\alpha q(t(\sigma)), {}_{t(\sigma)}D_{\sigma_b}^\beta q(t(\sigma))\right) t_\sigma^{'} d\sigma, \quad (2.61)
$$

where $t(\sigma_a) = a$, $t(\sigma_b) = b$,

$$
{}_{\sigma_a}D_{t(\sigma)}^\alpha q(t(\sigma))
$$
$$
= \frac{1}{\Gamma(n - \alpha)} \left(\frac{d}{dt(\sigma)}\right)^n \int_{\frac{a}{f(\lambda)}}^{\sigma f(\lambda)} (\sigma f(\lambda) - \theta)^{n-\alpha-1} q\left(\theta f^{-1}(\lambda)\right) d\theta
$$
$$
= \frac{(t_\sigma^{'})^{-\alpha}}{\Gamma(n - \alpha)} \left(\frac{d}{d\sigma}\right)^n \int_{\frac{a}{(t_\sigma^{'})^2}}^{\sigma} (\sigma - s)^{n-\alpha-1} q(s) ds
$$
$$
= (t_\sigma^{'})^{-\alpha} \, {}_{\frac{a}{(t_\sigma^{'})^2}}D_\sigma^\alpha q(\sigma),
$$

and, using the same reasoning,

$$
{}_{t(\sigma)}D_{\sigma_b}^\beta q(t(\sigma)) = (t_\sigma^{'})^{-\beta} \, {}_\sigma D_{\frac{b}{(t_\sigma^{'})^2}}^\beta q(\sigma).
$$

We have

$$\bar{I}[t(\cdot), q(t(\cdot))]$$

$$= \int_{\sigma_a}^{\sigma_b} L\left(t(\sigma), q(t(\sigma)), (t'_\sigma)^{-\alpha} \tfrac{a}{(t'_\sigma)^2} D^\alpha_\sigma q(\sigma), (t'_\sigma)^{-\beta} {}_\sigma D^\beta_{\tfrac{b}{(t'_\sigma)^2}} q(\sigma)\right) t'_\sigma d\sigma$$

$$\doteq \int_{\sigma_a}^{\sigma_b} \bar{L}_f\left(t(\sigma), q(t(\sigma)), t'_\sigma, \tfrac{a}{(t'_\sigma)^2} D^\alpha_\sigma q(t(\sigma)), {}_\sigma D^\beta_{\tfrac{b}{(t'_\sigma)^2}} q(t(\sigma))\right) d\sigma$$

$$= \int_a^b L\left(t, q(t), {}_a D^\alpha_t q(t), {}_t D^\beta_b q(t)\right) dt$$

$$= I[q(\cdot)].$$

By hypothesis, functional (2.61) is invariant under transformations (1.13), and it follows from Theorem 2.18 that

$$C_f\left(t(\sigma), q(t(\sigma)), t'_\sigma, \tfrac{a}{(t'_\sigma)^2} D^\alpha_\sigma q(t(\sigma)), {}_\sigma D^\beta_{\tfrac{b}{(t'_\sigma)^2}} q(t(\sigma))\right)$$

$$= \left(\partial_4 \bar{L}_f - \partial_5 \bar{L}_f\right) \cdot \xi + \frac{\partial}{\partial t'_\sigma} \bar{L}_f \tau \quad (2.62)$$

is a fractional conserved quantity. For $\lambda = 0$,

$$\tfrac{a}{(t'_\sigma)^2} D^\alpha_\sigma q(t(\sigma)) = {}_a D_t^{\,\alpha} q(t),$$

$$\sigma D^\beta_{\tfrac{b}{(t'_\sigma)^2}} q(t(\sigma)) = {}_t D_b^{\,\beta} q(t),$$

and we get

$$\partial_4 \bar{L}_f - \partial_5 \bar{L}_f = \partial_3 L - \partial_4 L, \quad (2.63)$$

and

$$\frac{\partial}{\partial t'_\sigma} \bar{L}_f = \partial_4 \bar{L}_f \cdot \frac{\partial}{\partial t'_\sigma}\left[\frac{(t'_\sigma)^{-\alpha}}{\Gamma(n-\alpha)}\left(\frac{d}{d\sigma}\right)^n \int_{\tfrac{a}{(t'_\sigma)^2}}^{\sigma} (\sigma - s)^{n-\alpha-1} q(s) ds\right] t'_\sigma$$

$$+ \partial_5 \bar{L}_f \cdot \frac{\partial}{\partial t'_\sigma}\left[\frac{(t'_\sigma)^{-\beta}}{\Gamma(n-\beta)}\left(-\frac{d}{d\sigma}\right)^n \int_\sigma^{\tfrac{b}{(t'_\sigma)^2}} (s - \sigma)^{n-\beta-1} q(s) ds\right] t'_\sigma + L$$

$$= \partial_4 \bar{L}_f \cdot \left[\frac{-\alpha(t'_\sigma)^{-\alpha-1}}{\Gamma(n-\alpha)}\left(\frac{d}{d\sigma}\right)^n \int_{\tfrac{a}{(t'_\sigma)^2}}^{\sigma} (\sigma - s)^{n-\alpha-1} q(s) ds\right]$$

$$+ \partial_5 \bar{L}_f \cdot \left[\frac{-\beta(t'_\sigma)^{-\beta-1}}{\Gamma(n-\beta)}\left(-\frac{d}{d\sigma}\right)^n \int_\sigma^{\tfrac{b}{(t'_\sigma)^2}} (s - \sigma)^{n-\beta-1} q(s) ds\right] + L$$

$$= -\alpha \partial_3 L \cdot {}_a D_t^\alpha q - \beta \partial_4 L \cdot {}_t D_b^\beta q + L. \quad (2.64)$$

Substituting (2.63) and (2.64) into equation (2.62), we obtain the fractional-conserved quantity (2.60). $\qquad\square$

In order to illustrate our results, we consider two problems with Lagrangians which are linear functions of the velocities.

Example 2.5. Let us consider the following fractional problem of the calculus of variations (2.51) with $n = 3$:

$$I[q(\cdot)] = I[q_1(\cdot), q_2(\cdot), q_3(\cdot)]$$

$$= \int_a^b \left(\left({}_aD_t^\alpha q_1 \right) q_2 - \left({}_aD_t^\alpha q_2 \right) q_1 - (q_1 - q_2)q_3 \right) dt \longrightarrow \min .$$

The problem is invariant under the transformations (1.13) with

$$(\tau, \xi_1, \xi_2, \xi_3) = (-ct, 0, 0, cq_3) ,$$

where c is an arbitrary constant. We obtain from our fractional Noether's theorem the following fractional-conserved quantity (2.60):

$$C_f \left(t, q, {}_aD_t^\alpha q \right) = \left[(1 - \alpha)\left(\left({}_aD_t^\alpha q_1 \right) q_2 - \left({}_aD_t^\alpha q_2 \right) q_1 \right) - (q_1 - q_2)q_3 \right] t . \tag{2.65}$$

We remark that the fractional-conserved quantity (2.65) depends only on the fractional derivatives of q_1 and q_2 if $\alpha \in (0, 1)$. For $\alpha = 1$, we obtain the classical result:

$$C(t, q) = (q_1 - q_2)\, q_3 t$$

is conserved for the problem of the calculus of variations

$$\int_a^b \left(\dot{q}_1 q_2 - \dot{q}_2 q_1 - (q_1 - q_2)q_3 \right) dt \longrightarrow \min .$$

Example 2.6. We consider now a variational functional (2.51) with $n = 4$:

$$I[q(\cdot)] = I[q_1(\cdot), q_2(\cdot), q_3(\cdot), q_4(\cdot)]$$

$$= - \int_a^b \left[\left({}_tD_b^\beta q_1 \right) q_2 + \left({}_tD_b^\beta q_3 \right) q_4 - \frac{1}{2} \left(q_4^2 - 2q_2q_3 \right) \right] dt .$$

The problem is invariant under (1.13) with

$$(\tau, \xi_1, \xi_2, \xi_3, \xi_4) = \left(\frac{2c}{3}t, cq_1, -cq_2, \frac{c}{3}q_3, -\frac{c}{3}q_4 \right) ,$$

where c is an arbitrary constant. We conclude from (2.60) that

$$C_f \left(t, q, {}_tD_b^\beta q \right) = \left[(\beta - 1) \left(\left({}_tD_b^\beta q_1 \right) q_2 + \left({}_tD_b^\beta q_3 \right) q_4 \right) \right.$$

$$\left. + \frac{1}{2} \left(q_4^2 - 2q_2q_3 \right) \right] \frac{2}{3}t + q_1q_2 + \frac{q_3q_4}{3} \tag{2.66}$$

is a fractional conserved quantity. In the particular case $\beta = 1$, the classical result follows from (2.66):

$$C(t, q) = q_1 q_2 + \frac{q_3 q_4}{3} + \frac{1}{3} \left(q_4^2 - 2 q_2 q_3 \right) t$$

is preserved along all the solutions of the Euler–Lagrange differential equations (1.11) of the problem

$$\int_a^b \dot{q}_1 q_2 + \dot{q}_3 q_4 + \frac{1}{2} \left(q_4^2 - 2 q_2 q_3 \right) dt \longrightarrow \min .$$

2.6 Multidimensional Fractional Noether's Theorem

In this section we present the Euler–Lagrange equations for fractional variational problems with multiple integrals. A fractional Noether-type theorem for conservative and non-conservative generalized physical systems is proved. We make use of partial fractional integrals and derivatives, which are a natural generalization of the corresponding one-dimensional fractional integrals and derivatives. For $(x_1, \ldots, x_n)$, $(\alpha_1, \ldots, \alpha_n)$, where $0 < \alpha_i < 1$, $i = 1, \ldots, n$ and $[a_1, b_1] \times \ldots \times [a_n, b_n]$, the partial Riemann–Liouville fractional integrals of order α_k with respect to x_k are defined by

$$_{a_k} I_{x_k}^{\alpha_k} f(x_1, \ldots, x_n)$$
$$= \frac{1}{\Gamma(\alpha_k)} \int_{a_k}^{x_k} (x_k - t_k)^{\alpha_k - 1} f(x_1, \ldots, x_{k-1}, t_k, x_{k+1}, \ldots, x_n) dt_k, \quad x_k > a_k,$$

$$_{x_k} I_{b_k}^{\alpha_k} f(x_1, \ldots, x_n)$$
$$= \frac{1}{\Gamma(\alpha_k)} \int_{x_k}^{b_k} (t_k - x_k)^{\alpha_k - 1} f(x_1, \ldots, x_{k-1}, t_k, x_{k+1}, \ldots, x_n) dt_k, \quad x_k < b_k.$$

Partial the Riemann–Liouville derivatives are defined by

$$_{a_k} D_{x_k}^{\alpha_k} f(x_1, \ldots, x_n)$$
$$= \frac{1}{\Gamma(1 - \alpha_k)} \frac{\partial}{\partial x_k} \int_{a_k}^{x_k} (x_k - t_k)^{-\alpha_k} f(x_1, \ldots, x_{k-1}, t_k, x_{k+1}, \ldots, x_n) dt_k,$$

$$_{x_k} D_{b_k}^{\alpha_k} f(x_1, \ldots, x_n)$$
$$= -\frac{1}{\Gamma(1 - \alpha_k)} \frac{\partial}{\partial x_k} \int_{x_k}^{b_k} (t_k - x_k)^{-\alpha_k} f(x_1, \ldots, x_{k-1}, t_k, x_{k+1}, \ldots, x_n) dt_k.$$

2.6.1   Multidimensional Euler–Lagrange Equations

Consider a physical system characterized by a set of functions

$$u_j(t, x_1, \ldots, x_n), \quad j = 1, \ldots, m, \tag{2.67}$$

depending on time t and the space coordinates $x_1, \ldots, x_n$. We can simplify the notation by interpreting (2.67) as a vector function $u = (u_1, \ldots, u_m)$ and writing $t = x_0$, $x = (x_0, x_1, \ldots, x_n)$, $dx = dx_0 dx_1 \cdots dx_n$. Then (2.67) becomes simply $u(x)$ and is called a (vector) field. We define the action functional in the form

$$\mathcal{J}(u) = \int_{a_0}^{b_0} dx_0 \int \ldots \int_R L(u, \nabla^\alpha u) dx_1 \cdots dx_n = \int_\Omega \mathcal{L}(u, \nabla^\alpha u) dx, \tag{2.68}$$

where $R = [a_1, b_1] \times \ldots \times [a_n, b_n]$, $\Omega = R \times [a_0, b_0]$, and ∇^α is the operator

$$\left({}_{a_0}D_{x_0}^{\alpha_0}, {}_{a_1}D_{x_1}^{\alpha_1}, \cdots, {}_{a_n}D_{x_n}^{\alpha_n} \right),$$

where $\alpha = (\alpha_0, \alpha_1, \ldots, \alpha_n)$, $0 < \alpha_i < 1$, $i = 0, \ldots, n$. The boundary conditions are specified by asking that $u(x) = \varphi(x)$ for $x \in \partial\Omega$, where $\varphi : \partial\Omega \to \mathbb{R}^m$ is a given function. The functions $L(u, \nabla^\alpha u)$ and $\mathcal{L}(u, \nabla^\alpha u)$ are called the Lagrangian and Lagrangian density of the field, respectively. Applying the principle of stationary action to (2.68) we obtain the multidimensional fractional Euler–Lagrange equations for the field.

We assume that:

(i) The set of admissible function consists of all function $u : \Omega \to \mathbb{R}^m$, such that ${}_{a_i}D_{x_i}^{\alpha_i} u_j \in C(\Omega, \mathbb{R})$, $i = 0, \ldots, n$, $j = 1, \ldots, m$, and $u|_{\partial\Omega} = \varphi$;

(ii) $\mathcal{L} \in C^1(\mathbb{R}^m \times \mathbb{R}^{m(n+1)}; \mathbb{R})$;

(iii) $x \to {}_{x_i}D_{b_i}^{\alpha_i} \frac{\partial \mathcal{L}}{\partial {}_{a_i}D_{x_i}^{\alpha_i} u_j}$, $i = 0, \ldots, n$, $j = 1, \ldots, m$ are continuous on $(a_0, b_0) \times \ldots \times (a_n, b_n)$ and integrable on Ω for all admissible functions.

Theorem 2.20. *If function u minimizes the action functional* (2.68), *then u satisfies the multidimensional fractional Euler–Lagrange differential equations:*

$$\frac{\partial \mathcal{L}}{\partial u_j} + \sum_{i=0}^{n} {}_{x_i}D_{b_i}^{\alpha_i} \frac{\partial \mathcal{L}}{\partial {}_{a_i}D_{x_i}^{\alpha_i} u_j} = 0, \quad j = 1, \ldots, m. \tag{2.69}$$

Proof.   The variations for this problem are functions $h : \Omega \to \mathbb{R}^m$ such that ${}_{a_i}D_{x_i}^{\alpha_i} h_j \in C(\Omega, \mathbb{R})$, $i = 0, \ldots, n$, $j = 1, \ldots, m$, and h vanishes on the boundary: $h(0) = 0$, $x \in \partial\Omega$. The condition that a function $u : \Omega \to \mathbb{R}^m$ be a critical point of the action $\mathcal{J}(u)$ is

$$\frac{d}{d\varepsilon} \mathcal{J}(u + \varepsilon h)\bigg|_{\varepsilon = 0} = 0$$

for all variations h with $h|_{\partial\Omega} = 0$. Differentiating under the integral sign we obtain

$$0 = \frac{d}{d\varepsilon}\mathcal{J}(u + \varepsilon h)\bigg|_{\varepsilon=0} = \int_\Omega \frac{d}{d\varepsilon}\mathcal{L}(u + \varepsilon h, \nabla^\alpha(u + \varepsilon h))\bigg|_{\varepsilon=0} dx$$

$$= \sum_{j=1}^m \int_\Omega \left(\frac{\partial\mathcal{L}}{\partial u_j} h_j + \sum_{i=0}^n \frac{\partial\mathcal{L}}{\partial_{a_i} D_{x_i}^{\alpha_i} u_j} {}_{a_i}D_{x_i}^{\alpha_i} h_j\right) dx.$$

The Fubini theorem allows us to rewrite integrals as the iterated integrals so that we can use the integration by parts formula (2.2):

$$\int_\Omega \left(\frac{\partial\mathcal{L}}{\partial u_j} + \sum_{i=0}^n {}_{x_i}D_{b_i}^{\alpha_i} \frac{\partial\mathcal{L}}{\partial_{a_i} D_{x_i}^{\alpha_i} u_j}\right) h_j dx, \quad j = 1, \ldots m$$

since $h|_{\partial\Omega} = 0$. Therefore,

$$0 = \sum_{j=1}^m \int_\Omega \left(\frac{\partial\mathcal{L}}{\partial u_j} + \sum_{i=0}^n {}_{x_i}D_{b_i}^{\alpha_i} \frac{\partial\mathcal{L}}{\partial_{a_i} D_{x_i}^{\alpha_i} u_j}\right) h_j dx.$$

It follows from the fundamental lemma of the calculus of variations that u satisfies (2.69). $\square$

In the following we present physical examples. The main aim is to illustrate what are the Euler–Lagrange equations when we have a Lagrangian density depending on fractional derivatives.

Example 2.7. Consider a minimizer of

$$\mathcal{J}(u) = \frac{1}{2} \int\int_R \sum_{i=1}^2 \left(\left({}_{a_i}D_{x_i}^{\alpha_i} u\right)^2 - fu\right) dx, \tag{2.70}$$

where $f : R \to \mathbb{R}$ is a given function, in a set of functions that satisfy condition $u = \varphi$ on ∂R, where φ is a given function defined on the boundary ∂R of $R = [a_1, b_1] \times [a_2, b_2]$. By Theorem 2.20 a minimizer of (2.70) satisfies the following equation:

$$\sum_{i=1}^2 {}_{x_i}D_{b_i}^{\alpha_i} {}_{a_i}D_{x_i}^{\alpha_i} u = f. \tag{2.71}$$

Observe that if α_i goes to 1, then the operator ${}_{a_i}D_{x_i}^{\alpha_i}$, $i = 1, 2$, can be replaced with $\frac{\partial}{\partial x_i}$, and the operator ${}_{x_i}D_{b_i}^{\alpha_i}$, $i = 1, 2$, can be replaced with $-\frac{\partial}{\partial x_i}$ (Podlubny, 1999). Thus, for $\alpha \to 1$ equation (2.71) becomes the Poisson equation that arises in the potential theory of electrostatic fields.

Example 2.8. Consider a motion of a medium whose displacement may be described by a scalar function $u(t, x)$, where $x = (x_1, x_2)$. For example, this function might represent the transverse displacement of a membrane. Suppose that the kinetic energy T and potential energy V of the medium are given by

$$T\left(\frac{\partial u}{\partial t}\right) = \frac{1}{2} \int \rho \left(\frac{\partial u}{\partial t}\right)^2 dx, \quad V(u) = \frac{1}{2} \int k\|\nabla u\|^2 dx,$$

where $\rho(x)$ is the mass density and $k(x)$ is the stiffness, both assume positive. Then, the classical action functional is

$$\mathcal{J}(u) = \frac{1}{2} \int \int_\Omega \left(\rho \left(\frac{\partial u}{\partial t}\right)^2 - k\|\nabla u\|^2\right) dxdt.$$

When we have the Lagrangian with the kinetic term depending on a fractional derivative, then the fractional action functional has the form

$$\mathcal{J}(u) = \frac{1}{2} \int \int_\Omega \left(\rho \left({}_0D_t^{\alpha_0} u\right)^2 - k\|\nabla u\|^2\right) dxdt. \qquad (2.72)$$

The fractional Euler–Lagrange equation satisfied by a stationary point of (2.72) is

$$\rho {}_tD_{t_1}^{\alpha_0} {}_0D_t^{\alpha_0} u - \nabla(k\nabla u) = 0.$$

If ρ and k are constants, then equation ${}_tD_{t_1}^{\alpha_0} {}_0D_t^{\alpha_0} u - c^2 \Delta u = 0$ where $c^2 = k/\rho$ can be called the time-fractional wave equation. In the case when the kinetic and potential energy depend on fractional derivatives, the action functional for the system has the form

$$\mathcal{J}(u) = \frac{1}{2} \int \int_\Omega \left[\rho \left({}_0D_t^{\alpha_0} u\right)^2 - k(({}_{a_1}D_{x_1}^{\alpha_1} u)^2 + ({}_{a_2}D_{x_2}^{\alpha_2} u)^2)\right] dxdt. \qquad (2.73)$$

The fractional Euler–Lagrange equation satisfied by a stationary point of (2.73) is

$$\rho {}_tD_{t_1}^{\alpha_0} {}_0D_t^{\alpha_0} u - k({}_{x_1}D_{b_1}^{\alpha_1} {}_{a_1}D_{x_1}^{\alpha_1} u + {}_{x_2}D_{b_2}^{\alpha_2} {}_{a_2}D_{x_2}^{\alpha_2} u) = 0.$$

If ρ and k are constants, then

$${}_tD_{t_1}^{\alpha_0} {}_0D_t^{\alpha_0} u - c^2({}_{x_1}D_{b_1}^{\alpha_1} {}_{a_1}D_{x_1}^{\alpha_1} u + {}_{x_2}D_{b_2}^{\alpha_2} {}_{a_2}D_{x_2}^{\alpha_2} u) = 0$$

can be called the space- and time-fractional wave equation. Observe that in the limit, $\alpha_0, \alpha_1, \alpha_2 \to 1$, we obtain the classical wave equation

$$\frac{\partial^2 u}{\partial t^2} - c^2 \Delta u = 0$$

with wave-speed c.

Remark 2.24. Many classical partial differential equations possess fractional analogues (see, e.g., Momani, 2006; Yakubovich, 2010), which are obtained by changing the classical partial derivatives by fractional ones. Usually, these generalizations are supported by physical arguments. On the other hand, one can start with a variational formulation of a physical process in which one modifies the Lagrangian density by replacing integer order derivatives with fractional ones. Then the action integral in the sense of Hamilton is minimized and the governing equation of a physical process is obtained. Here we use the second approach and derive the wave equation and the Poisson equation via fractional calculus of variations. It is worth pointing out that although functionals have been written only in terms of the left fractional partial derivatives, the necessary conditions contain also the right fractional partial derivatives. As we know, this is because of the integration by parts formula. For a gentle explanation via the notion of fractional embedding we refer the reader to Cresson, 2007, 2010.

2.6.2 *Fractional Noether-Type Theorem*

In this section we prove the fractional Noether-type theorem. We start by introducing the notion of variational invariance.

Definition 2.13. *Functional* (2.68) *is said to be invariant under an ε-parameter group of infinitesimal transformations* $\bar{u}(x) = u + \varepsilon\xi(x, u) + o(\varepsilon)$ *if*

$$\int_{\Omega^*} \mathcal{L}(u, \nabla^\alpha u)dx = \int_{\Omega^*} \mathcal{L}(\bar{u}, \nabla^\alpha \bar{u})dx \tag{2.74}$$

for any $\Omega^* \subseteq \Omega$.

The next theorem establishes a necessary condition of invariance.

Theorem 2.21. *If functional* (2.68) *is invariant in the sense of Definition 2.13, then*

$$\sum_{j=1}^{m} \frac{\partial \mathcal{L}}{\partial u_j}\xi_j + \sum_{j=1}^{m}\sum_{i=0}^{n} \frac{\partial \mathcal{L}}{\partial_{a_i} D_{x_i}^{\alpha_i} u_j} a_i D_{x_i}^{\alpha_i}\xi_j = 0. \tag{2.75}$$

Proof. Equation (2.74) is equivalent to

$$\mathcal{L}(u, \nabla^\alpha u) = \mathcal{L}(u + \varepsilon\xi(x, u) + o(\varepsilon), \nabla^\alpha(u + \varepsilon\xi(x, u) + o(\varepsilon))). \tag{2.76}$$

Differentiating both sides of (2.76) with respect to ε, then substituting $\varepsilon = 0$, and using the definitions and properties of the fractional partial derivatives, we obtain equation (2.75). $\qquad\square$

The following definition is similar to the one of Section 2.5 and is useful in order to formulate the fractional Noether-type theorem.

Definition 2.14. *For $0 < \gamma < 1$ and a given ordered pair of functions f and g $(f, g : \mathbb{R}^{n+1} \to \mathbb{R})$ we introduce the following operator:*

$$\mathcal{D}^\gamma (f, g) = f \cdot {}_{a_i}D_{x_i}^\gamma g - g \cdot {}_{x_i}D_{b_i}^\gamma f,$$

where $x_i \in [a_i, b_i]$, $i = 0, \ldots, n$.

We now prove the fractional Noether-type theorem.

Theorem 2.22. *If functional (2.68) is invariant in the sense of Definition 2.13, then*

$$\sum_{i=0}^n \sum_{j=1}^m \mathcal{D}^{\alpha_i} \left(\frac{\partial \mathcal{L}}{\partial_{a_i}D_{x_i}^{\alpha_i} u_j}, \xi_j \right) = 0 \tag{2.77}$$

along any solution of (2.69).

Proof. Using the fractional Euler–Lagrange equations (2.69), we have

$$\sum_{j=1}^m \frac{\partial \mathcal{L}}{\partial u_j} = -\sum_{j=1}^m \sum_{i=0}^n {}_{x_i}D_{b_i}^{\alpha_i} \frac{\partial \mathcal{L}}{\partial_{a_i}D_{x_i}^{\alpha_i} u_j}. \tag{2.78}$$

Substituting (2.78) in the necessary condition of invariance (2.75) we get

$$
\begin{aligned}
0 &= -\sum_{j=1}^m \sum_{i=0}^n {}_{x_i}D_{b_i}^{\alpha_i} \frac{\partial \mathcal{L}}{\partial_{a_i}D_{x_i}^{\alpha_i} u_j} \xi_j + \sum_{j=1}^m \sum_{i=0}^n \frac{\partial \mathcal{L}}{\partial_{a_i}D_{x_i}^{\alpha_i} u_j} {}_{a_i}D_{x_i}^{\alpha_i} \xi_j \\
&= \sum_{i=0}^n \left(\sum_{j=1}^m \frac{\partial \mathcal{L}}{\partial_{a_i}D_{x_i}^{\alpha_i} u_j} {}_{a_i}D_{x_i}^{\alpha_i} \xi_j - \sum_{j=1}^m {}_{x_i}D_{b_i}^{\alpha_i} \frac{\partial \mathcal{L}}{\partial_{a_i}D_{x_i}^{\alpha_i} u_j} \xi_j \right).
\end{aligned}
$$

By definition of operator $\mathcal{D}^\gamma$ we obtain equation (2.77). $\square$

Remark 2.25. In the particular case when α_i goes to 1, $i = 0, \ldots, n$, we get from the fractional conservation law (2.77) the classical Noether conservation law without transformation of the independent variables (that is, conservation of current):

$$\sum_{i=0}^n \frac{\partial}{\partial x_i} \sum_{j=1}^m \frac{\partial \mathcal{L}}{\partial \left(\frac{\partial u_j}{\partial x_i} \right)} \xi_j = 0 \tag{2.79}$$

along any extremal surface of

$$\mathcal{J}(u) = \int_\Omega \mathcal{L}(u, \nabla u) dx.$$

Example 2.9. Let us consider again the Lagrangian given in Example 2.8: $L = T - V$. Then the Hamiltonian, $H = T + V$, is conserved (see Goldstein, 1951). However, if the kinetic and potential energy depend on fractional derivatives, that is, the action functional has the form (2.73), then it can be shown that the Hamiltonian is not conserved (cf. Klimek, 2002; Riewe, 1996). Observe that functional (2.73) is invariant (in the sense of Definition 2.13) under transformations $\bar{u}(t, x) = u(t, x) + \varepsilon t^{\alpha_0 - 1}(x_1 - a_1)^{\alpha_1 - 1}(x_2 - a_2)^{\alpha_2 - 1}$. Therefore, by Theorem 2.22, we obtain:

$$\sum_{i=0}^{2} \mathcal{D}^{\alpha_i} \left(\frac{\partial \mathcal{L}}{\partial_{a_i} D_{x_i}^{\alpha_i} u}, x_0^{\alpha_0 - 1}(x_1 - a_1)^{\alpha_1 - 1}(x_2 - a_2)^{\alpha_2 - 1} \right) = 0, \qquad (2.80)$$

with $x_0 = t$ and $a_0 = 0$, along any extremal surface of (2.73). Note that when α_i goes to 1, $i = 0, 1, 2$, expression (2.80) has the form (2.79). This shows that the proved fractional Noether-type theorem can be applied for conservative and non-conservative physical systems.

2.7 Modified Optimal Energy and Initial Memory

In this section we study fractional-order systems using a vector state space representation, where we also include initialization. Our systems are described by initialized Riemann–Liouville derivatives. The initial condition problem for fractional linear systems is a subject under strong consideration. Proper initialization is crucial to the solution and understanding of fractional differential equations. An important tool in the solution of fractional differential or differintegral equations is provided by the Laplace transform. We consider the problem of solving initialized fractional differential systems using the Laplace transform. This can be interpreted as the initial memory of the system. Then we develop the idea by looking at a given system not in the perspective of the state but also taking into account memory, that can be of a different fractional order than the order of the system. We obtain explicit steering laws with respect to the values of the fractional integrals of the state variables. The classical controllability Gramian is generalized, and steering functions between memory values characterized.

2.7.1 *Linear Systems and the Controllability Gramian*

We consider an initialized fractional linear time-invariant control system

$$_0D_t^\alpha x(t) = Ax(t) + Bu(t) , \quad t \geq 0 , \qquad (\Sigma_{\alpha,\psi})$$

where $x(t) \in \mathbb{R}^n$, $u(t) \in \mathbb{R}^m$, and $\alpha \in (0,1]$ and matrices $A \in \mathbb{R}^{n \times n}$ and $B \in \mathbb{R}^{n \times m}$, together with an initialization vector $\psi(t)$, $t \leq 0$, are given. We consider piecewise constant controls $u(\cdot)$. The derivative notation is

$$_0D_t^\alpha x(t) = {_0d_t^\alpha} x(t) + \psi(t) ,$$

where $d_t^\alpha x(t)$ is the uninitialized fractional derivative starting at $t = 0$ and ψ is the initialization vector function, determined as

$$\psi(t) = \lim_{\omega \to -\infty} \frac{d}{dt} \left(\frac{1}{\Gamma(1-\alpha)} \int_\omega^0 \frac{x(\tau)}{(t-\tau)^\alpha} d\tau \right) .$$

We are involving the history from the left axis. In the particular case when $x(t) = a$ for $-\infty < t \leq 0$, we obtain the following:

$$\psi(t) = \lim_{\omega \to -\infty} \frac{d}{dt} \left(\frac{1}{\Gamma(1-\alpha)} \int_\omega^0 \frac{a}{(t-\tau)^\alpha} d\tau \right) = -\frac{at^{-\alpha}}{\Gamma(1-\alpha)} .$$

Proposition 2.7. *Let $0 < \alpha \leq 1$. The forward trajectory of the system $\Sigma_{\alpha,\psi}$ evaluated at $t > 0$ has the following form:*

$$\gamma(t, \psi, u) = \int_0^t S(t-\tau) \left(Bu(\tau) - \psi(\tau) \right) d\tau ,$$

where $S(t) = e_\alpha^{At}$. Moreover, for $x(t) = a$, $t \in (-\infty, 0]$, we have:

$$\gamma(t, a, u) = E_\alpha(At^\alpha)a + \int_0^t S(t-\tau)Bu(\tau)d\tau . \qquad (2.81)$$

Proof. We prove here the second part. We have

$$X(s) = (Is^\alpha - A)^{-1}BU(s) - (Is^\alpha - A)^{-1}\Psi(s) .$$

Applying the inverse Laplace transform for

$$-(Is^\alpha - A)^{-1}\Psi(s) = (Is^\alpha - A)^{-1}s^{\alpha-1}a$$

we get

$$\mathcal{L}^{-1}\left[-(Is^\alpha - A)^{-1}\Psi(s) \right](t) = E_\alpha(At^\alpha)a .$$

Hence the formula (2.81) holds. □

We introduce now the more general situation when one has some external measure of the memory. This is represented by the fractional integral of order β.

Definition 2.15. *Let $\beta \geq 1 - \alpha$. The memory of order β of the forward trajectory $\gamma(t, \psi, u)$ evaluated at time $t > 0$ is defined by*
$$M_\beta(t, \psi, u) := I_{0+}^\beta \gamma(t, \psi, u).$$
Additionally, we define the value of the memory of order β at $t = 0$ as $M_\beta(0, \psi, u) := \lim_{t \to 0+} I_{0+}^\beta \gamma(t, \psi, u)$.

Proposition 2.8. *Let $\beta \geq 1 - \alpha$. Then for the system $\Sigma_{\alpha, \psi}$ we have*
$$M_\beta(t, \psi, u) = \int_0^t \Phi_\beta(t - \tau)\left(Bu(\tau) - \psi(\tau)\right) d\tau \text{ for } t \geq 0,$$
where
$$\Phi_\beta(t) = t^{\alpha+\beta-1} E_{\alpha, \alpha+\beta}(At^\alpha) = t^{\alpha+\beta-1} \sum_{k=0}^\infty A^k \frac{t^{k\alpha}}{\Gamma((k+1)\alpha + \beta)}.$$
Moreover, for $\psi(t) = a$ for $t \in (-\infty, 0]$ we have
$$M_\beta(t, \psi = a, u) = t^\beta E_{\alpha, \beta}(At^\alpha) a + \int_0^t \Phi_\beta(t - \tau) Bu(\tau) d\tau.$$

Proof. Using the Laplace transformation we get
$$\mathcal{L}\left[M_\beta(t, \psi, u)\right](s) = \frac{X(s)}{s^\beta}$$
$$= -\sum_{k=0}^\infty A^k s^{-(k+1)\alpha-\beta} \Psi(s) + \sum_{k=0}^\infty s^{-(k+1)\alpha-\beta} A^k BU(s)).$$
The desired formula follows from the fact that
$$\Phi_\beta(t) = \mathcal{L}^{-1}\left[\sum_{k=0}^\infty A^k s^{-(k+1)\alpha-\beta}\right](t)$$
$$= t^{\alpha+\beta-1} \sum_{k=0}^\infty A^k \frac{t^{k\alpha}}{\Gamma((k+1)\alpha + \beta)}$$
$$= t^{\alpha+\beta-1} E_{\alpha, \alpha+\beta}(At^\alpha).$$
Moreover, for $\Psi(s) = -s^{\alpha-1} a = \mathcal{L}\left[-\frac{t^{-\alpha}}{\Gamma(1-\alpha)}\right](s)$ we arrive at
$$\mathcal{L}^{-1}\left[-\sum_{k=0}^\infty A^k s^{-(k+1)\alpha-\beta} \Psi(s)\right](t)$$
$$= \mathcal{L}^{-1}\left[\sum_{k=0}^\infty A^k s^{-(k+1)\alpha-\beta} a\right](t) = t^\beta \sum_{k=0}^\infty A^k \frac{t^{k\alpha}}{\Gamma(k\alpha + \beta)} a$$
$$= t^\beta E_{\alpha, \beta}(At^\alpha). \qquad \square$$

Definition 2.16. *Let $T > 0$. The system $\Sigma_{\alpha,\psi}$ is controllable with memory of order $\beta \geq 1 - \alpha$ on $[0,T]$ if there exists a control u defined on $[0,T]$ such that $M_\beta(t, \psi, u)_{|t=T} = b$.*

We denote by

$$Q_T = \int_0^T (T - t)^{2(1-\alpha-\beta)} \Phi_\beta(T - t) BB^* \Phi_\beta^*(T - t) dt$$

the β-controllability Gramian on the time interval $[0,T]$ corresponding to the system $\Sigma_{\alpha,\psi}$. As in the classical case, Q_T is symmetric and non-negative definite.

Theorem 2.23. *Let $T > 0$ and Q_T be nonsingular. Then,*

(i) for $b \in \mathbb{R}^n$ the control law

$$\overline{u}(t) = -(T - t)^{2(1-\alpha-\beta)} B^* \Phi_\beta^*(T - t) Q_T^{-1} f_T(\psi, b), \qquad (2.82)$$

where $f_T(\psi, b) = -b - \int_0^T \Phi_\beta(T - \tau)\psi(\tau)d\tau$, drives the system to b in time T.

(ii) Among all possible controls from $L_\alpha^2\left([0,T], \mathbb{R}^m\right)$ driving the system to b in time T, the control $\overline{u}$ defined by (2.82) minimizes the modified energy integral

$$\mathcal{E}^{\alpha,\beta}(u) := \int_0^T \left|(T - t)^{\alpha+\beta-1}u(t)\right|^2 dt.$$

Moreover,

$$\mathcal{E}^{\alpha,\beta}(\overline{u}) = \int_0^T |(T - t)^{\alpha+\beta-1}\overline{u}(t)|^2 dt = < Q_T^{-1} f_T(\psi, b), f_T(\psi, b) >,$$

where $< \cdot, \cdot >$ denotes the inner product, and

$$f_T(\psi, b) = -b - \int_0^T \Phi_\beta(T - \tau)\psi(\tau)d\tau.$$

Proof. From the form of $\overline{u}$ we directly have that

$$M_\beta(T, \psi, \overline{u}) = f_T(\psi, b) + b$$

$$- \left(\int_0^T (T - t)^{2(1-\alpha-\beta)} \Phi_\beta(T - t) BB^* \Phi_\beta^*(T - t) dt\right) Q_T^{-1} f_T(\psi, b)$$

$$= f_T(\psi, b) + b - Q_T Q_T^{-1} f_T(\psi, b)$$

$$= b.$$

Item (i) is proved. Similarly to the classical situation, to prove item (ii) we begin by noticing that

$$\int_0^T \left| (T-t)^{\alpha+\beta-1} \overline{u}(t) \right|^2 dt$$

$$= \int_0^T \left| (T-t)^{1-\alpha-\beta} B^* \Phi_\beta^* (T-t) Q_T^{-1} f_T(\psi, b) \right|^2 dt$$

$$= \int_0^T |T-t|^{2(1-\alpha-\beta)}$$

$$\times \left\langle B^* \Phi_\beta^* (t-t) Q_T^{-1} f_T(\psi, b), B^* \Phi_\beta^* (T-t) Q_T^{-1} f_T(\psi, b) \right\rangle dt$$

$$= \left\langle \int_0^T |T-t|^{2(1-\alpha-\beta)} \Phi_\beta (T-t) B B^* \Phi_\beta^* (T-t) dt, Q_T^{-1} f_T(\psi, b) \right\rangle$$

$$= \left\langle Q_T Q_T^{-1} f_T(\psi, b), Q_T^{-1} f_T(\psi, b) \right\rangle$$

$$= \left\langle f_T(\psi, b), Q_T^{-1} f_T(\psi, b) \right\rangle .$$

Let us take another control u for which $(T-t)^{\alpha+\beta-1} u(t)$ is square integrable on $[0, T]$ and $M_\beta(T, \psi, u) = b$. Then,

$$\int_0^T (T-t)^{2(\alpha+\beta-1)} \left\langle u(t), \overline{u}(t) \right\rangle dt$$

$$= -\int_0^T (T-t)^{2(\alpha+\beta-1)}$$

$$\times \left\langle u(t), (T-t)^{2(1-\alpha-\beta)} B^* \Phi_\beta^* (T-t) Q_T^{-1} f_T(\psi, b) \right\rangle dt$$

$$= -\int_0^T \left\langle u(t), B^* \Phi_\beta^* (T-t) Q_T^{-1} f_T(\psi, b) \right\rangle dt$$

$$= \left\langle f_T(\psi, b), Q_T^{-1} f_T(\psi, b) \right\rangle .$$

Hence,

$$\int_0^T (T-t)^{2(1-\alpha-\beta)} \left\langle u(t), \overline{u}(t) \right\rangle dt = \int_0^T (T-t)^{2(1-\alpha-\beta)} \left\langle \overline{u}(t), \overline{u}(t) \right\rangle dt$$

and we obtain

$$\int_0^T (T-t)^{2(1-\alpha-\beta)} |u(t)|^2 dt$$

$$= \int_0^T (T-t)^{2(1-\alpha-\beta)} |\overline{u}(t)|^2 dt + \int_0^T (T-t)^{2(1-\alpha-\beta)} |u(t) - \overline{u}(t)|^2 dt ,$$

which gives the intended minimality property for the integral. $\qquad \square$

Example 2.10. Let $\Sigma_{\alpha=0.5,\psi}$ be the following system in $\mathbb{R}^2$:

$$\begin{cases} D_{0+}^{0.5} x_1(t) = x_2(t), \\ D_{0+}^{0.5} x_2(t) = u(t). \end{cases}$$

Let us take $b = \begin{pmatrix} 0 \\ 0 \end{pmatrix}$. Since $A = \begin{pmatrix} 0 & 1 \\ 0 & 0 \end{pmatrix}$, $B = \begin{pmatrix} 0 \\ 1 \end{pmatrix}$, and $A^2 = \mathbf{0}$, we obtain for $t_0 = 0$ the formula for the solution with the initialization ψ in the following form:

$$\gamma(t, \psi, u) = \int_0^t \begin{pmatrix} 1 \\ \frac{1}{\sqrt{\pi(t-\tau)}} \end{pmatrix} (u(\tau) - \psi(\tau)) \, d\tau.$$

Let $\beta \geq 0.5$. Then,

$$\Phi_\beta(t) = \begin{pmatrix} \frac{t^{\beta-0.5}}{\Gamma(0.5+\beta)} & \frac{t^\beta}{\Gamma(1+\beta)} \\ 0 & \frac{t^{\beta-0.5}}{\Gamma(0.5+\beta)} \end{pmatrix}.$$

For simplicity, let us take $\beta = 0.5$. Then,

$$\Phi_{\beta=0.5}(t) = \begin{pmatrix} 1 & \frac{2\sqrt{t}}{\sqrt{\pi}} \\ 0 & 1 \end{pmatrix}$$

and the Gramian is the following:

$$Q_{T,\beta=0.5} = \begin{pmatrix} \frac{2T^2}{\pi} & \frac{4T^{1.5}}{3\sqrt{\pi}} \\ \frac{4T^{1.5}}{3\sqrt{\pi}} & T \end{pmatrix}.$$

Thus, $\bar{u}(t) = \left[\frac{\sqrt{T-t}}{\Gamma(3/2)}, 1 \right] Q_T^{-1} f_T(\psi, b)$. Let us take the end-point b equal to the zero vector and $\psi_2(t) \equiv 0$. Then, $\bar{u}(t) = \left(\frac{6\sqrt{\pi}}{T^{3/2} - \frac{9\sqrt{\pi}\sqrt{T-t}}{T^2}} \right) \int_0^T \psi_1(\tau) d\tau$, and the norm

$$\varepsilon^{0.5,0.5}(\bar{u}) = \left(\int_0^T \psi_1(\tau) d\tau \right)^2 \int_0^T \left(\frac{6\sqrt{\pi}}{T^{3/2}} - \frac{9\sqrt{\pi}\sqrt{T-t}}{T^2} \right) dt$$

with $\bar{u}(t) = \frac{9\pi}{2T^2} \left(\int_0^T \psi_1(\tau) d\tau \right)^2$.

2.7.2   Steering Laws

In the classical theory it is quite easy to explain why the rank condition of matrix B equal to n is sufficient for controllability. In the case of fractional order systems we need to use a special function.

Proposition 2.9. *Let* $rank\, B = n$, *matrix* B^+ *be such that* $BB^+ = I$, *and* $g_\beta(\cdot)$ *be the matrix function such that* $\Phi_\beta(t)g_\beta(t) = I$, $t > 0$, *and* $\lim\limits_{t \to 0+} \Phi_\beta(t)g_\beta(t) = I$. *Then the control*

$$\widehat{u}(t) = \frac{\Gamma(\alpha + \beta)}{T} B^+ g(T - t) \left(b + \int_0^T \Phi_\beta(T - \tau)\psi(\tau)d\tau \right), \quad t \in [0, T],$$

transfers system $\Sigma_{\alpha,\psi}$ *to* b *in time* $T > 0$.

Proof. It follows by direct calculation:

$$\gamma(T, \psi, \widehat{u}) = -\int_0^T \Phi_\beta(T - \tau)\psi(\tau)d\tau$$

$$+\frac{1}{T}\int_0^T \Phi_\beta(T - t)BB^+ g_\beta(T - t) \left(b + \int_0^T \Phi_\beta(T - \tau)\psi(\tau)d\tau \right) dt = b. \qquad \square$$

Example 2.11. Let us consider the system

$$\Sigma_{\alpha,\psi} : \quad D_{0+}^\alpha x(t) = u(t),$$

where $D_{0+}^\alpha x(t) = {}_0 D_t^\alpha x(t) + \psi(t)$. Let $b \in \mathbb{R}$, $T > 0$. Then,

$$\Phi_\beta(t) = \frac{t^{\alpha+\beta-1}}{\Gamma(\alpha + \beta)}, \quad B^+ = B = 1,$$

$$M(T, \psi, u) = \frac{1}{\Gamma(\alpha + \beta)}\int_0^T (T - t)^{\alpha+\beta-1}\left(u(t) - \psi(t) \right) dt.$$

The steering law

$$\widehat{u}(t) = \frac{\Gamma(\alpha + \beta)}{T}(T - t)^{1-\alpha-\beta}\left(b + \int_0^T \Phi_\beta(T - \tau)\psi(\tau)d\tau \right)$$

transfers the system to b.

If the rank condition is satisfied, the control $\overline{u}$ given by (2.82) drives the system to b at time T. Our goal now is to find another formula for the steering control by using the matrix $[A|B]$ instead of the controllability matrix Q_T. It is a classical result that if rank $[A|B] = n$, then there exists a matrix $K \in M(mn, n)$ such that $[A|B]\, K = I \in M(n, n)$ or, equivalently, there are matrices $K_1, K_2, \ldots K_n \in M(m, n)$ such that

$$BK_1 + ABK_2 + \cdots + A^{n-1}BK_n = I.$$

Let $R_{0+}^{\alpha,0}\mu(t) = \mu(t)$. We define, recursively, $R_{0+}^{\alpha,j+1}\mu(t) := {}_0 D_t^\alpha\left(R_{0+}^{\alpha,j}\mu(t) \right)$, $j \in \mathbb{N}$.

Theorem 2.24. *Let* $\beta \geq 0$ *and rank* $[A|B] = rank\, \left[B, AB, \ldots, A^{n-1}B \right] = n$ *for system* $\Sigma_{\alpha,\psi}$. *Let* p *be such that* $\Phi_\beta(T - t) \in {}_t I_T^\alpha(L_p)$ *and let* φ *be a real function given on* $[0, T]$ *such that* *(i)* $\int_0^T \varphi(t)dt = 1$;

(ii) $R_{0+}^{\alpha,j}\mu(t) \in {}_0I_t^\alpha(L_q)$ *for* $j = 0,\ldots,n-1$, *where*

$$\mu(t) = g_\beta(t)\left(b + \int_0^T \Phi_\beta(T-\tau)\psi(\tau)d\tau\right)\varphi(t),$$

and $\Phi_\beta(T-t)g_\beta(t) = I_n$ *for* $t \in [0,T]$, *while* $1/p + 1/q \le 1 + \alpha$.

Then, the control

$$\hat{u}(t) = K_1\,\mu(t) + K_2\,{}_0D_t^\alpha\mu(t) + \cdots + K_n\,R_{0+}^{\alpha,n-1}\mu(t), \quad t \in [0,T],$$

transfers the system $\Sigma_{\alpha,\psi}$ *to the value* b *at time* $T > 0$.

Proof.   Using $j-1$ times the formula (2.2) of integration by parts and Lemma 2.1, we get for $j = 1,\ldots,n$ that

$$\int_0^T \Phi_\beta(T-t)BK_j R_{0+}^{\alpha,j-1}\mu(t)dt = \int_0^T \Phi_\beta(T-t)A^{j-1}BK_j\mu(t)dt.$$

Then,

$$\int_0^T \Phi_\beta(T-t)B\hat{u}(t)dt = \int_0^T \Phi_\beta(T-t)B\mu(t)dt$$

and finally

$$\gamma(T,\psi,\hat{u}) = -\int_0^T \Phi_\beta(T-\tau)\psi(\tau)d\tau$$

$$+ \int_0^T \Phi_\beta g_\beta(t)\left(b + \int_0^T \Phi_\beta(T-\tau)\psi(\tau)d\tau\right)\varphi(t)dt = b. \qquad \square$$

2.8   Fractional Conservation Laws in Optimal Control

In this section, we prove a Noether-like theorem in the more general context of the fractional optimal control. As a corollary, it follows that in the fractional case the autonomous Hamiltonian no longer defines a conservation law. Instead, it is proved that the fractional conservation law adds to the Hamiltonian a new term which depends on the fractional-order of differentiation, the generalized momentum, and the fractional derivative of the state variable.

Let us consider the following fractional problem of the calculus of variations: to find function $q(\cdot)$ that minimizes the integral functional

$$I[q(\cdot)] = \int_a^b L\left(t, q(t), {}_aD_t^\alpha q(t)\right)dt, \tag{2.83}$$

where the Lagrangian $L : [a, b] \times \mathbb{R}^n \times \mathbb{R}^n \to \mathbb{R}$ is a C^2 function with respect to all its arguments, and $0 < \alpha < 1$. Observe that, functional (2.83) is a particular case of functional (2.51). Also, for the convenience of the reader, we state the necessary facts from Section 2.5 in the context of functional (2.83).

Definition 2.17 (cf. Definition 2.11). *We say that $C_f (t, q, {}_aD_t^\alpha q)$ is a fractional conservation law if, and only if, it is possible to write C_f in the form of a sum of products,*

$$C_f (t, q, d) = \sum_{i=1}^{r} C_i^1 (t, q, d) \cdot C_i^2 (t, q, d) \tag{2.84}$$

for some $r \in \mathbb{N}$, and for each $i = 1, \ldots, r$ the pair C_i^1 and C_i^2 satisfy one of the following relations:

$$\mathcal{D}_t^\alpha \left(C_i^1 (t, q, {}_aD_t^\alpha q), C_i^2 (t, q, {}_aD_t^\alpha q) \right) = 0 \tag{2.85}$$

or

$$\mathcal{D}_t^\alpha \left(C_i^2 (t, q, {}_aD_t^\alpha q), C_i^1 (t, q, {}_aD_t^\alpha q) \right) = 0 \tag{2.86}$$

along all the fractional Euler–Lagrange extremals, i.e., along all the solutions of the fractional Euler–Lagrange equations

$$\partial_2 L (t, q, {}_aD_t^\alpha q) + {}_tD_b^\alpha \partial_3 L (t, q, {}_aD_t^\alpha q) = 0. \tag{2.87}$$

We then write $\mathcal{D} \{ C_f (t, q, {}_aD_t^\alpha q) \} = 0$.

Remark 2.26. For $\alpha = 1$ (2.85) and (2.86) coincide, and

$$\mathcal{D} \{ C (t, q, {}_aD_t^\alpha q) \} = 0$$

is reduced to

$$\frac{d}{dt} \{ C (t, q(t), \dot{q}(t)) \} = 0 \Leftrightarrow C (t, q(t), \dot{q}(t)) \equiv \text{constant},$$

which is the standard meaning of a conservation law, i.e., a function $C (t, q, \dot{q})$ preserved along all the Euler–Lagrange extremals $q(t)$, $t \in [a, b]$, of the problem. This implies that if $(p(t), q(t))$ is a solution to the classical Hamilton–Jacobi equations of motion, then C defines a conservation law of the Hamiltonian equations with Hamiltonian H if $\{H, C\} = 0$ or $\{C, H\} = 0$, where $\{\cdot, \cdot\}$ denotes the canonical Poisson bracket operator.

Definition 2.18 (cf. Definition 2.12). *Functional* (2.83) *is said to be invariant under the one-parameter group of infinitesimal transformations*

$$\begin{cases} \bar{t} = t + \varepsilon\tau(t,q) + o(\varepsilon)\,, \\ \bar{q}(\bar{t}) = q(t) + \varepsilon\xi(t,q) + o(\varepsilon)\,, \end{cases} \tag{2.88}$$

if, and only if,

$$\int_{t_a}^{t_b} L\left(t, q(t), {}_{t_a}D_t^\alpha q(t)\right) dt = \int_{\bar{t}(t_a)}^{\bar{t}(t_b)} L\left(\bar{t}, \bar{q}(\bar{t}), {}_{\bar{t}_a}D_{\bar{t}}^\alpha \bar{q}(\bar{t})\right) d\bar{t} \tag{2.89}$$

for any subinterval $[t_a, t_b] \subseteq [a, b]$.

Remark 2.27. Having in mind that condition (2.89) is to be satisfied for any subinterval $[t_a, t_b] \subseteq [a, b]$, we can remove the integral signs in (2.89). This is done in the new Definition 2.21.

The next theorem provides an extension of the classical Noether's theorem to fractional problems of the calculus of variations.

Theorem 2.25 (cf. Theorem 2.19). *If functional* (2.83) *is invariant under* (2.88), *then*

$$\left[L\left(t, q, {}_aD_t^\alpha q\right) - \alpha\partial_3 L\left(t, q, {}_aD_t^\alpha q\right) \cdot {}_aD_t^\alpha q\right]\tau(t,q)$$
$$+ \partial_3 L\left(t, q, {}_aD_t^\alpha q\right) \cdot \xi(t,q)$$

is a fractional conservation law (cf. Definition 2.17).

Using Theorem 2.25, we obtain here a Noether's theorem for the following fractional optimal control problem:

$$I[q(\cdot), u(\cdot)] = \int_a^b L\left(t, q(t), u(t)\right) dt \longrightarrow \min\,, \tag{2.90}$$
$$ {}_aD_t^\alpha q(t) = \varphi\left(t, q(t), u(t)\right)\,,$$

together with the initial condition $q(a) = q_a$. In problem (2.90), the Lagrangian $L : [a,b] \times \mathbb{R}^n \times \mathbb{R}^m \to \mathbb{R}$ and the velocity vector $\varphi : [a,b] \times \mathbb{R}^n \times \mathbb{R}^m \to \mathbb{R}^n$ are assumed to be C^1 functions with respect to all the arguments. In agreement with the calculus of variations, we also assume that the admissible control functions take values on an open set of $\mathbb{R}^m$.

Definition 2.19. *A pair* $(q(\cdot), u(\cdot))$ *satisfying the fractional control system* ${}_aD_t^\alpha q(t) = \varphi\left(t, q(t), u(t)\right)$ *of problem* (2.90), $t \in [a,b]$, *is called a process.*

Theorem 2.26. *If* $(q(\cdot), u(\cdot))$ *is an optimal process for problem* (2.90), *then there exists a co-vector function* $p(\cdot)$ *such that the following conditions hold:*
(i) the Hamiltonian system

$$\begin{cases} {}_aD_t^\alpha q(t) & = \partial_4 \mathcal{H}(t, q(t), u(t), p(t)), \\ {}_tD_b^\alpha p(t) & = \partial_2 \mathcal{H}(t, q(t), u(t), p(t)); \end{cases}$$

(ii) and the stationary condition

$$\partial_3 \mathcal{H}(t, q(t), u(t), p(t)) = 0;$$

with the Hamiltonian $\mathcal{H}$ *defined by*

$$\mathcal{H}(t, q, u, p) = L(t, q, u) + p \cdot \varphi(t, q, u). \tag{2.91}$$

Remark 2.28. In classical mechanics, the Lagrange multiplier p is called the generalized momentum. In the language of optimal control, p is known as the adjoint variable.

Definition 2.20. *Any triplet* $(q(\cdot), u(\cdot), p(\cdot))$ *satisfying the conditions of Theorem 2.26 will be called a fractional Pontryagin extremal.*

For the fractional problem of the calculus of variations (2.83) one has $\varphi(t, q, u) = u \Rightarrow \mathcal{H} = L + p \cdot u$, and we obtain from Theorem 2.26 that

$$_aD_t^\alpha q = u,$$
$$_tD_b^\alpha p = \partial_2 L,$$
$$\partial_3 \mathcal{H} = 0 \Leftrightarrow p = -\partial_3 L \Rightarrow {}_tD_b^\alpha p = -{}_tD_b^\alpha \partial_3 L.$$

Comparing the two expressions for ${}_tD_b^\alpha p$, one arrives at the Euler–Lagrange differential equations (2.87): $\partial_2 L = -{}_tD_b^\alpha \partial_3 L$.

We define the notion of invariance for problem (2.90) in terms of the Hamiltonian, by introducing the augmented functional

$$J[q(\cdot), u(\cdot), p(\cdot)] = \int_a^b [\mathcal{H}(t, q(t), u(t), p(t)) - p(t) \cdot {}_aD_t^\alpha q(t)] \, dt, \tag{2.92}$$

where $\mathcal{H}$ is given by (2.91).

Remark 2.29. Theorem 2.26 is easily obtained by applying the necessary optimality condition (2.87) to problem (2.92).

Definition 2.21. *A fractional optimal control problem* (2.90) *is said to be invariant under the ε-parameter local group of transformations*

$$\begin{cases} \bar{t} = t + \varepsilon\tau(t, q(t), u(t), p(t)) + o(\varepsilon) \,, \\ \bar{q}(\bar{t}) = q(t) + \varepsilon\xi(t, q(t), u(t), p(t)) + o(\varepsilon) \,, \\ \bar{u}(\bar{t}) = u(t) + \varepsilon\sigma(t, q(t), u(t), p(t)) + o(\varepsilon) \,, \\ \bar{p}(\bar{t}) = p(t) + \varepsilon\zeta(t, q(t), u(t), p(t)) + o(\varepsilon) \,, \end{cases} \qquad (2.93)$$

if, and only if,

$$[\mathcal{H}(\bar{t}, \bar{q}(\bar{t}), \bar{u}(\bar{t}), \bar{p}(\bar{t})) - \bar{p}(\bar{t}) \cdot {}_{\bar{a}}D_{\bar{t}}^{\alpha}\bar{q}(\bar{t})] \, d\bar{t}$$
$$= [\mathcal{H}(t, q(t), u(t), p(t)) - p(t) \cdot {}_{a}D_{t}^{\alpha}q(t)] \, dt \,. \quad (2.94)$$

Theorem 2.27 (Fractional Noether's theorem). *If the fractional optimal control problem* (2.90) *is invariant under* (2.93), *then*

$$[\mathcal{H} - (1 - \alpha)\, p(t) \cdot {}_{a}D_{t}^{\alpha}q(t)]\, \tau - p(t) \cdot \xi \qquad (2.95)$$

is a fractional conservation law, that is,

$$\mathcal{D}\left\{[\mathcal{H} - (1 - \alpha)\, p(t) \cdot {}_{a}D_{t}^{\alpha}q(t)]\, \tau - p(t) \cdot \xi\right\} = 0$$

along all the fractional Pontryagin extremals.

Proof. The fractional conservation law (2.95) is obtained applying Theorem 2.25 to the augmented functional (2.92). ☐

Remark 2.30. For $\alpha \to 1$, the fractional optimal control problem (2.90) is reduced to the classical optimal control problem

$$I[q(\cdot), u(\cdot)] = \int_{a}^{b} L\left(t, q(t), u(t)\right) dt \longrightarrow \min \,,$$
$$\dot{q}(t) = \varphi\left(t, q(t), u(t)\right) \,,$$

and we obtain from Theorem 2.27 the optimal control version of Noether's theorem: invariance under a one-parameter group of transformations (2.93) implies that

$$C(t, q, u, p) = \mathcal{H}(t, q, u, p)\tau - p \cdot \xi \qquad (2.96)$$

is constant along any Pontryagin extremal (one obtains (2.96) from (2.95) setting $\alpha = 1$).

Theorem 2.27 provides a new interesting insight to the fractional autonomous variational problems. Let us consider the autonomous fractional optimal control problem, i.e., the situation when the Lagrangian L and the fractional velocity vector φ do not depend explicitly on time t:

$$I[q(\cdot), u(\cdot)] = \int_a^b L(q(t), u(t))\, dt \longrightarrow \min,$$

$$_aD_t^\alpha q(t) = \varphi(q(t), u(t)).$$

(2.97)

Corollary 2.1. *For the autonomous problem* (2.97) *the following fractional conservation law holds:*

$$\mathcal{D}\{\mathcal{H} - (1-\alpha)\, p(t) \cdot {_aD_t^\alpha q(t)}\} = 0.$$

(2.98)

Proof. The Hamiltonian $\mathcal{H}$ does not depend explicitly on time, and it is easy to check that (2.97) is invariant under time-translations: invariance condition (2.94) is satisfied with $\bar{t} = t + \varepsilon$, $\bar{q}(\bar{t}) = q(t)$, $\bar{u}(\bar{t}) = u(t)$ and $\bar{p}(\bar{t}) = p(t)$. In fact, given that $d\bar{t} = dt$, (2.94) holds trivially, proving that $_{\bar{a}}D_{\bar{t}}^\alpha \bar{q}(\bar{t}) = {_aD_t^\alpha q(t)}$:

$$_{\bar{a}}D_{\bar{t}}^\alpha \bar{q}(\bar{t})$$

$$= \frac{1}{\Gamma(n-\alpha)} \left(\frac{d}{d\bar{t}}\right)^n \int_{\bar{a}}^{\bar{t}} (\bar{t}-\theta)^{n-\alpha-1} \bar{q}(\theta)\, d\theta$$

$$= \frac{1}{\Gamma(n-\alpha)} \left(\frac{d}{dt}\right)^n \int_{a+\varepsilon}^{t+\varepsilon} (t+\varepsilon-\theta)^{n-\alpha-1} \bar{q}(\theta)\, d\theta$$

$$= \frac{1}{\Gamma(n-\alpha)} \left(\frac{d}{dt}\right)^n \int_a^t (t-s)^{n-\alpha-1} \bar{q}(s+\varepsilon)\, ds$$

$$= {_aD_t^\alpha \bar{q}(t+\varepsilon)} = {_aD_t^\alpha \bar{q}(\bar{t})}$$

$$= {_aD_t^\alpha q(t)}.$$

Using the notation in (2.93), one has $\tau = 1$ and $\xi = \sigma = \zeta = 0$. Conclusion (2.98) follows from Theorem 2.27. $\square$

Remark 2.31. In the classical framework of optimal control theory one has $\alpha = 1$ and our operator $\mathcal{D}$ coincides with $\frac{d}{dt}$. We then get from (2.98) the classical result: the Hamiltonian $\mathcal{H}$ is a preserved quantity along any Pontryagin extremal of the problem.

We begin by illustrating our results with two Lagrangians that do not depend explicitly on the time variable t. Here, we use our Corollary 2.1 to obtain new fractional conservation laws.

Example 2.12. We begin by considering a simple fractional problem of the calculus of variations:

$$I[q(\cdot)] = \frac{1}{2} \int_0^1 \left({}_0D_1^\alpha q(t) \right)^2 dt \longrightarrow \min, \quad \alpha > \frac{1}{2}. \tag{2.99}$$

Equation (2.91) takes the form

$$\mathcal{H} = -\frac{1}{2}p^2. \tag{2.100}$$

We conclude from Corollary 2.1 that

$$\frac{p^2}{2}(1 - 2\alpha) \tag{2.101}$$

is a fractional constant of motion.

Example 2.13. Let us now consider the following fractional optimal control problem:

$$I[q(\cdot)] = \frac{1}{2} \int_0^1 \left[q^2(t) + u^2(t) \right] dt \longrightarrow \min, \tag{2.102}$$

$${}_0D_1^\alpha q(t) = -q(t) + u(t),$$

under the initial condition $q(0) = 1$. The Hamiltonian $\mathcal{H}$ (2.91) has the form

$$\mathcal{H} = \frac{1}{2} \left(q^2 + u^2 \right) + p(-q + u).$$

From Corollary 2.1 it follows that

$$\frac{1}{2} \left(q^2 + u^2 \right) + \alpha p(-q + u) \tag{2.103}$$

defines a fractional conservation law.

For $\alpha = 1$, the fractional constants of motion (2.101) and (2.103) give conservation of energy.

Finally, we give an example of an optimal control problem with three state variables and two controls $(n = 3, m = 2)$.

Example 2.14. Consider the following fractional optimal control problem:

$$\int_a^b \left(u_1(t)^2 + u_2(t)^2 \right) dt \longrightarrow \min, \tag{2.104}$$

$$\begin{cases} {}_aD_t^\alpha q_1(t) = u_1(t)\cos(q_3(t)), \\ {}_aD_t^\alpha q_2(t) = u_1(t)\sin(q_3(t)), \\ {}_aD_t^\alpha q_3(t) = u_2(t). \end{cases} \tag{2.105}$$

For $\alpha = 1$ the control system (2.105) serves as a model for the kinematics of a car. From Corollary 2.1 one gets that

$$u_1^2 + u_2^2 + p_1 \left(u_1 \cos(q_3) - (1 - \alpha)_a D_t^\alpha q_1\right)$$
$$+ p_2 \left(u_1 \sin(q_3) - (1 - \alpha)_a D_t^\alpha q_2\right) + p_3 \left(u_2 - (1 - \alpha)_a D_t^\alpha q_3\right)$$

is a fractional conservation law.

The main difficulty of our approach is related to the computation of the invariance transformations. To illustrate this issue, let us consider problem (2.90) with

$$L(t, q, u) = L(t, u), \quad \varphi(t, q, u) = \varphi(t, u).$$

In the classical case, since q does not appear both in L and φ, such a problem is trivially invariant under translations on the variable q, i.e., condition (2.94) is verified for $\alpha = 1$ with $\bar{t} = t$, $\bar{q}(t) = q(t) + \varepsilon$, $\bar{u}(\bar{t}) = u(t)$ and $\bar{p}(\bar{t}) = p(t)$. In the fractional case this is not generally true: we have $d\bar{t} = dt$, but condition (2.94) is not satisfied since $_a D_{\bar{t}}^\alpha \bar{q}(\bar{t}) = {_a}D_t^\alpha q(t) + {_a}D_t^\alpha \varepsilon$ and the second term on the right-hand side is, in general, not equal to zero.

2.9 Fractional Version of Leitmann's Direct Method

Applications of necessary optimality conditions for fractional problems of the calculus of variations involve the solution of fractional differential equations, which is a difficult task to accomplish analytically. Often, numerical methods or truncated Taylor series are used (Pooseh, Almeida and Torres, 2012a,b). In this section, we give examples of fractional problems of the calculus of variations and optimal control with a known exact expression for the global minimizers. Such minimizers are found utilizing Leitmann's principle.

Consider the following fractional problem of the calculus of variations: to minimize

$$\mathfrak{J}(y) = \int_a^b F(x, y(x), {_a}D_x^\alpha y(x))\, dx \qquad (2.106)$$

under the constraint

$$_a I_b^{1-\alpha} y(b) = y_b, \qquad (2.107)$$

where y_b is a fixed real. Here we consider continuous functions $y : [a, b] \to \mathbb{R}$ such that $_a D_x^\alpha y(x)$ exists on $[a, b]$. We also emphasize that the condition

$_aI_a^{1-\alpha}y(a) = 0$ appears implicitly, since function y is continuous. If we allow $\alpha = 1$, functional (2.106) becomes the standard one,

$$\mathfrak{J}(y) = \int_a^b F(x, y(x), y'(x))\, dx\,, \tag{2.108}$$

and equality (2.107) reads as $y(b) = y_b$, that is, we obtain a classical problem of the calculus of variations: to minimize (2.108) under given boundary conditions $y(a) = 0$ and $y(b) = y_b$. The next result is obtained by following a method introduced by Leitmann.

Theorem 2.28. *Let $y(x) = z(x, \tilde{y}(x))$ be a continuous transformation having a unique inverse $\tilde{y}(x) = \tilde{z}(x, y(x))$, such that there exists an one-to-one correspondence*

$$y(x) \Leftrightarrow \tilde{y}(x),$$

for every function y satisfying

$$_aI_b^{1-\alpha}y(b) = y_b \tag{2.109}$$

and for every function $\tilde{y}$ satisfying

$$_aI_b^{1-\alpha}\tilde{y}(b) = {}_aI_b^{1-\alpha}\tilde{z}(x, y(x))|_{x=b}. \tag{2.110}$$

In addition assume that there exists a C^1 function $H : [a, b] \times \mathbb{R} \to \mathbb{R}$ such that the relation

$$F(x, y(x), {}_aD_x^\alpha y(x)) - F(x, \tilde{y}(x), {}_aD_x^\alpha \tilde{y}(x)) = \frac{d}{dx} H(x, {}_aI_x^{1-\alpha}\tilde{y}(x))$$

holds. Then there exists a one-to-one correspondence between the minimizers $y^*(x)$ of $\mathfrak{J}$ verifying (2.109) and the minimizers $\tilde{y}^*(x) = \tilde{z}(x, y^*(x))$ of $\mathfrak{J}$ verifying (2.110).

Proof. This is obvious from

$$\begin{aligned}
\mathfrak{J}(y^*) - \mathfrak{J}(\tilde{y}^*) &= \int_a^b F(x, y^*(x), {}_aD_x^\alpha y^*(x)) - F(x, \tilde{y}^*(x), {}_aD_x^\alpha \tilde{y}^*(x))\, dx \\
&= \int_a^b \frac{d}{dx} H(x, {}_aI_x^{1-\alpha}\tilde{y}^*(x)) \\
&= H(b, {}_aI_b^{1-\alpha}\tilde{y}^*(b)) - H(a, 0)
\end{aligned}$$

and the fact that the right-hand side is a constant. $\square$

We illustrate the application of Leitmann's method with examples from three different classes of problems: Proposition 2.10 gives an exact solution for a standard fractional variational problem (2.106)–(2.107); Proposition 2.11 gives the global minimizer for a family of problems whose Lagrangian depends simultaneously on a fractional derivative and a fractional integral of y; Proposition 2.12 gives the exact solution of a fractional optimal control problem.

Proposition 2.10. *Let α be an arbitrary real between 0 and 1. The global minimizer of the fractional problem*

$$\mathfrak{J}(y) = \int_0^1 ({}_0D_x^\alpha y(x))^2 \, dx \longrightarrow \min \tag{2.111}$$

under the constraint

$$_0I_1^{1-\alpha} y(1) = c \tag{2.112}$$

is

$$y(x) = \frac{c}{\alpha\Gamma(\alpha)} x^\alpha. \tag{2.113}$$

Remark 2.32. We note that for $\alpha = 1$ problem (2.111)–(2.112) reduces to the standard problem of the calculus of variations

$$\int_0^1 (y'(x))^2 \, dx \longrightarrow \min$$

$$y(0) = 0, \quad y(1) = c. \tag{2.114}$$

The global minimizer to problem (2.114) is $y(x) = cx$ (cf. p. 509 of Silva and Torres, 2006), which coincides with our solution (2.113) for $\alpha = 1$.

Proof (of Proposition 2.10). Let $y = \tilde{y} + f$, where f is a function such that $_0D_x^\alpha f(x) = K$, for all $x \in [0,1]$ (K to be specified later). We consider functions $\tilde{y}$ satisfying the equation

$$_0I_1^{1-\alpha} \tilde{y}(x) = 0. \tag{2.115}$$

Since $_0D_x^\alpha f = K$, then

$$_0I_x^\alpha \, _0D_x^\alpha f(x) = {}_0I_x^\alpha K$$

and the continuity on f implies that $_aI_a^{1-\alpha} f(a) = 0$. Consequently

$$f(x) = \frac{K}{\alpha\Gamma(\alpha)} x^\alpha.$$

Moreover,

$$\int_0^1 ({}_0D_x^\alpha y)^2 \, dx = \int_0^1 ({}_0D_x^\alpha \tilde{y} + {}_0D_x^\alpha f)^2 \, dx$$

$$= \int_0^1 ({}_0D_x^\alpha \tilde{y})^2 \, dx + \int_0^1 (2K {}_0D_x^\alpha \tilde{y} + K^2) \, dx$$

$$= \int_0^1 ({}_0D_x^\alpha \tilde{y})^2 \, dx + \int_0^1 \frac{d}{dx} (2K {}_0I_x^{1-\alpha} \tilde{y} + K^2 x) \, dx.$$

Observe that $\int_0^1 ({}_0D_x^\alpha \tilde{y})^2 \geq 0$ and $\int_0^1 {}_0D_x^\alpha \tilde{y}\, dx = 0$ if $\tilde{y}(x) \equiv 0$. Therefore, $\tilde{y}(x) \equiv 0$ is a minimizer of $\mathfrak{J}$, and satisfies equation (2.115). Since the second integral of the last sum is constant (actually, is equal to K^2), it follows that

$$y(x) = \frac{K}{\alpha \Gamma(\alpha)} x^\alpha$$

is a solution to problem (2.111)–(2.112). Since

$${}_0I_x^{1-\alpha} \frac{K}{\alpha \Gamma(\alpha)} x^\alpha = Kx,$$

it follows that ${}_0I_1^{1-\alpha}y(1) = K$, and so $K = c$. $\square$

We now study a variational problem where not only a fractional derivative of y but also a fractional integral of y appears in the Lagrangian. We will show how Leitmann's direct method can also be used to solve such problems.

Proposition 2.11. *Let g be a given function of class C^1 with $g(x) \neq 0$ on $[0, 1]$, and α and ξ be real numbers with α between zero and one. The global minimizer of the fractional variational problem*

$$\int_0^1 \left[{}_0D_x^\alpha y(x) \cdot g(x) + ({}_0I_x^{1-\alpha}y(x) + 1)g'(x) \right]^2 dx \longrightarrow \min \tag{2.116}$$

$${}_0I_1^{1-\alpha}y(1) = \xi$$

is given by the function

$$y(x) = {}_0D_x^{1-\alpha} \left(\frac{[g(1)(\xi + 1) - g(0)]x + g(0)}{g(x)} - 1 \right). \tag{2.117}$$

Remark 2.33. For $\alpha = 1$ problem (2.116) coincides with the classical problem of the calculus variations

$$\int_0^1 \left[y'(x) \cdot g(x) + (y(x) + 1)g'(x) \right]^2 dx \longrightarrow \min$$

$$y(0) = 0, \quad y(1) = \xi$$

that has been studied by Leitmann in 1967. For the integer-order case $\alpha = 1$, our function (2.117) reduces to $y(x) = \frac{[g(1)(\xi+1)-g(0)]x+g(0)}{g(x)} - 1$, which coincides with the results of Leitmann.

Proof (**of Proposition 2.11**). To begin, observe that

$$\frac{d}{dx}\left[({}_0I_x^{1-\alpha}y(x) + 1)g(x) \right] = {}_0D_x^\alpha y(x) \cdot g(x) + ({}_0I_x^{1-\alpha}y(x) + 1)g'(x).$$

Let y be an admissible function to problem (2.116), and consider the transformation $y(x) = \tilde{y}(x) + f(x)$, with f to be determined later. Then

$$\left(\frac{d}{dx}\left[\left(_0I_x^{1-\alpha}\tilde{y}(x) + {_0I_x^{1-\alpha}}f(x) + 1\right)g(x)\right]\right)^2 - \left(\frac{d}{dx}\left[\left(_0I_x^{1-\alpha}\tilde{y}(x) + 1\right)g(x)\right]\right)^2$$

$$= 2\frac{d}{dx}\left[\left(_0I_x^{1-\alpha}\tilde{y}(x) + 1\right)g(x)\right] \cdot \frac{d}{dx}\left[_0I_x^{1-\alpha}f(x) \cdot g(x)\right]$$

$$+ \left(\frac{d}{dx}\left[_0I_x^{1-\alpha}f(x) \cdot g(x)\right]\right)^2$$

$$= \frac{d}{dx}\left[_0I_x^{1-\alpha}f(x) \cdot g(x)\right] \cdot \frac{d}{dx}\left[\left(2\left(_0I_x^{1-\alpha}\tilde{y}(x) + 1\right) + {_0I_x^{1-\alpha}}f(x)\right)g(x)\right].$$

Let us determine f in such a way that

$$\frac{d}{dx}\left[_0I_x^{1-\alpha}f(x) \cdot g(x)\right] = const.$$

Integrating, we deduce that

$$_0I_x^{1-\alpha}f(x) \cdot g(x) = Ax + B,$$

i.e.,

$$f(x) = {_0D_x^{1-\alpha}}\left(\frac{Ax + B}{g(x)}\right).$$

Consider now the new problem

$$\int_0^1 \left(_0D_x^{\alpha}\tilde{y}(x) \cdot g(x) + \left(_0I_x^{1-\alpha}\tilde{y}(x) + 1\right)g'(x)\right)^2 dx \longrightarrow \min$$

$$_0I_1^{1-\alpha}\tilde{y}(1) = \frac{1}{g(1)} - 1.$$

(2.118)

It is easy to see, and a trivial exercise to check, that

$$\tilde{y}(x) = {_0D_x^{1-\alpha}}\left(\frac{1}{g(x)} - 1\right)$$

is a solution to problem (2.118). Therefore,

$$y(x) = {_0D_x^{1-\alpha}}\left(\frac{Ax + C}{g(x)} - 1\right), \quad \text{with } C = B + 1,$$

is a solution to (2.116). Using the boundary conditions

$$_0I_0^{1-\alpha}y(0) = 0 \quad \text{and} \quad {_0I_1^{1-\alpha}}y(1) = \xi,$$

we obtain the values for the constants A and C: $A = g(1)(\xi + 1) - g(0)$ and $C = g(0)$. □

Proposition 2.12 deals with a fractional optimal control problem. Using Leitmann's method we solve it.

Proposition 2.12. *Let $\alpha \in (0,1)$ be a real number. Consider the following fractional optimal control problem:*

$$\mathfrak{J}(u_1, u_2) = \int_0^1 \left[(u_1(x))^2 + (u_2(x))^2 \right] dx \longrightarrow \min \qquad (2.119)$$

subject to the fractional control system

$$\begin{cases} {}_0D_x^\alpha y_1(x) = \exp(u_1(x)) + u_1(x) + u_2(x) \\ {}_0D_x^\alpha y_2(x) = u_2(x) \end{cases} \qquad (2.120)$$

and fractional boundary conditions

$$\begin{cases} {}_0I_1^{1-\alpha} y_1(1) = 2 \\ {}_0I_1^{1-\alpha} y_2(1) = 1 \, . \end{cases} \qquad (2.121)$$

The global minimizer to problem (2.119)–(2.121) *is*

$$(u_1(x), u_2(x)) \equiv (0,1), \quad (y_1(x), y_2(x)) = \left(\frac{2x^\alpha}{\alpha\Gamma(\alpha)}, \frac{x^\alpha}{\alpha\Gamma(\alpha)} \right) . \qquad (2.122)$$

Proof. We use the following change of variables:

$$\begin{cases} \tilde{y}_1(x) = y_1(x) - {}_0I_x^\alpha 1 \\ \tilde{y}_2(x) = y_2(x) - {}_0I_x^\alpha 1 \\ \tilde{u}_1(x) = u_1(x) \\ \tilde{u}_2(x) = u_2(x) - 1 \, . \end{cases}$$

For the new variables, we compute

$$\begin{aligned} {}_0D_x^\alpha \tilde{y}_1 &= {}_0D_x^\alpha y_1 - {}_0D_x^\alpha {}_0I_x^\alpha 1 \\ &= \exp(u_1) + u_1 + u_2 - 1 \\ &= \exp(\tilde{u}_1) + \tilde{u}_1 + \tilde{u}_2 \end{aligned}$$

and

$$\begin{aligned} {}_0D_x^\alpha \tilde{y}_2 &= {}_0D_x^\alpha y_2 - {}_0D_x^\alpha {}_0I_x^\alpha 1 \\ &= \tilde{u}_2 \, . \end{aligned}$$

Therefore, $(\tilde{y}_1, \tilde{y}_2, \tilde{u}_1, \tilde{u}_2)$ satisfies (2.120). System (2.121) becomes

$${}_0I_1^{1-\alpha} \tilde{y}_1(1) = 1 \quad \text{and} \quad {}_0I_1^{1-\alpha} \tilde{y}_2(1) = 0 \qquad (2.123)$$

since ${}_0I_x^{1-\alpha} {}_0I_x^\alpha 1 = {}_0I_x^1 1 = x$.

Consider a new problem, which we label as $(\tilde{P})$: minimize $\mathfrak{J}(\cdot, \cdot)$ subject to system (2.120) and conditions (2.123). Evaluating the functional $\mathfrak{J}(\cdot, \cdot)$ at $(\tilde{u}_1, \tilde{u}_2)$, we obtain

$$
\begin{aligned}
\mathfrak{J}(\tilde{u}_1, \tilde{u}_2) &= \int_0^1 \left[(\tilde{u}_1(x))^2 + (\tilde{u}_2(x))^2 \right] dx \\
&= \int_0^1 \left[(u_1(x))^2 + (u_2(x))^2 - 2u_2(x) + 1 \right] dx \\
&= \mathfrak{J}(u_1, u_2) + \int_0^1 \left(-2{}_0D_x^\alpha y_2(x) + 1 \right) dx \\
&= \mathfrak{J}(u_1, u_2) + \int_0^1 \frac{d}{dx} \left(-2{}_0I_x^{1-\alpha} y_2(x) + x \right) dx.
\end{aligned}
$$

Observe that the second term of the last expression is constant. Because $\tilde{u}_1(x) \equiv \tilde{u}_2(x) \equiv 0$ is a solution to the problem $(\tilde{P})$, we get

$$
\begin{cases}
{}_0D_x^\alpha \tilde{y}_1 = 1 \\
{}_0D_x^\alpha \tilde{y}_2 = 0 \,,
\end{cases}
$$

i.e.,

$$
\begin{cases}
\tilde{y}_1(x) = {}_0I_x^\alpha 1 = \frac{x^\alpha}{\alpha \Gamma(\alpha)} \\
\tilde{y}_2(x) = {}_0I_x^\alpha 0 = 0 \,.
\end{cases}
$$

Therefore, the solution to problem (2.119)–(2.121) is given by

$$
(u_1(x), u_2(x)) = (0, 1)
$$

and

$$
(y_1(x), y_2(x)) = \left(\frac{2x^\alpha}{\alpha \Gamma(\alpha)}, \frac{x^\alpha}{\alpha \Gamma(\alpha)} \right).
$$

$\square$

Chapter 3

Fractional Calculus of Variations via Caputo Operators

This chapter is dedicated to the fractional calculus of variations with Caputo derivatives. We begin with the definitions and main properties of Caputo fractional operators (Section 3.1). The basic problem of the calculus of variations is presented in Section 3.2; transversality conditions for variable end-points problems are discussed in Section 3.3; while in Section 3.4 we study isoperimetric problems. Section 3.5 deals with sufficient conditions of optimality under an appropriate convexity assumption, and in Section 3.6 we discuss fractional control systems. Section 3.7 presents a Noether-type theorem for fractional optimal control problems with Caputo derivatives, while in Section 3.8 we prove a Noether's theorem for fractional variational problems in the Riesz–Caputo sense. Section 3.9 introduces the multiple dimensional Caputo fractional variational calculus. In Section 3.10 we extend the second Noether theorem to fractional variational problems that are invariant under infinitesimal transformations that depend upon r arbitrary functions and their fractional derivatives in the sense of Caputo. The result is illustrated using the fractional Lagrangian density of the electromagnetic field. Finally, in Section 3.11 we introduce a new combined fractional Caputo derivative and study variational problems with this more general fractional operator; in Section 3.11.3 we obtain necessary optimality conditions for variational problems with a Lagrangian depending on a Caputo fractional derivative, a fractional and an indefinite integral, and we discuss a numerical scheme for solving the proposed fractional variational problems. The chapter is based on original results from the papers by Almeida and Torres, 2011a; Frederico and Torres, 2008b, 2010; Malinowska and Torres, 2010b,c, 2011; Malinowska, 2012b; Malinowska and Torres, 2012a; Mozyrska and Torres, 2010; Odzijewicz, Malinowska and Torres, 2012b; Almeida, Pooseh and Torres, 2012.

3.1   Caputo Fractional Derivatives

In this section we collect the definitions of fractional derivatives in the sense of Caputo and their main properties. For a full treatment, we refer the reader to Kilbas, Srivastava and Trujillo, 2006; Podlubny, 1999; Samko, Kilbas and Marichev, 1993.

Definition 3.1. *The left and right Caputo fractional derivatives of order* $\alpha \geq 0$ *on* $[t_0, t_1]$, *denoted by* ${}^C_{t_0}D^\alpha_t$ *and* ${}^C_t D^\alpha_{t_1}$, *respectively, are defined via the Riemann–Liouville fractional derivatives by*

$$
{}^C_{t_0}D^\alpha_t \varphi(t) := {}_{t_0}D^\alpha_t \left(\varphi(t) - \sum_{k=0}^{n-1} \frac{\varphi^{(k)}(t_0)(t - t_0)^k}{k!} \right) \tag{3.1}
$$

and

$$
{}^C_t D^\alpha_{t_1} \varphi(t) := {}_t D^\alpha_{t_1} \left(\varphi(t) - \sum_{k=0}^{n-1} \frac{\varphi^{(k)}(t_1)(t_1 - t)^k}{k!} \right), \tag{3.2}
$$

where $n = [\alpha] + 1$ *for* $\alpha \notin \mathbb{N}_0$, *and* $n = \alpha$ *for* $\alpha \in \mathbb{N}_0$.

When $0 < \alpha < 1$, the relations (3.1) and (3.2) take the following form:

$$
{}^C_{t_0}D^\alpha_t \varphi(t) = {}_{t_0}D^\alpha_t \left(\varphi(t) - \varphi(t_0) \right), \quad {}^C_t D^\alpha_{t_1} \varphi(t) = {}_t D^\alpha_{t_1} \left(\varphi(t) - \varphi(t_1) \right).
$$

Let $AC[t_0, t_1]$ be the space of functions that are absolutely continuous on $[t_0, t_1]$, and $AC^n[t_0, t_1]$ denote the space of functions φ that have continuous derivatives up to order $n - 1$ on $[t_0, t_1]$ and such that $\varphi^{(n-1)} \in AC[a, b]$.

Proposition 3.1 (Kilbas, Srivastava and Trujillo, 2006). *Let* $\alpha \geq 0$, $n = [\alpha] + 1$ *if* $\alpha \notin \mathbb{N}_0$ *and* $n = \alpha$ *if* $\alpha \in \mathbb{N}_0$. *If* $\varphi \in AC^n[t_0, t_1]$, *then the Caputo fractional derivatives* ${}^C_{t_0}D^\alpha_t \varphi(t)$ *and* ${}^C_t D^\alpha_{t_1} \varphi(t)$ *exist almost everywhere on* $[t_0, t_1]$. *Moreover,*

(a) If $\alpha \notin \mathbb{N}_0$, *then*

$$
{}^C_{t_0}D^\alpha_t \varphi(t) = \frac{1}{\Gamma(n - \alpha)} \int_{t_0}^t \varphi^{(n)}(\tau)(t - \tau)^{n-\alpha-1} d\tau = {}_{t_0}I^{n-\alpha}_t \left(\varphi^{(n)} \right)(t),
$$

$$
{}^C_t D^\alpha_{t_1} \varphi(t) = \frac{(-1)^n}{\Gamma(n - \alpha)} \int_t^{t_1} \varphi^{(n)}(\tau)(\tau - t)^{n-\alpha-1} d\tau
$$

$$
= (-1)^n {}_t I^{n-\alpha}_{t_1} \left(\varphi^{(n)} \right)(t).
$$

(b) If $\alpha = n \in \mathbb{N}_0$, *then* ${}^C_{t_0}D^n_t \varphi(t) = \varphi^{(n)}(t)$ *and* ${}^C_t D^n_{t_1} \varphi(t) = (-1)^n \varphi^{(n)}(t)$.

If the Riemann–Liouville and the Caputo fractional derivatives exist, then they are connected with each other by the following relations:

$$
{}_a^C D_x^\alpha f(x) = {}_a D_x^\alpha f(x) - \sum_{k=0}^{n-1} \frac{f^{(k)}(a)}{\Gamma(k - \alpha + 1)}(x - a)^{k-\alpha}
$$

and

$$
{}_x^C D_b^\alpha f(x) = {}_x D_b^\alpha f(x) - \sum_{k=0}^{n-1} \frac{f^{(k)}(b)}{\Gamma(k - \alpha + 1)}(b - x)^{k-\alpha}.
$$

Therefore,

$$
\text{if } f(a) = f'(a) = \ldots = f^{(n-1)}(a) = 0, \text{ then } {}_a^C D_x^\alpha f(x) = {}_a D_x^\alpha f(x)
$$

and

$$
\text{if } f(b) = f'(b) = \ldots = f^{(n-1)}(b) = 0, \text{ then } {}_x^C D_b^\alpha f(x) = {}_x D_b^\alpha f(x).
$$

Proposition 3.2. *Let f and g be two functions defined on $[a, b]$ such that ${}_a^C D_x^\alpha f$ and ${}_a^C D_x^\alpha g$ exist almost everywhere. Moreover, let $c_1, c_2 \in \mathbb{R}$. Then, ${}_a^C D_x^\alpha(c_1 f + c_2 g)$ exists almost everywhere, and*

$$
{}_a^C D_x^\alpha(c_1 f + c_2 g) = c_1 {}_a^C D_x^\alpha f + c_2 {}_a^C D_x^\alpha g.
$$

Similar results hold for the right Caputo fractional derivative.

Proposition 3.3. *Let $\alpha > 0$ and $n = [\alpha] + 1$. Also let $f \in C^n[a, b]$. Then, the Caputo fractional derivatives ${}_a^C D_x^\alpha f$ and ${}_x^C D_b^\alpha g$ are continuous on $[a, b]$.*

Some properties valid for integer differentiation and integer integration remain valid for fractional differentiation and fractional integration. Namely, the Caputo fractional derivatives and the Riemann–Liouville fractional integrals are inverse operations:

(a) If $f \in L_\infty(a, b)$ and $\alpha > 0$, then

$$
{}_a^C D_x^\alpha \, {}_a I_x^\alpha f(x) = f(x) \quad \text{and} \quad {}_x^C D_b^\alpha \, {}_x I_b^\alpha f(x) = f(x).
$$

(b) If $f \in AC^n[a, b]$ and $\alpha > 0$, then

$$
{}_a I_x^\alpha \, {}_a^C D_x^\alpha f(x) = f(x) - \sum_{k=0}^{n-1} \frac{f^{(k)}(a)}{k!}(x - a)^k
$$

and

$$
{}_x I_b^\alpha \, {}_x^C D_b^\alpha f(x) = f(x) - \sum_{k=0}^{n-1} \frac{(-1)^k f^{(k)}(b)}{k!}(b - x)^k.
$$

We need for our purposes the following integration by parts formulas. For $\alpha > 0$ one has

$$\int_a^b g(x) \cdot {}_a^C D_x^\alpha f(x) dx$$

$$= \int_a^b f(x) \cdot {}_x D_b^\alpha g(x) dx + \sum_{j=0}^{n-1} \left[{}_x D_b^{\alpha+j-n} g(x) \cdot {}_x D_b^{n-1-j} f(x) \right]_a^b$$

and

$$\int_a^b g(x) \cdot {}_x^C D_b^\alpha f(x) dx$$

$$= \int_a^b f(x) \cdot {}_a D_x^\alpha g(x) dx + \sum_{j=0}^{n-1} \left[(-1)^{n+j} {}_a D_x^{\alpha+j-n} g(x) \cdot {}_a D_x^{n-1-j} f(x) \right]_a^b,$$

where ${}_a D_x^k g(x) = {}_a I_x^{-k} g(x)$ and ${}_x D_b^k g(x) = {}_x I_b^{-k} g(x)$ if $k < 0$. Therefore, if $0 < \alpha < 1$, we obtain that

$$\int_a^b g(x) \cdot {}_a^C D_x^\alpha f(x) dx = \int_a^b f(x) \cdot {}_x D_b^\alpha g(x) dx + \left[{}_x I_b^{1-\alpha} g(x) \cdot f(x) \right]_a^b \quad (3.3)$$

and

$$\int_a^b g(x) \cdot {}_x^C D_b^\alpha f(x) dx = \int_a^b f(x) \cdot {}_a D_x^\alpha g(x) dx - \left[{}_a I_x^{1-\alpha} g(x) \cdot f(x) \right]_a^b. \quad (3.4)$$

Moreover, if f is a function such that $f(a) = f(b) = 0$, we have simpler formulas:

$$\int_a^b g(x) \cdot {}_a^C D_x^\alpha f(x) dx = \int_a^b f(x) \cdot {}_x D_b^\alpha g(x) dx \quad (3.5)$$

and

$$\int_a^b g(x) \cdot {}_x^C D_b^\alpha f(x) dx = \int_a^b f(x) \cdot {}_a D_x^\alpha g(x) dx. \quad (3.6)$$

Remark 3.1. Observe that the left-hand side of equations (3.5) and (3.6) contain a Caputo fractional derivative, while the right-hand side contains a Riemann–Liouville fractional derivative. However, since there exists a relation between the two derivatives, we can present such formulas with only one type of fractional derivative, although in this case the resulting equation will contain some extra terms.

For a function $x : [0, T] \to \mathbb{R}^n$ we use a similar notation to that in the classical case:

$$
{}_0^C D_t^\alpha x(t) = {}_0^C D_t^\alpha \begin{pmatrix} x_1(t) \\ \vdots \\ x_n(t) \end{pmatrix} = \begin{pmatrix} {}_0^C D_t^\alpha x_1(t) \\ \vdots \\ {}_0^C D_t^\alpha x_n(t) \end{pmatrix}.
$$

We recall that when for each component we use the same fractional order α of differentiation (in the Riemann–Liouville or Caputo sense), the fractional derivative is said to be of commensurate order.

Proposition 3.4. *For $\alpha > 0$ the following holds:*

(i) ${}_{t_0}^C D_t^\alpha E_\alpha(A(t - t_0)^\alpha) = A E_\alpha(A(t - t_0)^\alpha);$

(ii) ${}_{t_0} D_t^\alpha e_\alpha^{A(t-t_0)} = A e_\alpha^{A(t-t_0)};$

(iii) ${}_t D_T^\alpha S(T - \tau) = A S(T - \tau),$ *where* $S(t) = e_\alpha^{At}.$

Proof. (i) Directly from the definition of the classical Mittag–Leffler function E_α, and from the formula of the Caputo derivative of a power function, one has:

$$
{}_{t_0}^C D_t^\alpha (t - t_0)^\beta = \frac{\Gamma(\beta + 1)}{\Gamma(\beta - \alpha + 1)} (t - t_0)^{\beta - \alpha}, \quad \beta \neq 0.
$$

Since the Caputo derivative of a constant function is zero,

$$
{}_{t_0}^C D_t^\alpha E_\alpha(A(t - t_0)^\alpha) = {}_{t_0}^C D_t^\alpha \sum_{k=0}^\infty A^k \frac{(t - t_0)^{k\alpha}}{\Gamma(k\alpha + 1)} = \sum_{k=1}^\infty A^k \frac{(t - t_0)^{(k-1)\alpha}}{\Gamma((k - 1)\alpha + 1)}
$$

$$
= A E_\alpha(A(t - t_0)^\alpha).
$$

(ii) As the formula for the α-exponential function does not consist of constant terms, and since the formula for the Riemann–Liouville derivative of a power function is the same as for the Caputo derivative, we have:

$$
{}_{t_0} D_t^\alpha e_\alpha^{A(t-t_0)} = {}_{t_0} D_t^\alpha \sum_{k=0}^\infty A^k \frac{(t - t_0)^{(k+1)\alpha - 1}}{\Gamma[(k + 1)\alpha]}
$$

$$
= \sum_{k=1}^\infty A^k \frac{(t - t_0)^{k\alpha - 1}}{\Gamma(k\alpha)}
$$

$$
= A e_\alpha^{A(t-t_0)},
$$

where we used the fact that $\lim_{\alpha \to 0} \frac{1}{\Gamma(\alpha)} = 0$.

(iii) Using the formulas

$$
{}_t D_T^\alpha (T - \tau)^{\beta - 1} = \frac{\Gamma(\beta)}{\Gamma(\beta - \alpha)} (T - \tau)^{\beta - \alpha - 1}, \quad \lim_{\beta \to \alpha} D_{T-}^\beta \frac{(T - \tau)^{\alpha - 1}}{\Gamma(\alpha)} = 0
$$

we get:

$$
\begin{aligned}
{}_tD_T^\alpha S(T-\tau) &= {}_tD_T^\alpha \left(I\frac{1}{\Gamma(\alpha)}(T-\tau)^{\alpha-1} + A\frac{(T-\tau)^{2\alpha-1}}{\Gamma 2\alpha} + \cdots \right) \\
&= A\frac{(T-\tau)^{\alpha-1}}{\Gamma(\alpha)} + A^2\frac{(T-\tau)^{2\alpha-1}}{\Gamma(2\alpha)} + \cdots \\
&= (T-\tau)^{\alpha-1}A\sum_{k=0}^\infty A^k\frac{(T-\tau)^{k\alpha}}{\Gamma[(k+1)\alpha]} \\
&= AS(T-\tau)\,.
\end{aligned}
$$

$\square$

Proposition 3.5. *For $\alpha > 0$ the following relation holds:*

$$
E_\alpha(A(t-t_0)^\alpha) = I + \int_{t_0}^t Ae_\alpha^{A(t-\tau)}d\tau\,.
$$

Proof.   Follows by direct calculation of the integral:

$$
\begin{aligned}
\int_{t_0}^t Ae_\alpha^{A(t-\tau)}d\tau &= \int_{t_0}^t \sum_{k=0}^\infty A^{k+1}\frac{(t-\tau)^{(k+1)\alpha-1}}{\Gamma[(k+1)\alpha]}d\tau \\
&= \sum_{k=1}^\infty A^k\frac{(t-t_0)^{k\alpha}}{\Gamma(k\alpha+1)} = E_\alpha(A(t-t_0)^\alpha) - I\,.
\end{aligned}
$$

$\square$

Since $e_\alpha^{At} = t^{\alpha-1}E_{\alpha,\alpha}(At^\alpha)$ and each Mittag–Leffler function $E_{\alpha,\alpha}(az^\alpha)$, $\alpha > 0$, is an entire function on the complex plane, we can state the following.

Proposition 3.6. *Let $\alpha > 0$. There is a uniquely determined function $g(t) = t^{1-\alpha}G(t)$ such that $e_\alpha^{At}g(t) = E_{\alpha,\alpha}(At^\alpha)G(t) = I$, for $t \neq 0$, and $\lim_{t\to 0} e_\alpha^{At}g(t) = I$.*

Lemma 3.1. *Let $\alpha > 0$ and $\psi(t) \in {}_0I_t^\alpha(L_q)$, where $1/p + 1/q \leq 1 + \alpha$ for p such that all components of $S(T-t)$ belong to ${}_tI_T^\alpha(L_p)$. Then,*

$$
\int_0^T S(T-\tau){}_0D_t^\alpha\psi(\tau)d\tau = \int_0^T AS(T-\tau)\psi(\tau)d\tau\,.
$$

Proof.   Taking into account (2.2) and item (iii) of Proposition 3.4, we have

$$
\int_0^T S(T-\tau){}_0D_t^\alpha\psi(\tau)d\tau = \int_0^T \psi(\tau){}_tD_T^\alpha S(T-\tau)d\tau = \int_0^T AS(T-\tau)\psi(\tau)d\tau\,.
$$

$\square$

Definition 3.2. *Let f be a continuous function in the interval $[a, b]$. For $t \in [a, b]$, the Riesz fractional integral ${}_{a}^{R}I_{b}^{\alpha}$ of order α, $\alpha > 0$, is defined by*

$$
{}_{a}^{R}I_{b}^{\alpha}f(t) := \frac{1}{2\Gamma(\alpha)} \int_{a}^{b} |t - \theta|^{\alpha-1} f(\theta)d\theta. \tag{3.7}
$$

Remark 3.2. From definitions of the Riemann–Liouville and the Riesz fractional integrals it follows that

$$
{}_{a}^{R}I_{b}^{\alpha}f(t) = \frac{1}{2} \left[{}_{a}I_{t}^{\alpha}f(t) +_{t} I_{b}^{\alpha}f(t) \right]. \tag{3.8}
$$

Definition 3.3. *Let f be a continuous function in $[a, b]$. For $t \in [a, b]$, the Riesz fractional derivative ${}_{a}^{R}D_{b}^{\alpha}$ and the Riesz–Caputo fractional derivative ${}_{a}^{RC}D_{b}^{\alpha}$ of order α are defined by*

$$
{}_{a}^{R}D_{b}^{\alpha}f(t) := D^{n} {}_{a}^{R}I_{t}^{n-\alpha}f(t) = \frac{1}{\Gamma(n-\alpha)} \left(\frac{d}{dt}\right)^{n} \int_{a}^{b} |t - \theta|^{n-\alpha-1} f(\theta)d\theta,
$$
$$\tag{3.9}$$

$$
{}_{a}^{RC}D_{b}^{\alpha}f(t) := {}_{a}^{R}I_{t}^{n-\alpha}D^{n}f(t) = \frac{1}{\Gamma(n-\alpha)} \int_{a}^{b} |t - \theta|^{n-\alpha-1} \left(\frac{d}{d\theta}\right)^{n} f(\theta)d\theta,
$$
$$\tag{3.10}$$

where $n \in \mathbb{N}$ is such that $n - 1 \leq \alpha < n$.

Therefore,

$$
{}_{a}^{R}D_{b}^{\alpha}f(t) = \frac{1}{2} \left[{}_{a}D_{t}^{\alpha}f(t) + (-1)^{n} {}_{t}D_{b}^{\alpha}f(t) \right]
$$

and

$$
{}_{a}^{RC}D_{b}^{\alpha}f(t) = \frac{1}{2} \left[{}_{a}^{C}D_{t}^{\alpha}f(t) + (-1)^{n} {}_{t}^{C}D_{b}^{\alpha}f(t) \right].
$$

In the particular case $0 < \alpha < 1$, we have

$$
{}_{a}^{R}D_{b}^{\alpha}f(t) = \frac{1}{2} \left[{}_{a}D_{t}^{\alpha}f(t) - {}_{t}D_{b}^{\alpha}f(t) \right] \tag{3.11}
$$

and

$$
{}_{a}^{RC}D_{b}^{\alpha}f(t) = \frac{1}{2} \left[{}_{a}^{C}D_{t}^{\alpha}f(t) - {}_{t}^{C}D_{b}^{\alpha}f(t) \right]. \tag{3.12}
$$

If $\alpha = 1$, we obtain that

$$
{}_{a}^{R}D_{b}^{1}f(t) = {}_{a}^{RC}D_{b}^{1}f(t) = \frac{d}{dt}f(t).
$$

3.2 Fundamental Problem of the Calculus of Variations

In this section the Euler–Lagrange equation is proved for functionals where the lower and upper bounds of the integral are distinct from the bounds of the Caputo derivative. As a corollary we obtain an equation for variational functionals containing left and right Caputo fractional derivatives. Fix $0 < \alpha, \beta \leq 1$. To simplify, we denote

$$[y](x) = (x, y(x), {}^C_a D^\alpha_x y(x), {}^C_x D^\beta_b y(x)).$$

Let

$${}^\alpha_a E^\beta_b = \left\{ y : [a, b] \to \mathbb{R} \mid {}^C_a D^\alpha_x y \text{ and } {}^C_x D^\beta_b y \text{ exist and are continuous on } [a, b] \right\}.$$

Definition 3.4. *The space of admissible variations* ${}^C Var(a, b)$ *for the Caputo derivatives is defined by*

$${}^C Var(a, b) = \left\{ h \in {}^\alpha_a E^\beta_b \mid h(a) = h(b) = 0 \right\}.$$

We consider the functional

$$J^*(y) = \int_A^B L(x, y(x), {}^C_a D^\alpha_x y(x), {}^C_x D^\beta_b y(x)) dx, \qquad (3.13)$$

where $[A, B] \subset [a, b]$. We assume that the map $(x, y, u, v) \to L(x, y, u, v)$ is a function of class C^1. Denoting by $\partial_i L$ the partial derivative of L with respect to the ith variable, $i = 1, 2, 3, 4$, we also assume that $\partial_3 L$ has continuous right Riemann–Liouville fractional derivative of order α and $\partial_4 L$ has continuous left Riemann–Liouville fractional derivative of order β.

Theorem 3.1 (Fractional Euler–Lagrange equation). *If y is a local minimizer of J^* given by (3.13), satisfying the boundary conditions $y(a) = y_a$ and $y(b) = y_b$, then y satisfies the system*

$$\begin{cases} \partial_2 L + {}_x D^\alpha_B \partial_3 L + {}_A D^\beta_x \partial_4 L = 0 & \text{for all } x \in [A, B] \\ {}_x D^\alpha_B \partial_3 L - {}_x D^\alpha_A \partial_3 L = 0 & \text{for all } x \in (a, A) \\ {}_A D^\beta_x \partial_4 L - {}_B D^\beta_x \partial_4 L = 0 & \text{for all } x \in (B, b). \end{cases}$$

Proof. Let y be a minimizer and let $\hat{y} = y + \epsilon \eta$ be a variation of y, $\eta \in {}^C Var(a, b)$, such that $\eta(A) = \eta(B) = 0$. Define the new function $\hat{J}^*(\epsilon) = J^*(y + \epsilon \eta)$. By hypothesis, y is a local extremizer of J^* and so $\hat{J}^*$

has a local extremum at $\epsilon = 0$. Therefore, the following holds:

$$0 = \int_A^B \left[\partial_2 L\, \eta + \partial_3 L\, {}_a^C D_x^\alpha \eta + \partial_4 L\, {}_x^C D_b^\beta \eta \right] dx$$

$$= \int_A^B \partial_2 L\, \eta dx + \left[\int_a^B \partial_3 L\, {}_a^C D_x^\alpha \eta dx - \int_a^A \partial_3 L\, {}_a^C D_x^\alpha \eta dx \right]$$

$$+ \left[\int_A^b \partial_4 L\, {}_x^C D_b^\beta \eta dx - \int_B^b \partial_4 L\, {}_x^C D_b^\beta \eta dx \right].$$

Integrating by parts the four last terms gives

$$0 = \int_A^B \partial_2 L\, \eta dx + \left[\int_a^B \eta\, {}_x D_B^\alpha \partial_3 L dx - \int_a^A \eta\, {}_x D_A^\alpha \partial_3 L dx \right]$$

$$+ \left[\int_A^b \eta\, {}_A D_x^\beta \partial_4 L dx - \int_B^b \eta\, {}_B D_x^\beta \partial_4 L dx \right]$$

$$= \int_A^B \partial_2 L\, \eta dx$$

$$+ \left[\int_a^A \eta\, {}_x D_B^\alpha \partial_3 L dx + \int_A^B \eta\, {}_x D_B^\alpha \partial_3 L dx - \int_a^A \eta\, {}_x D_A^\alpha \partial_3 L dx \right]$$

$$+ \left[\int_A^B \eta\, {}_A D_x^\beta \partial_4 L dx + \int_B^b \eta\, {}_A D_x^\beta \partial_4 L dx - \int_B^b \eta\, {}_B D_x^\beta \partial_4 L dx \right]$$

$$= \int_a^A \left[{}_x D_B^\alpha \partial_3 L - {}_x D_A^\alpha \partial_3 L \right] \eta dx + \int_A^B \left[\partial_2 L + {}_x D_B^\alpha \partial_3 L + {}_A D_x^\beta \partial_4 L \right] \eta dx$$

$$+ \int_B^b \left[{}_A D_x^\beta \partial_4 L - {}_B D_x^\beta \partial_4 L \right] \eta dx.$$

Since η is an arbitrary function, we can assume that $\eta(x) = 0$ for all $x \in [A, b]$ and so, by the fundamental lemma of the calculus of variations,

$$ {}_x D_B^\alpha \partial_3 L - {}_x D_A^\alpha \partial_3 L = 0, \text{ for all } x \in (a, A).$$

Similarly, one proves the other two equations:

$$\partial_2 L + {}_x D_B^\alpha \partial_3 L + {}_A D_x^\beta \partial_4 L = 0, \text{ for all } x \in [A, B],$$

and

$$ {}_A D_x^\beta \partial_4 L - {}_B D_x^\beta \partial_4 L = 0, \text{ for all } x \in (B, b). \qquad \square$$

We now present first order necessary optimality conditions for functionals defined on ${}_a^\alpha E_b^\beta$, of the type

$$J(y) = \int_a^b L[y](x) dx. \tag{3.14}$$

Theorem 3.2 (Fractional Euler–Lagrange equation). *Let* J *be the functional as in* (3.14) *and* y *a local minimizer of* J *satisfying the boundary conditions* $y(a) = y_a$ *and* $y(b) = y_b$. *Then,* y *satisfies the Euler–Lagrange equation*

$$\partial_2 L + {}_x D_b^\alpha \partial_3 L + {}_a D_x^\beta \partial_4 L = 0. \tag{3.15}$$

When the term ${}_x^C D_b^\beta y$ is not present in the function L, then equation (3.15) reduces to a simpler one:

$$\partial_2 L + {}_x D_b^\alpha \partial_3 L = 0.$$

Moreover, if we allow $\alpha = 1$, and since, in that case, the right Riemann–Liouville fractional derivative is equal to $-d/dx$, we obtain the classical Euler–Lagrange equation:

$$\partial_2 L - \frac{d}{dx} \partial_3 L = 0.$$

3.3   Variable End-points

The purpose of this section is to study general transversality conditions for fractional variational problems defined in terms of the Caputo derivative. Subsection 3.3.1 is devoted to more general problems of the fractional calculus of variations with a Lagrangian that may also depend on the unspecified end-points $y(a)$ and $y(b)$. In Subsection 3.3.2 we consider the Bolza-type fractional variational problem and develop the transversality conditions appropriate to various types of variable end-points. Subsection 3.3.3 provides the necessary optimality conditions for fractional variational problems with a Lagrangian that may also depend on the unspecified end-point $\varphi(b)$, where $x = \varphi(t)$ is a given curve. Finally, in Subsection 3.3.4 we present a necessary optimality condition for the infinite horizon fractional variational problem.

3.3.1   *Generalized Natural Boundary Conditions*

Let us consider the following problem:

$$\mathcal{J}(y) = \int_a^b L(x, y(x), {}_a^C D_x^\alpha y(x), {}_x^C D_b^\beta y(x), y(a), y(b)) \, dx \longrightarrow \text{extr} \tag{3.16}$$

$$(y(a) = y_a), \quad (y(b) = y_b).$$

Using parentheses around the end-point conditions means that the conditions may or may not be present. We assume that:

(a) $L \in C^1([a,b] \times \mathbb{R}^5; \mathbb{R})$;

(b) $x \to \partial_3 L(x, y(x), {}_a^C D_x^\alpha y(x), {}_x^C D_b^\beta y(x), y(a), y(b))$ has continuous right RLFD of order α, where $\alpha \in (0, 1)$;

(c) $x \to \partial_4 L(x, y(x), {}_a^C D_x^\alpha y(x), {}_x^C D_b^\beta y(x), y(a), y(b))$ has continuous left RLFD of order β, where $\beta \in (0, 1)$.

Remark 3.3. We are assuming that the admissible functions y are such that ${}_a^C D_x^\alpha y(x)$ and ${}_x^C D_b^\beta y(x)$ exist on the closed interval $[a, b]$.

We denote by $\partial_i L$, $i = 1, \ldots, 6$, the partial derivative of function L with respect to its ith argument. For simplicity of notation we introduce the operator $\{\cdot\}$ defined by

$$\{y\}(x) = (x, y(x), {}_a^C D_x^\alpha y(x), {}_x^C D_b^\beta y(x), y(a), y(b)).$$

The next theorem gives the necessary optimality conditions for problem (3.16).

Theorem 3.3. *Let y be a local extremizer to problem* (3.16). *Then, y satisfies the fractional Euler–Lagrange equation*

$$\partial_2 L\{y\}(x) + {}_x D_b^\alpha \partial_3 L\{y\}(x) + {}_a D_x^\beta \partial_4 L\{y\}(x) = 0 \qquad (3.17)$$

for all $x \in [a, b]$. Moreover, if $y(a)$ is not specified, then

$$\int_a^b \partial_5 L\{y\}(x)\, dx = \left[{}_x I_b^{1-\alpha} \partial_3 L\{y\}(x) - {}_a I_x^{1-\beta} \partial_4 L\{y\}(x) \right]\Big|_{x=a} \qquad (3.18)$$

if $y(b)$ is not specified, then

$$\int_a^b \partial_6 L\{y\}(x)\, dx = \left[{}_a I_x^{1-\beta} \partial_4 L\{y\}(x) - {}_x I_b^{1-\alpha} \partial_3 L\{y\}(x) \right]\Big|_{x=b}. \qquad (3.19)$$

Proof. Suppose that y is an extremizer of $\mathcal{J}$. We can proceed as Lagrange did, by considering the value of $\mathcal{J}$ at a nearby function $\tilde{y} = y + \varepsilon h$, where $\varepsilon \in \mathbb{R}$ is a small parameter and h is an arbitrary admissible function. We do not require $h(a) = 0$ or $h(b) = 0$ in case $y(a)$ or $y(b)$, respectively, is free (it is possible that both are free). Let

$$\varphi(\varepsilon) = \int_a^b L\{y + \varepsilon h\}(x)\, dx.$$

Since ${}_a^C D_x^\alpha$ and ${}_x^C D_b^\beta$ are linear operators, it follows that

$${}_a^C D_x^\alpha(y(x) + \varepsilon h(x)) = {}_a^C D_x^\alpha y(x) + \varepsilon\, {}_a^C D_x^\alpha h(x)$$
$${}_x^C D_b^\beta(y(x) + \varepsilon h(x)) = {}_x^C D_b^\beta y(x) + \varepsilon\, {}_x^C D_b^\beta h(x).$$

A necessary condition for y to be an extremizer is given by

$$\frac{d\varphi}{d\varepsilon}\Big|_{\varepsilon=0} = 0$$

$$\Leftrightarrow \int_a^b \Big[\partial_2 L(\{y\}(x))h(x) + \partial_3 L(\{y\}(x))_a^C D_x^\alpha h(x) + \partial_4 L(\{y\}(x))_x^C D_b^\beta h(x)$$

$$+ \partial_5 L(\{y\}(x))h(a) + \partial_6 L(\{y\}(x))h(b)\Big]dx = 0. \quad (3.20)$$

Using formulae (3.3)–(3.4) for integration by parts, the second and the third integral can be written as

$$\int_a^b \partial_3 L\{y\}(x)_a^C D_x^\alpha h(x)dx$$

$$= \int_a^b h(x)_x D_b^\alpha \partial_3 L\{y\}(x)dx + {}_x I_b^{1-\alpha} \partial_3 L\{y\}(x)h(x)\Big|_{x=a}^{x=b},$$

$$\int_a^b \partial_4 L\{y\}(x)_x^C D_b^\beta h(x)dx \qquad (3.21)$$

$$= \int_a^b h(x)_a D_x^\beta \partial_4 L\{y\}(x)dx - {}_a I_x^{1-\beta} \partial_4 L\{y\}(x)h(x)\Big|_{x=a}^{x=b}.$$

Substituting (3.21) into (3.20), we get

$$\int_a^b \Big[\partial_2 L\{y\}(x) + {}_x D_b^\alpha \partial_3 L\{y\}(x) + {}_a D_x^\beta \partial_4 L\{y\}(x)\Big]h(x)$$

$$+ {}_x I_b^{1-\alpha} \partial_3 L\{y\}(x)h(x)\Big|_{x=a}^{x=b} - {}_a I_x^{1-\beta} \partial_4 L\{y\}(x)h(x)\Big|_{x=a}^{x=b}$$

$$+ \int_a^b \big(\partial_5 L\{y\}(x)h(a) + \partial_6 L\{y\}(x)h(b)\big)\,dx = 0. \quad (3.22)$$

We first consider functions h such that $h(a) = h(b) = 0$. Then, by the fundamental lemma of the calculus of variations, we deduce that

$$\partial_2 L\{y\}(x) + {}_x D_b^\alpha \partial_3 L\{y\}(x) + {}_a D_x^\beta \partial_4 L\{y\}(x) = 0 \qquad (3.23)$$

for all $x \in [a, b]$. Therefore, in order for y to be an extremizer to the problem (3.16), y must be a solution of the fractional Euler–Lagrange equation. But if y is a solution of (3.23), the first integral in expression (3.22) vanishes, and then the condition (3.20) takes the form

$$h(b) \left\{ \int_a^b \partial_6 L\{y\}(x)dx - \Big[{}_a I_x^{1-\beta} \partial_4 L\{y\}(x) - {}_x I_b^{1-\alpha} \partial_3 L\{y\}(x)\Big]\Big|_{x=b} \right\}$$

$$+ h(a) \left\{ \int_a^b \partial_5 L\{y\}(x)dx - \Big[{}_x I_b^{1-\alpha} \partial_3 L\{y\}(x) - {}_a I_x^{1-\beta} \partial_4 L\{y\}(x)\Big]\Big|_{x=a} \right\}$$

$$= 0.$$

If $y(a) = y_a$ and $y(b) = y_b$ are given in the formulation of problem (3.16), then the latter equation is trivially satisfied since $h(a) = h(b) = 0$. When $y(a)$ is free, then

$$\int_a^b \partial_5 L\{y\}(x)dx - \left[{}_xI_b^{1-\alpha}\partial_3 L\{y\}(x) - {}_aI_x^{1-\beta}\partial_4 L\{y\}(x) \right]\big|_{x=a} = 0,$$

when $y(b)$ is free, then

$$\int_a^b \partial_6 L\{y\}(x)dx - \left[{}_aI_x^{1-\beta}\partial_4 L\{y\}(x) - {}_xI_b^{1-\alpha}\partial_3 L\{y\}(x) \right]\big|_{x=b} = 0$$

since $h(a)$ or $h(b)$ is, respectively, arbitrary. $\qquad\qquad\square$

In the case where L does not depend on $y(a)$ and $y(b)$, by Theorem 3.3 we obtain the following result.

Corollary 3.1. *If y is a local extremizer to problem*

$$\mathcal{J}(y) = \int_a^b L(x, y(x), {}_a^C D_x^\alpha y(x), {}_x^C D_b^\beta y(x))\, dx \longrightarrow extr,$$

then y satisfies the fractional Euler–Lagrange equation

$$\partial_2 L(x, y(x), {}_a^C D_x^\alpha y(x), {}_x^C D_b^\beta y(x)) + {}_x D_b^\alpha \partial_3 L(x, y(x), {}_a^C D_x^\alpha y(x), {}_x^C D_b^\beta y(x))$$
$$+ {}_a D_x^\beta \partial_4 L(x, y(x), {}_a^C D_x^\alpha y(x), {}_x^C D_b^\beta y(x)) = 0$$

for all $x \in [a, b]$. Moreover, if $y(a)$ is not specified, then

$$\left[{}_xI_b^{1-\alpha}\partial_3 L(x, y(x), {}_a^C D_x^\alpha y(x), {}_x^C D_b^\beta y(x)) \right.$$
$$\left. - {}_aI_x^{1-\beta}\partial_4 L(x, y(x), {}_a^C D_x^\alpha y(x), {}_x^C D_b^\beta y(x)) \right]\bigg|_{x=a} = 0,$$

if $y(b)$ is not specified, then

$$\left[{}_aI_x^{1-\beta}\partial_4 L(x, y(x), {}_a^C D_x^\alpha y(x), {}_x^C D_b^\beta y(x)) \right.$$
$$\left. - {}_xI_b^{1-\alpha}\partial_3 L(x, y(x), {}_a^C D_x^\alpha y(x), {}_x^C D_b^\beta y(x)) \right]\bigg|_{x=b} = 0.$$

We note that the generalized Euler–Lagrange equation contains both the left and the right fractional derivative. The generalized natural conditions contain also the left and the right fractional integral. Although the functional has been written only in terms of the Caputo fractional derivatives, necessary conditions (3.17)–(3.19) contain Caputo fractional derivatives, Riemann–Liouville fractional derivatives, and Riemann–Liouville fractional integrals.

Observe that if α goes to 1, then the operators $_a^C D_x^\alpha$ and $_a D_x^\alpha$ can be replaced with $\frac{d}{dx}$ and the operators $_x^C D_b^\alpha$ and $_x D_b^\alpha$ can be replaced with $-\frac{d}{dx}$. Thus, if the $_x^C D_b^\beta y$ term is not present in (3.16), then for $\alpha \to 1$ we obtain a corresponding result in the classical context of the calculus of variations (see Cruz, Torres and Zinober, 2010 and Corollary 1 of Malinowska and Torres, 2010a).

Corollary 3.2. *If y is a local extremizer for*

$$\mathcal{J}(y) = \int_a^b L(x, y(x), y'(x), y(a), y(b))\, dx \longrightarrow extr$$

$$(y(a) = y_a), \quad (y(b) = y_b),$$

then

$$\frac{d}{dx}\partial_3 L(x, y(x), y'(x), y(a), y(b)) = \partial_2 L(x, y(x), y'(x), y(a), y(b))$$

for all $x \in [a, b]$. Moreover, if $y(a)$ is free, then

$$\partial_3 L(a, y(a), y'(a), y(a), y(b)) = \int_a^b \partial_5 L(x, y(x), y'(x), y(a), y(b)) dx;$$

if $y(b)$ is free, then

$$\partial_3 L(b, y(b), y'(b), y(a), y(b)) = -\int_a^b \partial_6 L(x, y(x), y'(x), y(a), y(b)) dx.$$

Example 3.1. Consider the following problem:

$$\mathcal{J}(y) = \frac{1}{2}\int_0^1 \left[(_0^C D_x^\alpha y(x))^2 - 2y(x) + \gamma y^2(0) \right] dx \longrightarrow \min,$$

where $\gamma \in \mathbb{R}^+$. For this problem, the fractional Euler–Lagrange equation and the natural boundary conditions are given, respectively, as

$$_x D_1^\alpha{_0^C} D_x^\alpha y(x) = 1,$$

$$\int_0^1 \gamma y(0) dx = \left._x I_1^{1-\alpha}{_0^C} D_x^\alpha y(x)\right|_{x=0},$$

$$0 = -\left._x I_1^{1-\alpha}{_0^C} D_x^\alpha y(x)\right|_{x=1}.$$

Therefore, we obtain

$$y(x) = \frac{1}{\gamma\Gamma(2)} + \frac{1}{\Gamma(\alpha+1)}\int_0^x (x-t)^{\alpha-1}(1-t)^\alpha dt$$

as a candidate solution.

3.3.2 Transversality Conditions I

Let us introduce the linear space

$$D = \left\{ (x,t) \in C^1([a,b]) \times [a,b] \mid x(a) = x_a \right\}$$

endowed with the norm

$$\|(x,t)\|_{1,\infty} := \max_{a \le t \le b} |x(t)| + \max_{a \le t \le b} \left| {}_a^C D_t^\alpha x(t) \right| + |t|.$$

For simplicity of notation, we introduce operator $[\cdot]$ defined by

$$[x](t) = (t, x(t), {}_a^C D_t^\alpha x(t)).$$

We consider the following type of functionals:

$$J(x,T) = \int_a^T L[x](t)\,dt + \phi(T, x(T)), \tag{3.24}$$

where the Lagrangian $L : [a,b] \times \mathbb{R}^2 \to \mathbb{R}$ and the terminal cost function $\phi : [a,b] \times \mathbb{R} \to \mathbb{R}$ are at least of class C^1. Observe that we have a free end-point T and no constraint on $x(T)$. Therefore, they become a part of the optimal choice process. We address the problem of finding a pair (x, T) which minimizes (or maximizes) the functional J on D, i.e., there exists $\delta > 0$ such that $J(x,T) \le J(\bar{x}, t)$ (or $J(x,T) \ge J(\bar{x}, t)$) for all $(\bar{x}, t) \in D$ with $\|(\bar{x} - x, t - T)\|_{1,\infty} < \delta$.

Theorem 3.4. *Consider the functional given by* (3.24). *Suppose that* (x, T) *minimizes (or maximizes) functional* (3.24) *on* D. *Then the pair* (x, T) *is a solution of the fractional Euler–Lagrange equation*

$$\partial_2 L[x](t) + {}_tD_T^\alpha(\partial_3 L[x](t)) = 0 \tag{3.25}$$

on the interval $[a, T]$, *and satisfies the transversality conditions*

$$\begin{cases} L[x](T) + \partial_1\phi(T, x(T)) - x'(T) \left[{}_tI_T^{1-\alpha}\partial_3 L[x](t) \right]_{t=T} = 0 \\ \left[{}_tI_T^{1-\alpha}\partial_3 L[x](t) \right]_{t=T} + \partial_2\phi(T, x(T)) = 0. \end{cases} \tag{3.26}$$

Proof. Let us consider a variation $(x(t) + \epsilon h(t), T + \epsilon \triangle T)$, where $h \in C^1([a,b])$, $\triangle T \in \mathbb{R}$ and $\epsilon \in \mathbb{R}$ with $|\epsilon| \ll 1$. The constraint on $x = a$ implies that all admissible variations must fulfill the condition $h(a) = 0$. Define $j(\cdot)$ on a neighborhood of zero by

$$\begin{aligned} j(\epsilon) &= J(x + \epsilon h, T + \epsilon \triangle T) \\ &= \int_a^{T+\epsilon\triangle T} L[x + \epsilon h](t)\,dt + \phi(T + \epsilon\triangle T, (x + \epsilon h)(T + \epsilon\triangle T)). \end{aligned}$$

If (x, T) minimizes (or maximizes) functional (3.24) on D, then $j'(0) = 0$. Therefore, one has

$$0 = \int_a^T \left[\partial_2 L[x](t) h(t) + \partial_3 L[x](t)_a^C D_t^\alpha h(t) \right] dt + L[x](T) \triangle T$$
$$+ \partial_1 \phi(T, x(T)) \triangle T + \partial_2 \phi(T, x(T))[h(T) + x'(T) \triangle T].$$

Integrating by parts (cf. equations (3.3)–(3.4)), and since $h(a) = 0$, we get

$$0 = \int_a^T \left[\partial_2 L[x](t) + {}_t D_T^\alpha (\partial_3 L[x](t)) \right] h(t) \, dt + \left[{}_t I_T^{1-\alpha} (\partial_3 L[x](t)) h(t) \right]_{t=T}$$
$$+ L[x](T) \triangle T + \partial_1 \phi(T, x(T)) \triangle T + \partial_2 \phi(T, x(T))[h(T) + x'(T) \triangle T]$$
$$= \int_a^T \left[\partial_2 L[x](t) + {}_t D_T^\alpha (\partial_3 L[x](t)) \right] h(t) \, dt$$
$$+ \triangle T \left[L[x](T) + \partial_1 \phi(T, x(T)) - x'(T) \left[{}_t I_T^{1-\alpha} \partial_3 L[x](t) \right]_{t=T} \right]$$
$$+ \left[\left[{}_t I_T^{1-\alpha} \partial_3 L[x](t) \right]_{t=T} + \partial_2 \phi(T, x(T)) \right] [h(T) + x'(T) \triangle T].$$
$$(3.27)$$

As h and $\triangle T$ are arbitrary we can choose $h(T) = 0$ and $\triangle T = 0$. Then, by the fundamental lemma of the calculus of variations, we deduce equation (3.25). But if x is a solution of (3.25), then the condition (3.27) takes the form

$$0 = \triangle T \left[L[x](T) + \partial_1 \phi(T, x(T)) - x'(T) \left[{}_t I_T^{1-\alpha} \partial_3 L[x](t) \right]_{t=T} \right]$$
$$+ \left[\left[{}_t I_T^{1-\alpha} \partial_3 L[x](t) \right]_{t=T} + \partial_2 \phi(T, x(T)) \right] [h(T) + x'(T) \triangle T].$$

Restricting ourselves to those h for which $h(T) = -x'(T) \triangle T$, we get the first equation of (3.26). Analogously, considering those variations for which $\triangle T = 0$ we get the second equation of (3.26). $\square$

In the case when α goes to 1, by Theorem 3.4 we obtain the following result.

Corollary 3.3. *If (x, T) is an extremizer of*

$$J(x, T) = \int_a^T L(t, x(t), x'(t)) \, dt + \phi(T, x(T))$$

on the set

$$\left\{ x \in C^1([a, b]) \mid x(a) = x_a \right\} \times [a, b],$$

then it is a solution of the three following equations:

$$\partial_2 L(t, x(t), x'(t)) - \frac{d}{dt} \partial_3 L(t, x(t), x'(t)) = 0, \quad \text{for all } t \in [a, T],$$

$$L(T, x(T), x'(T)) + \partial_1 \phi(T, x(T)) - x'(T) \partial_3 L(T, x(T), x'(T)) = 0,$$

and

$$\partial_3 L(T, x(T), x'(T)) + \partial_2 \phi(T, x(T)) = 0.$$

Example 3.2. Let

$$J(x,T) = \int_0^T \left[t^2 - 1 + ({}_a^C D_t^\alpha x(t))^2\right] dt, \quad T \in [0,10].$$

It easy to verify that a constant function $x(t) = K$ and the end-point $T = 1$ satisfies the necessary conditions of optimality of Theorem 3.4, with the value of K being determined by the initial-point $x(0)$.

Now we shall rewrite the necessary conditions (3.26) in terms of the increment on time $\triangle T$ and on the consequent increment on x, $\triangle x_T$. To start, we fix $\epsilon = 1$ and variation functions h satisfying the additional condition $h'(T) = 0$. Define the total increment by

$$\triangle x_T = (x + h)(T + \triangle T) - x(T).$$

Doing Taylor's expansion up to the first order, for a small $\triangle T$, we have

$$(x + h)(T + \triangle T) - (x + h)(T) = x'(T)\triangle T + O(\triangle T)^2,$$

and so we can write $h(T)$ in terms of $\triangle T$ and $\triangle x_T$:

$$h(T) = \triangle x_T - x'(T)\triangle T + O(\triangle T)^2. \tag{3.28}$$

If (x,T) is an extremizer of J, then

$$\partial_2 L[x](t) + {}_t D_T^\alpha(\partial_3 L[x](t)) = 0 \quad \text{holds for all } t \in [a,T].$$

Therefore, substituting (3.28) into equation (3.27) we obtain

$$\triangle T \left[L[x](T) + \partial_1 \phi(T, x(T)) - x'(T)\left[{}_t I_T^{1-\alpha}\partial_3 L[x](t)\right]_{t=T}\right]$$
$$+ \triangle x_T \left[\left[{}_t I_T^{1-\alpha}\partial_3 L[x](t)\right]_{t=T} + \partial_2 \phi(T, x(T))\right] + O(\triangle T)^2 = 0. \tag{3.29}$$

We remark that the above equation is evaluated in one single point $x = T$. Equation (3.29) replaces the missing terminal condition in the problem.

For special values of $\triangle T$ and $\triangle x_T$, we obtain some particular but important results on fractional transversality conditions.

A. Vertical terminal line

In this case the upper bound T is fixed and consequently the variation $\triangle T = 0$. Therefore, equation (3.29) becomes

$$\triangle x_T \left[\left[{}_t I_T^{1-\alpha}\partial_3 L[x](t)\right]_{t=T} + \partial_2 \phi(T, x(T))\right] = 0,$$

and by the arbitrariness of $\triangle x_T$, we deduce that

$$\left[{}_t I_T^{1-\alpha}\partial_3 L[x](t)\right]_{t=T} + \partial_2 \phi(T, x(T)) = 0.$$

When $\phi \equiv 0$, we get the natural boundary condition:

$$\left[{}_t I_T^{1-\alpha}\partial_3 L[x](t)\right]_{t=T} = 0.$$

B. Horizontal terminal line

Now we have $\triangle x_T = 0$ but $\triangle T$ is arbitrary. Hence, equation (3.29) implies

$$L[x](T) + \partial_1 \phi(T, x(T)) - x'(T) \left[{}_t I_T^{1-\alpha} \partial_3 L[x](t) \right]_{t=T} = 0.$$

C. Terminal curve

In this case the terminal point is described by a given function ψ, in the sense that $x(T) = \psi(T)$, where $\psi : [a, b] \to \mathbb{R}$ is a prescribed differentiable curve. For a small $\triangle T$, from Taylor's formula, one has

$$\begin{aligned} \triangle x_T &= \psi(T + \triangle T) - \psi(T) \\ &= \psi'(T)\triangle T + O(\triangle T)^2. \end{aligned} \tag{3.30}$$

Substituting (3.30) into (3.29) yields

$$(\psi'(T) - x'(T)) \left\{ \left[{}_t I_T^{1-\alpha} \partial_3 L[x](t) \right]_{t=T} + \partial_2 \phi(T, x(T)) \right\} + L[x](T) \\ + \partial_1 \phi(T, x(T)) = 0. \tag{3.31}$$

D. Truncated vertical terminal line

We now consider the case where $\triangle T = 0$ and $x(T) \geq x_{min}$. Here x_{min} is a minimum permissible level of x. By the Kuhn–Tucker conditions, we obtain

$$\left[{}_t I_T^{1-\alpha} \partial_3 L[x](t) \right]_{t=T} + \partial_2 \phi(T, x(T)) \leq 0, \quad x(T) \geq x_{min},$$

and $(x(T) - x_{min}) \left\{ \left[{}_t I_T^{1-\alpha} \partial_3 L[x](t) \right]_{t=T} + \partial_2 \phi(T, x(T)) \right\} = 0$

for the maximization problem; and

$$\left[{}_t I_T^{1-\alpha} \partial_3 L[x](t) \right]_{t=T} + \partial_2 \phi(T, x(T)) \geq 0, \quad x(T) \geq x_{min},$$

and $(x(T) - x_{min}) \left\{ \left[{}_t I_T^{1-\alpha} \partial_3 L[x](t) \right]_{t=T} + \partial_2 \phi(T, x(T)) \right\} = 0$

if x is a minimizer of J.

E. Truncated horizontal terminal line

In this situation, the constraints are $\triangle x_T = 0$ and $T \leq T_{max}$. With an analogous discussion as for the previous case, we obtain:

if x is a maximizer of J, then

$$L[x](T) + \partial_1 \phi(T, x(T)) - x'(T) \left[{}_t I_T^{1-\alpha} \partial_3 L[x](t) \right]_{t=T} \geq 0, \quad T \leq T_{max},$$

and $(T - T_{max}) \left[L[x](T) + \partial_1 \phi(T, x(T)) - x'(T) \left[{}_t I_T^{1-\alpha} \partial_3 L[x](t) \right]_{t=T} \right] = 0;$

and if x is a minimizer of J, then

$$L[x](T) + \partial_1 \phi(T, x(T)) - x'(T) \left[{}_t I_T^{1-\alpha} \partial_3 L[x](t) \right]_{t=T} \leq 0, \quad T \leq T_{max},$$

and $(T - T_{max}) \left[L[x](T) + \partial_1 \phi(T, x(T)) - x'(T) \left[{}_t I_T^{1-\alpha} \partial_3 L[x](t) \right]_{t=T} \right] = 0.$

3.3.3 Transversality Conditions II

For simplicity of notation, we introduce the operator $\{\cdot, \cdot\}$ defined by

$$\{x, \varphi\}(t, T) = (t, x(t), {}^C_a D^\alpha_t x(t), \varphi(T)).$$

Let us consider now the following variational problem:

$$J(x, T) = \int_a^T L\{x, \varphi\}(t, T)dt \longrightarrow \text{extr}$$
$$(x, T) \in \text{D} \tag{3.32}$$
$$x(T) = \varphi(T),$$

where $L : [a, b] \times \mathbb{R}^3 \to \mathbb{R}$ and $\varphi : [a, b] \to \mathbb{R}$ are at least of class C^1. Here $x = \varphi(t)$ is a specified curve.

Theorem 3.5. *Suppose that (x, T) is a solution to problem (3.32). Then the pair (x, T) is a solution of the Euler–Lagrange equation*

$$\partial_2 L\{x, \varphi\}(t, T) + {}_t D^\alpha_T(\partial_3 L\{x, \varphi\}(t, T)) = 0 \tag{3.33}$$

on the interval $[a, T]$, and satisfies the transversality condition

$$(\varphi'(T) - x'(T)) \left[{}_t I^{1-\alpha}_T \partial_3 L\{x, \varphi\}(t, T)\right]_{t=T}$$
$$+ \varphi'(T) \int_a^T \partial_4 L\{x, \varphi\}(t, T)dt + L\{x, \varphi\}(T, T) = 0. \tag{3.34}$$

Proof. Suppose that (x, T) is a solution to problem (3.32) and consider the value of J at an admissible variation $(x(t) + \epsilon h(t), T + \epsilon \triangle T)$, where $\epsilon \in \mathbb{R}$ is a small parameter, $\triangle T \in \mathbb{R}$, and $h \in C^1([a, b])$ with $h(a) = 0$. Let

$$j(\epsilon) = J(x + \epsilon h, T + \epsilon \triangle T) = \int_a^{T+\epsilon \triangle T} L\{x + \epsilon h, \varphi\}(t, T + \epsilon \triangle T)dt.$$

Then, a necessary condition for (x, T) to be a solution to problem (3.32) is given by

$$j'(0) = 0 \Leftrightarrow \int_a^T \left[\partial_2 L\{x, \varphi\}(t, T)h(t) + \partial_3 L\{x, \varphi\}(t, T){}^C_a D^\alpha_t h(t)\right.$$
$$\left. + \partial_4 L\{x, \varphi\}(t, T)\varphi'(T)\triangle T\right]dt + L\{x, \varphi\}(T, T)\triangle T = 0.$$

Integrating by parts (cf. equations (3.3)–(3.4)), and since $h(a) = 0$, we get

$$0 = \int_a^T \left[\partial_2 L\{x, \varphi\}(t, T) + {}_t D^\alpha_T(\partial_3 L\{x, \varphi\}(t, T))\right] h(t)\, dt$$
$$+ \left[{}_t I^{1-\alpha}_T(\partial_3 L\{x, \varphi\}(t, T))h(t)\right]_{t=T} + L\{x, \varphi\}(T, T)\triangle T \tag{3.35}$$
$$+ \int_a^T \left[\partial_4 L\{x, \varphi\}(t, T)\varphi'(T)\triangle T\right]dt.$$

As h and $\triangle T$ are arbitrary, first we consider h and $\triangle T$ such that $h(T) = 0$ and $\triangle T = 0$. Then, by the fundamental lemma of the calculus of variations, we deduce equation (3.33) for all $t \in [a, T]$. Therefore, in order for (x, T) to be a solution to problem (3.32), x must be a solution of the fractional Euler–Lagrange equation. But if x is a solution of (3.33), the first integral in expression (3.35) vanishes, and then the condition (3.35) takes the form

$$\left[{}_tI_T^{1-\alpha}(\partial_3 L\{x, \varphi\}(t, T))h(t) \right]_{t=T} + L\{x, \varphi\}(T, T)\triangle T$$

$$+ \int_a^T [\partial_4 L\{x, \varphi\}(t, T)\varphi'(T)\triangle T]\, dt = 0. \quad (3.36)$$

Since the right-hand point of x lies on the curve $z = \varphi(t)$, we have

$$x(T + \epsilon\triangle T) + \epsilon h(T + \epsilon\triangle T) = \varphi(T + \epsilon\triangle T).$$

Hence, differentiating with respect to ϵ and setting $\epsilon = 0$, we get

$$h(T) = (\varphi'(T) - x'(T))\triangle T. \quad (3.37)$$

Substituting (3.37) into (3.36) yields

$$\triangle T(\varphi'(T) - x'(T)) \left[{}_tI_T^{1-\alpha}\partial_3 L\{x, \varphi\}(t, T) \right]_{t=T}$$

$$+ \varphi'(T)\triangle T \int_a^T \partial_4 L\{x, \varphi\}(t, T)dt + L\{x, \varphi\}(T, T)\triangle T = 0.$$

Since $\triangle T$ can take any value, we obtain condition (3.34).   □

In the case when α goes to 1, by Theorem 3.5 we obtain the following result.

Corollary 3.4. *If (x, T) is an extremizer of*

$$J(x, T) = \int_a^T L(t, x(t), x'(t), \varphi(T))\, dt$$

on the set

$$\left\{ x \in C^1([a, b]) \mid x(a) = x_a \right\} \times [a, b],$$

then it is a solution of the two following equations:

$$\partial_2 L(t, x(t), x'(t), \varphi(T)) - \frac{d}{dt}\partial_3 L(t, x(t), x'(t), \varphi(T)) = 0, \quad \textit{for all } t \in [a, T],$$

$$(\varphi'(T) - x'(T))\partial_3 L(T, x(T), x'(T), \varphi(T))$$

$$+ \varphi'(T) \int_a^T \partial_4 L(t, x(t), x'(t), \varphi(T))\, dt + L(T, x(T), x'(T), \varphi(T)) = 0.$$

Example 3.3. Consider the following problem

$$J(x,T) = \int_0^T \left[({}_0^C D_t^\alpha x(t))^2 + \varphi^2(T) \right] dt \longrightarrow \min$$

$$x(T) = \varphi(T) = T,$$

$$x(0) = 1.$$

(3.38)

For this problem, the fractional Euler–Lagrange equation and the transversality condition (see Theorem 3.5) are given, respectively, as

$$_tD_T^\alpha \left({}_0^C D_t^\alpha x(t) \right) = 0,$$

(3.39)

$$2(1 - x'(T)) \left[{}_tI_T^{1-\alpha} {}_0^C D_t^\alpha x(t) \right] |_{t=T} + 2T^2 + ({}_0^C D_t^\alpha x(T))^2 + T^2 = 0. \quad (3.40)$$

Note that it is a difficult task to solve the above fractional equations. For $0 < \alpha < 1$ a numerical method should be used. When α goes to 1, problem (3.38) becomes

$$J(x,T) = \int_0^T \left[(x'(t))^2 + \varphi^2(T) \right] dt \longrightarrow \min$$

$$x(T) = \varphi(T) = T,$$

$$x(0) = 1,$$

(3.41)

and equations (3.39)–(3.40) are replaced by

$$x''(t) = 0,$$

(3.42)

$$2(1 - x'(T))x'(T) + 2T^2 + (x'(T))^2 + T^2 = 0.$$

(3.43)

Solving equations (3.42)–(3.43) we obtain that

$$\tilde{x}(t) = \frac{\sqrt{-6 + 6\sqrt{13}} - 6}{\sqrt{-6 + 6\sqrt{13}}} t + 1, \quad T = \frac{1}{6}\sqrt{-6 + 6\sqrt{13}}$$

is an extremal to problem (3.41).

3.3.4 *Infinite Horizon Fractional Variational Problems*

Starting with Ramsey's pioneering work, infinite horizon variational problems have been widely used in economics. One may assume that, due to some constraints of an economical nature, the infinite horizon variational problem does not depend on the usual derivative but on the left Caputo fractional derivative. In this case one has to consider the following problem:

$$J(x) = \int_a^{+\infty} L[x](t)dt \longrightarrow \max$$

$$x \in D_\infty,$$

(3.44)

where $L : [a, +\infty] \times \mathbb{R}^2 \to \mathbb{R}$ is at least of class C^1 and

$$D_\infty = \left\{ x \in C^1([a, +\infty]) \mid x(a) = x_a \right\}.$$

Since the integral in problem (3.44) may not converge, an extension of the definition of optimality (used in Section 3.3.2 and Section 3.3.3) is needed. We follow Brock's notion of optimality: a function $x \in D_\infty$ is said to be weakly maximal to problem (3.44) if

$$\lim_{T \to +\infty} \inf_{T' \geq T} \int_a^{T'} [L[\overline{x}](t) - L[x](t)] \, dt \leq 0$$

for all $\overline{x} \in D_\infty$.

Lemma 3.2. *If g is continuous on $[a, +\infty)$ and*

$$\lim_{T \to +\infty} \inf_{T' \geq T} \int_a^{T'} g(t) h(t) \, dt = 0$$

for all continuous functions $h : [a, +\infty) \to \mathbb{R}$ with $h(a) = 0$, then $g(t) = 0$ for all $t \geq a$.

Proof.   Suppose that $g \not\equiv 0$. With any loss of generality, we may assume that there exists some $t_0 \in (a, +\infty)$ such that $g(t_0) > 0$. By continuity, $g(t) > 0$ on some neighborhood of t_0, say (t_1, t_2). Let h be a continuous function such that $h(t) = 0$ on $[a, t_1] \cup [t_2, +\infty)$ and positive elsewhere. Then, for T' sufficiently large,

$$\int_a^{T'} g(t) h(t) \, dt = \int_{t_1}^{t_2} g(t) h(t) \, dt > 0$$

and so

$$\lim_{T \to +\infty} \inf_{T' \geq T} \int_a^{T'} g(t) h(t) \, dt > 0,$$

which is a contradiction.   □

Theorem 3.6. *Suppose that x is weakly maximal to problem (3.44). Let $h \in C^1([a, +\infty])$ and $\epsilon \in \mathbb{R}$. Define*

$$A(\epsilon, T') = \int_a^{T'} \frac{L[x + \epsilon h](t) - L[x](t)}{\epsilon} \, dt;$$

$$V(\epsilon, T) = \inf_{T' \geq T} \int_a^{T'} [L[x + \epsilon h](t) - L[x](t)] \, dt;$$

$$V(\epsilon) = \lim_{T \to +\infty} V(\epsilon, T).$$

Suppose that

(i) $\lim_{\epsilon \to 0} \dfrac{V(\epsilon, T)}{\epsilon}$ *exists for all T;*

(ii) $\lim_{T \to +\infty} \dfrac{V(\epsilon, T)}{\epsilon}$ *exists uniformly for ϵ;*

(iii) *For every $T' > a$, $T > a$, and $\epsilon \in \mathbb{R} \setminus \{0\}$, there exists a sequence $(A(\epsilon, T'_n))_{n \in \mathbb{N}}$ such that*

$$\lim_{n \to +\infty} A(\epsilon, T'_n) = \inf_{T' \geq T} A(\epsilon, T')$$

uniformly for ϵ.

Then x is a solution of the fractional Euler–Lagrange equation

$$\partial_2 L[x](t) + {}_t D_T^\alpha (\partial_3 L[x](t)) = 0$$

for all $t \in [a, +\infty)$, and for all $T > t$. Moreover, it satisfies the transversality condition

$$\lim_{T \to +\infty} \inf_{T' \geq T} {}_t I_{T'}^{1-\alpha} \partial_3 L[x](t)]_{t=T'} x(T') = 0.$$

Proof. Observe that $V(\epsilon) \leq 0$ for every $\epsilon \in \mathbb{R}$, and $V(0) = 0$. Thus $V'(0) = 0$, i.e.,

$$0 = \lim_{\epsilon \to 0} \frac{V(\epsilon)}{\epsilon}$$
$$= \lim_{T \to +\infty} \inf_{T' \geq T} \int_a^{T'} \lim_{\epsilon \to 0} \frac{L[x + \epsilon h](t) - L[x](t)}{\epsilon} dt$$
$$= \lim_{T \to +\infty} \inf_{T' \geq T} \int_a^{T'} \left[\partial_2 L[x](t) h(t) + \partial_3 L[x](t) {}_a^C D_t^\alpha h(t) \right] dt.$$

Thus, integrating by parts, and since $h(a) = 0$, we get

$$\lim_{T \to +\infty} \inf_{T' \geq T} \left\{ \int_a^{T'} [\partial_2 L[x](t) + {}_t D_{T'}^\alpha (\partial_3 L[x](t))] h(t) dt \right.$$

$$\left. + [{}_t I_{T'}^{1-\alpha} \partial_3 L[x](t) h(t)]_{t=T'} \right\} = 0. \quad (3.45)$$

By the arbitrariness of h, we may assume that $h(T') = 0$, and so

$$\lim_{T \to +\infty} \inf_{T' \geq T} \int_a^{T'} [\partial_2 L[x](t) + {}_t D_{T'}^\alpha (\partial_3 L[x](t))] h(t) dt = 0. \quad (3.46)$$

Then, applying Lemma 3.2, we obtain

$$\partial_2 L[x](t) + {}_t D_T^\alpha (\partial_3 L[x](t)) = 0$$

for all $t \in [a, +\infty)$, and for all $T > t$. Substituting equation (3.46) into equation (3.45) we get

$$\lim_{T \to +\infty} \inf_{T' \geq T} [{}_t I_{T'}^{1-\alpha} \partial_3 L[x](t)]_{t=T'} h(T') = 0. \qquad (3.47)$$

To eliminate h from equation (3.47), consider the particular case $h(t) = \chi(t)x(t)$, where $\chi : [a, +\infty) \to \mathbb{R}$ is a function of class C^1 such that $\chi(a) = 0$ and $\chi(t) = const \neq 0$, for all $t > t_0$, for some $t_0 > a$. We deduce that

$$\lim_{T \to +\infty} \inf_{T' \geq T} [{}_t I_{T'}^{1-\alpha} \partial_3 L[x](t)]_{t=T'} x(T') = 0. \qquad \square$$

3.4   The Fractional Isoperimetric Problem

We state the fractional isoperimetric problem as follows. Given a functional J as in (3.14) in which functions y minimize (or maximize) J subject to given boundary conditions

$$y(a) = y_a, \ y(b) = y_b, \qquad (3.48)$$

and an integral constraint

$$I(y) = \int_a^b g[y]dx = l. \qquad (3.49)$$

Here, as before, we consider a function g of class C^1, such that $\partial_3 g$ has continuous right Riemann–Liouville fractional derivative of order α and $\partial_4 g$ has continuous left Riemann–Liouville fractional derivative of order β, and functions $y \in {}_a^\alpha E_b^\beta$, where ${}_a^\alpha E_b^\beta$ is defined as in Section 3.2. A function $y \in {}_a^\alpha E_b^\beta$ that satisfies (3.48) and (3.49) is called admissible.

Definition 3.5. *An admissible function y is an extremal for I in (3.49) if it satisfies the equation*

$$\partial_2 g[y](x) + {}_x D_b^\alpha \partial_3 g[y](x) + {}_a D_x^\beta \partial_4 g[y](x) = 0, \ \textit{for all } x \in [a, b].$$

Observe that Definition 3.5 makes sense by Theorem 3.2. To solve the isoperimetric problem, the idea is to consider a new extended function. The exact formula for such an extended function depends on y being or not being an extremal for the integral functional $I(y)$ (cf. Theorems 3.7 and 3.8).

Theorem 3.7. *Let y be a local minimizer for J given by (3.14), subject to the conditions (3.48) and (3.49). If y is not an extremal for the functional I, then there exists a constant λ such that y satisfies*

$$\partial_2 F + {}_x D_b^\alpha \partial_3 F + {}_a D_x^\beta \partial_4 F = 0 \qquad (3.50)$$

for all $x \in [a, b]$, where $F = L + \lambda g$.

Proof. Is similar to that of Theorem 2.6 and can be found in Almeida and Torres, 2011a. □

Example 3.4. Let $\overline{y}(x) = E_\alpha(x^\alpha)$, $x \in [0,1]$, where E_α is the Mittag–Leffler function:

$$E_\alpha(x) = \sum_{k=0}^{\infty} \frac{x^k}{\Gamma(\alpha k + 1)}, \quad x \in \mathbb{R}, \quad \alpha > 0.$$

When $\alpha = 1$, the Mittag–Leffler function is the exponential function: $E_1(x) = e^x$. The left Caputo fractional derivative of $\overline{y}$ is $\overline{y}$ (cf. p. 98 of Kilbas, Srivastava and Trujillo, 2006):

$$_0^C D_x^\alpha \overline{y} = \overline{y}.$$

Consider the following fractional variational problem:

$$\begin{cases} J(y) = \displaystyle\int_0^1 (_0^C D_x^\alpha y)^2 \, dx \quad \to \quad \text{extr}, \\[2mm] I(y) = \displaystyle\int_0^1 \overline{y} \, _0^C D_x^\alpha y \, dx = \int_0^1 (\overline{y})^2 \, dx, \\[2mm] y(0) = 1 \quad \text{and} \quad y(1) = E_\alpha(1). \end{cases} \quad (3.51)$$

The augmented function is

$$F(x, y, {}_0^C D_x^\alpha y, {}_x^C D_1^\beta y, \lambda) = (_0^C D_x^\alpha y)^2 + \lambda \overline{y} \, _0^C D_x^\alpha y,$$

and the fractional Euler–Lagrange equation is

$$\partial_2 F + {}_x D_1^\alpha \partial_3 F + {}_x D_1^\beta \partial_4 F = 0$$

i.e.,

$$_x D_1^\alpha (2 \, _0^C D_x^\alpha y + \lambda \overline{y}) = 0.$$

A solution of this problem is $\lambda = -2$ and $y = \overline{y}$.

Observe that, as $\alpha \to 1$, the variational problem (3.51) becomes

$$\int_0^1 y'^2 \, dx \quad \to \quad \text{extr},$$

$$\int_0^1 \overline{y} \, y' \, dx = \frac{1}{2}(e^2 - 1),$$

$$y(0) = 1 \quad \text{and} \quad y(1) = e,$$

and the Euler–Lagrange equation is

$$\partial_2 F - \frac{d}{dx}\partial_3 F = 0 \Leftrightarrow -\frac{d}{dx}(2y' - 2\overline{y}) = 0, \quad (3.52)$$

where $F = y'^2 - 2\overline{y}y'$. Also, for $\alpha = 1$, $\overline{y}(x) = e^x$, which is obviously a solution of the differential equation (3.52) (cf. Fig. 3.1).

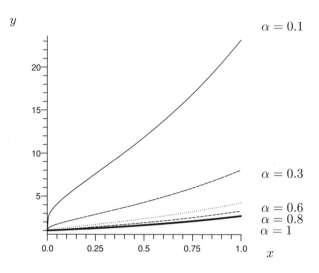

Fig. 3.1   Solutions to problem (3.51).

We now study the case when y is an extremal of I (the so called abnormal case).

Theorem 3.8.   *Let y be a local minimizer to J (3.14), subject to the conditions (3.48) and (3.49). Then, there exist two constants λ_0 and λ, with $(\lambda_0, \lambda) \neq (0, 0)$, such that*

$$\partial_2 K + {}_x D_b^\alpha \partial_3 K + {}_a D_x^\beta \partial_4 K = 0,$$

where $K = \lambda_0 L + \lambda g$.

Proof.   Is similar to that of Theorem 2.7 and can be found in Almeida and Torres, 2011a.   □

An extension

We now present a solution to the isoperimetric problem for functionals of type (3.13). Similarly, one has an integral constraint, but this time of the form

$$I^*(y) = \int_A^B g(x, y(x), {}_a^C D_x^\alpha y(x), {}_x^C D_b^\beta y(x)) dx = l. \tag{3.53}$$

Again, we need the concept of extremal for a functional of type (3.53).

Definition 3.6.   *A function y is called extremal for I^* given by (3.53) if*

$$\partial_2 g[y](x) + {}_x D_B^\alpha \partial_3 g[y](x) + {}_A D_x^\beta \partial_4 g[y](x) = 0, \text{ for all } x \in [A, B].$$

Theorem 3.9. *If y is a local minimizer to J^* given by (3.13), when restricted to the conditions $y(a) = y_a$, $y(b) = y_b$, and (3.53), and if y is not an extremal for I^*, then there exists a constant λ such that*

$$\begin{cases} \partial_2 F + {}_xD_B^\alpha \partial_3 F + {}_AD_x^\beta \partial_4 F = 0 & \text{for all } x \in [A, B] \\ {}_xD_B^\alpha \partial_3 F - {}_xD_A^\alpha \partial_3 F = 0 & \text{for all } x \in (a, A) \\ {}_AD_x^\beta \partial_4 F - {}_BD_x^\beta \partial_4 F = 0 & \text{for all } x \in (B, b), \end{cases} \tag{3.54}$$

where $F = L + \lambda g$.

Remark 3.4. In the case $[A, B] = [a, b]$, Theorem 3.9 is reduced to Theorem 3.7.

Proof. Is similar to that of Theorem 2.11 and can be found in Almeida and Torres, 2011a. □

The following result generalizes Theorem 3.8 and can be proved in a similar way.

Theorem 3.10. *If y is a local minimizer to J^* given by (3.13), subject to the boundary conditions (3.48) and the integral constraint (3.53), then there exist two constants λ_0 and λ, not both zero, such that*

$$\begin{cases} \partial_2 K + {}_xD_B^\alpha \partial_3 K + {}_AD_x^\beta \partial_4 K = 0 & \text{for all } x \in [A, B] \\ {}_xD_B^\alpha \partial_3 K - {}_xD_A^\alpha \partial_3 K = 0 & \text{for all } x \in (a, A) \\ {}_AD_x^\beta \partial_4 K - {}_BD_x^\beta \partial_4 K = 0 & \text{for all } x \in (B, b), \end{cases}$$

where $K = \lambda_0 L + \lambda g$.

3.5 Sufficient Conditions for Optimality

We are now interested in finding sufficient conditions for J to attain extremes. Typically, some conditions of convexity over the Lagrangian are needed.

Theorem 3.11. *Suppose that the function $L(\underline{x}, y, u, v)$ is jointly convex in (y, u, v). Then each solution y_0 of the fractional Euler–Lagrange equation (3.15) minimizes*

$$J(y) = \int_a^b L[y]dx$$

when restricted to the boundary conditions $y(a) = y_0(a)$ and $y(b) = y_0(b)$.

Proof.   Is similar to that of Theorem 2.14 and can be found in Almeida and Torres, 2011a.   □

The same procedure can be easily adapted for the isoperimetric problem.

Theorem 3.12. *Suppose that, for some constant* λ, *the functions* $L(\underline{x}, y, u, v)$ *and* $\lambda g(\underline{x}, y, u, v)$ *are jointly convex in* (y, u, v). *Let* $F = L + \lambda g$. *Then each solution* y_0 *of the fractional equation*

$$\partial_2 F + {}_x D_b^\alpha \partial_3 F + {}_a D_x^\beta \partial_4 F = 0$$

minimizes

$$J(y) = \int_a^b L[y] dx$$

under the constraints $y(a) = y_0(a)$, $y(b) = y_0(b)$, *and*

$$I(y) = \int_a^b g[y] dx = l, \quad l \in \mathbb{R}.$$

Proof.   Let us prove that y_0 minimizes

$$\widetilde{F}(y) = \int_a^b \left(L[y] + \lambda g[y] \right) dx.$$

First, it is easy to prove that $L(\underline{x}, y, u, v) + \lambda g(\underline{x}, y, u, v)$ is convex. Let $\eta \in {}_a^\alpha E_b^\beta$ (where ${}_a^\alpha E_b^\beta$ is defined as in Section 3.2) be such that $\eta(a) = \eta(b) = 0$. Then, by Theorem 3.11, $\widetilde{F}(y_0 + \eta) \geq \widetilde{F}(y_0)$. In other words, if $y \in {}_a^\alpha E_b^\beta$ is any function such that $y(a) = y_0(a)$ and $y(b) = y_0(b)$, then

$$\int_a^b L[y] \, dx + \int_a^b \lambda g[y] \, dx \geq \int_a^b L[y_0] \, dx + \int_a^b \lambda g[y_0] \, dx.$$

If we restrict the integral constraint, we obtain

$$\int_a^b L[y] \, dx + l \geq \int_a^b L[y_0] \, dx + l,$$

and so

$$\int_a^b L[y] \, dx \geq \int_a^b L[y_0] \, dx,$$

proving the desired result.   □

Example 3.5. Recall Example 3.4.   Since $L(\underline{x}, y, u, v) = u^2$ and $\lambda g(\underline{x}, y, u, v) = -2\overline{y}u$ are both convex, we conclude that $\overline{y}$ is actually a minimizer for the fractional variational problem (3.51).

The next theorem presents a sufficient optimality condition for problem (3.16).

Theorem 3.13. *Let $L(\underline{x}, y, z, t, u, v)$ be jointly convex (concave) in (y, z, t, u, v). If y_0 satisfies conditions (3.17)-(3.19), then y_0 is a global minimizer (maximizer) to problem (3.16).*

Proof. Is similar to that of Theorem 2.14 and can be found in Malinowska and Torres, 2010b.   □

Example 3.6. Consider the following problem:

$$\mathcal{J}(y) = \frac{1}{2} \int_0^1 \left[\left({}_0^C D_x^\alpha y(x) \right)^2 + \gamma y^2(0) + \lambda (y(1) - 1)^2 \right] dx \longrightarrow \min \quad (3.55)$$

where $\gamma, \lambda \in \mathbb{R}^+$. For (3.55), the generalized Euler–Lagrange equation and the natural boundary conditions (see Theorem 3.3) are given, respectively, as

$$_x D_1^\alpha \left({}_0^C D_x^\alpha y(x) \right) = 0, \quad (3.56)$$

$$\int_0^1 \gamma y(0) dx = {}_x I_1^{1-\alpha} \left({}_0^C D_x^\alpha y(x) \right) \big|_{x=0}, \quad (3.57)$$

$$\int_0^1 \lambda (y(1) - 1) dx = -{}_x I_1^{1-\alpha} \left({}_0^C D_x^\alpha y(x) \right) \big|_{x=1}. \quad (3.58)$$

Note that it is difficult to solve the above fractional equations. For $0 < \alpha < 1$ a numerical method should be used. When α goes to 1, problem (3.55) tends to

$$\mathcal{J}(y) = \frac{1}{2} \int_0^1 \left[(y'(x))^2 + \gamma y^2(0) + \lambda (y(1) - 1)^2 \right] dx \longrightarrow \min \quad (3.59)$$

and equations (3.56)–(3.58) could be replaced with

$$y''(x) = 0, \quad (3.60)$$

$$\gamma y(0) = y'(0), \quad (3.61)$$

$$\lambda (y(1) - 1) = -y'(1). \quad (3.62)$$

Solving equations (3.60)–(3.62) we obtain that

$$\bar{y}(x) = \frac{\gamma \lambda}{\gamma \lambda + \lambda + \gamma} x + \frac{\lambda}{\gamma \lambda + \lambda + \gamma}$$

is a candidate for a minimizer. Observe that problem (3.55) satisfies assumptions of Theorem 3.13. Therefore, $\bar{y}$ is a global minimizer to problem (3.59). Expression $\gamma y^2(0) + \lambda (y(1) - 1)^2$ added to the Lagrangian $(y'(x))^2$ works like a penalty function when γ and λ go to infinity. The penalty function itself grows, and forces the merit function to increase in value when the constraints $y(0) = 0$ and $y(1) = 1$ are violated, and causes no growth when constraints are fulfilled.

3.6   Minimal Modified Energy Control

In this section fractional control systems with the Caputo derivative are considered. The modified controllability Gramian and the minimum energy optimal control problem are investigated. Construction of minimizing steering controls for the modified energy functional are proposed.

3.6.1   *Fractional Linear Control Systems and the Gramian*

We consider the following linear time-invariant control system of order $\alpha \in (0, 1]$, denoted by Σ:

$$_0^C D_t^\alpha x(t) = Ax(t) + Bu(t) \,, \quad y = Cx(t) \,,$$

where $x(t) \in \mathbb{R}^n$, $u(t) \in \mathbb{R}^m$, matrix $A \in M_{n \times n}(\mathbb{R})$, $B \in M_{n \times m}(\mathbb{R})$, $C \in M_{p \times n}(\mathbb{R})$, and $_0^C D_t^\alpha$ indicates the fractional Caputo derivative of commensurate order α. The forward trajectory of the system Σ, starting at $t_0 = 0$ and evaluated at $t \geq 0$, is the solution of the initial value problem $_0^C D_t^\alpha x(t) = Ax(t) + Bu(t)$, $x(0) = a \in \mathbb{R}^n$ (Kilbas, Srivastava and Trujillo, 2006):

$$\gamma(t, a, u) = \left(I + \int_0^t S(\tau)A d\tau \right) a + \int_0^t S(t - \tau)Bu(\tau)d\tau \,, \qquad (3.63)$$

where $S(t) = e_\alpha^{At}$. Moreover, we can represent (3.63) in the following way:

$$\gamma(t, a, u) = S_0(t)a + \int_0^t S(t - \tau)Bu(\tau)d\tau \,,$$

where $S_0(t) = E_\alpha(At^\alpha) = I + \int_0^t S(\tau)A d\tau$. Taking into account the output of Σ, the forward output trajectory is then defined by values evaluated at $t \geq 0$:

$$\eta(t, a, u) = C\gamma(t, a, u) = C\left(I + \int_0^t S(\tau)A d\tau \right) a + C \int_0^t S(t - \tau)Bu(\tau)d\tau \,.$$

For system Σ we define the notion of controllability in the standard manner.

Definition 3.7. *Let $T > 0$. The system Σ is controllable on $[0, T]$ if for any $a \in \mathbb{R}^n$ and $b \in \mathbb{R}^n$ there is a control $u(\cdot)$ defined on $[0, T]$ which steers the initial state $\gamma(0, a, u) = a$ to the final state $\gamma(T, a, u) = b$.*

We denote by

$$Q_T = \int_0^T S(T-t)BB^*S^*(T-t)(T-t)^{2(1-\alpha)}dt$$

the controllability Gramian of fractional order α, on the time interval $[0,T]$, corresponding to the system Σ. As in the classical case, Q_T is symmetric and non-negative definite. The term $(T-t)^{2(1-\alpha)}$ under the integral is called a neutralizer of the singularity at $t = T$. It is needed in order to ensure the convergence of the integral.

Let $T > 0$. By $L^2_\alpha([0,T],\mathbb{R}^m)$ we denote the set of functions $\varphi : [0,T] \to \mathbb{R}^m$ such that $\widetilde{\varphi}$ defined by $\widetilde{\varphi}(t) = (T-t)^{\alpha-1}\varphi(t)$ is square integrable on $[0,T]$.

Theorem 3.14. *Let $T > 0$ and Q_T be nonsingular. Then,*

(i) for any states a, $b \in \mathbb{R}^n$ the control law

$$\overline{u}(t) = -(T-t)^{2(1-\alpha)}B^*S^*(T-t)Q_T^{-1}f_T(a,b)\,, \quad t \in [0,T), \quad (3.64)$$

where

$$f_T(a,b) = \left(I + \int_0^T S(t)A\,dt\right)a - b = -b + S_0(T)a$$

and $\overline{u}(T) = 0$, drives point a to point b in time T;

(ii) among all possible controls from $L^2_\alpha([0,T],\mathbb{R}^m)$ driving a to b in time T, the control $\overline{u}$ defined by (3.64) minimizes the integral

$$\int_0^T |(T-t)^{\alpha-1}u(t)|^2dt\,. \quad (3.65)$$

Moreover,

$$\int_0^T |(T-t)^{\alpha-1}\overline{u}(t)|^2dt = <Q_T^{-1}f_T(a,b), f_T(a,b)>\,,$$

where $< \cdot, \cdot >$ denotes the inner product.

Proof. Is similar to that of Theorem 2.23 and can be found in Mozyrska and Torres, 2010. □

As in the classical case, the operator

$$u \mapsto \gamma(T,0,u) = \int_0^T S(t)Bu(T-t)dt$$

is a linear operator from the space $L^2_\alpha([0,T],\mathbb{R}^m)$ into $\mathbb{R}^n$. Hence, the classical result can be replaced by the following proposition.

Proposition 3.7. *If any state $b \in \mathbb{R}^n$ is attainable from $a = 0$, then the matrix Q_T is nonsingular for any arbitrary $T > 0$.*

Example 3.7. Let Σ be the following system evaluated on $\mathbb{R}^2$:

$$\Sigma: \quad \begin{cases} {}^C D_{0+}^{0.5} x_1(t) = x_2(t), \\ {}^C D_{0+}^{0.5} x_2(t) = u(t). \end{cases}$$

Let us take $a = (1,0)^*$ and $b = (0,0)^*$. Since $A = \begin{pmatrix} 0 & 1 \\ 0 & 0 \end{pmatrix}$ and $B = \begin{pmatrix} 0 \\ 1 \end{pmatrix}$,

we obtain that $S(t) = \begin{pmatrix} \frac{1}{\sqrt{\pi t}} & 1 \\ 0 & \frac{1}{\sqrt{\pi t}} \end{pmatrix}$ while $S_0(t) = \begin{pmatrix} 1 & \frac{2\sqrt{t}}{\sqrt{\pi}} \\ 0 & 1 \end{pmatrix}$. Hence, the

formula for the solution with the initial condition $\gamma(0, a, u) = a$ is

$$\gamma(t, a, u) = \begin{pmatrix} 1 & \frac{2\sqrt{t}}{\sqrt{\pi}} \\ 0 & 1 \end{pmatrix} a + \int_0^t \begin{pmatrix} \frac{1}{\sqrt{\pi(t-\tau)}} & 1 \\ 0 & \frac{1}{\sqrt{\pi(t-\tau)}} \end{pmatrix} Bu(\tau) d\tau.$$

Let us take $u(t) \equiv 1$. Then, for the given a, $\gamma(t, a, u) = \left(1 + t \quad 2\frac{\sqrt{t}}{\sqrt{\pi}} \right)^*$.
From the last expression we see that using constant $u(\cdot) \equiv 1$ for $t > 0$ we are not able to steer the given initial point a to the origin.

Now let $f_T(a, b) = S_0(T)a - b = a$. The Gramian has the form

$$Q_T = \begin{pmatrix} \frac{T^2}{2} & \frac{2T^{3/2}}{3\sqrt{\pi}} \\ \frac{2T^{3/2}}{3\sqrt{\pi}} & \frac{T}{\pi} \end{pmatrix}$$

and the control

$$\overline{u}(t) = -\frac{18(T-t)}{T^2} + \frac{12\sqrt{T-t}}{T^{3/2}}$$

drives a to b with the modified energy

$$m = \int_0^T |(T-t)^{-0.5}\overline{u}(t)|^2 dt = \frac{18}{T^2}.$$

Example 3.8. Let $\alpha \in (0,1)$. Consider the following fractional system Σ on $\mathbb{R}^2$:

$$\Sigma: \quad \begin{cases} {}_0^C D_t^\alpha x_1(t) = x_2(t), \\ {}_0^C D_t^\alpha x_2(t) = -x_1(t) + u(t). \end{cases}$$

The matrix A is now skew-symmetric, and thus

$$A^0 = I, \quad A = \begin{pmatrix} 0 & 1 \\ -1 & 0 \end{pmatrix}, \quad A^2 = -I.$$

Hence, $A^k = I$ if $k = 0, 4, 8, \ldots$; $A^k = A$ if $k = 1, 5, 9, \ldots$; $A^k = -I$ if $k = 2, 6, 10, \ldots$; and $A^k = -A$ if $k = 3, 7, 11, \ldots$. Moreover,

$$S(t) = t^{\alpha-1}\left(I\frac{1}{\Gamma(\alpha)} + A\frac{t^\alpha}{\Gamma(2\alpha)} - I\frac{t^{2\alpha}}{\Gamma(3\alpha)} - A\frac{t^{3\alpha}}{\Gamma(4\alpha)} + \cdots\right)$$

$$= I\left(\frac{t^{\alpha-1}}{\Gamma(\alpha)} - \frac{t^{3\alpha-1}}{\Gamma(3\alpha)} + \cdots\right) + A\left(\frac{t^{2\alpha-1}}{\Gamma(2\alpha)} - \frac{t^{4\alpha-1}}{\Gamma(4\alpha)} + \cdots\right).$$

Using the notation

$$\sin_\alpha t = \sum_{k=0}^{\infty}(-1)^k\frac{t^{2(k+1)\alpha-1}}{\Gamma[2(k+1)\alpha]}, \quad \cos_\alpha t = \sum_{k=0}^{\infty}(-1)^k\frac{t^{(2k+1)\alpha-1}}{\Gamma[(2k+1)\alpha]},$$

we can write $S(t) = \begin{pmatrix} \cos_\alpha t & \sin_\alpha t \\ -\sin_\alpha t & \cos_\alpha t \end{pmatrix}$. As $B = \begin{pmatrix} 0 \\ 1 \end{pmatrix}$, we have

$$Q_T = \int_0^T (T-t)^{2(1-\alpha)} \cdot M_\alpha(t)dt$$

with

$$M_\alpha(t) = \begin{pmatrix} \sin_\alpha^2(T-t) & \sin_\alpha(T-t)\cos_\alpha(T-t) \\ \sin_\alpha(T-t)\cos_\alpha(T-t) & \cos_\alpha^2(T-t) \end{pmatrix}.$$

To get an exact formula for Q_T and Q_T^{-1} is difficult. We can, however, easily obtain approximations with the desired precision for both matrices, and an approximate formula for the optimal control. Let us consider a concrete situation. Let $\alpha = \frac{1}{2}$. Then $\sin_{\frac{1}{2}} t = e^{-t}$ and

$$\cos_{\frac{1}{2}} t = \frac{1}{\sqrt{\pi t}}\left(1 - \sum_{k=1}^{+\infty}\frac{2^k t^{2k-1}}{\prod_{i=1}^k(2i-1)}\right) = \frac{1}{\sqrt{\pi t}}(1 - 2t + \cdots).$$

We choose to approximate $\cos_{\frac{1}{2}} t$ by functions

$$c_L(t) = \frac{1}{\sqrt{\pi t}}\left(1 - \sum_{k=1}^{L}\frac{2^k t^{2k-1}}{\prod_{i=1}^k(2i-1)}\right), \quad L = 1, 2, \ldots$$

In this way we can approximate Q_T by

$$\int_0^T (T-t)\begin{pmatrix} e^{-2(T-t)} & c_L(T-t)e^{-(T-t)} \\ c_L(T-t)e^{-(T-t)} & c_L^2(t) \end{pmatrix} dt.$$

Let $T = 10$, the starting point to be $a = (0 \quad 1)^*$, and the final point to be the origin, i.e., $b = (0 \quad 0)^*$. Applying formula (3.64) for the steering control $\overline{u}$ with the approximate expressions for $S^*(T-t)$ and Q_T^{-1}, we check that we are very close to the final goal b. Taking more terms in the functions c_L, the value of the modified energy tends to the minimum. For example, for $L = 1$ we have $c_1(t) = \frac{1}{\sqrt{\pi t}}(1 - 2t)$ and the approximate value of energy is $m_1 = 1.02$; for $L = 11$ we have $m_{11} = 0.0921$; and for $L = 12$ we get $m_{12} = 0.0911$. These numbers were calculated using MATLAB with a final precision of four digits.

3.6.2 Rank Conditions and Steering Controls

In the classical theory it is quite easy to explain why the condition rank $B = n$ is sufficient for controllability. Roughly speaking, one needs to construct a special control $u(\cdot)$. In the case of fractional order systems we use the special function given by Proposition 3.6.

Proposition 3.8. *Let $rank\, B = n$ and B^+ be such that $BB^+ = I$. Let $g(\cdot)$ be the matrix function defined by Proposition 3.6. Then the control*

$$\widehat{u}(t) = \frac{1}{T} B^+ g(T - t)(b - S_0(T)a)\,, \quad t \in [0, T]\,,$$

transfers a to b in time $T > 0$.

Proof. The proof follows by a direct calculation:

$$\gamma(T, a, \widehat{u}) = S_0(T)a + \frac{1}{T} \int_0^T S(T - t)BB^+ g(T - t)(b - S_0(T)a)\, dt = b\,. \quad \square$$

Example 3.9. Let Σ be the following fractional system evaluated on $\mathbb{R}$:

$$\Sigma: \quad {}^C_0 D_t^\alpha x(t) = u(t)\,,$$

where $\alpha \in (0, 1]$. Let us take $a, b \in \mathbb{R}$, $T > 0$. Then,

$$S(t) = \frac{t^{\alpha - 1}}{\Gamma(\alpha)}\,, \quad S_0(t) = 1\,, \quad B = B^+ = 1\,,$$

and

$$\gamma(T, a, u) = a + \frac{1}{\Gamma(\alpha)} \int_0^T (T - t)^{\alpha - 1} u(t) dt\,.$$

Consider, according to Proposition 3.6, $g(t) = t^{1-\alpha}\Gamma(\alpha)$. From Proposition 3.8,

$$\widehat{u}(t) = \frac{\Gamma(\alpha)}{T}(T - t)^{1-\alpha}(b - a)$$

transfers a to b: $\gamma(T, a, \widehat{u}) = b$ with the modified energy (3.65) of value

$$m = \frac{\Gamma^2(\alpha)(b - a)^2}{T}\,.$$

It is also the minimum energy as $\widehat{u}(t) = \overline{u}(t)$.

According to Theorem 3.14 and Proposition 3.7, we can state:

Theorem 3.15. *The following conditions are equivalent:*

(i) An arbitrary state $b \in \mathbb{R}^n$ is attainable from 0.

(ii) System Σ is controllable.

(iii) System Σ is controllable at a given time $T > 0$.

(iv) Matrix Q_T is nonsingular for some $T > 0$.

(v) Matrix Q_T is nonsingular for an arbitrary $T > 0$.

(vi) rank $[A|B] = $ rank $\left[B, AB, \ldots, A^{n-1}B\right] = n$.

If the rank condition is satisfied, then the control $\bar{u}(\cdot)$ given by (3.64) steers a to b at time T. Our goal now is to find another formula for the steering control by using the matrix $[A|B]$ instead of the controllability matrix Q_T. It is a classical result that if rank $[A|B] = n$, then there exists a matrix $K \in M(mn, n)$ such that $[A|B] K = I \in M(n, n)$ or, equivalently, there are matrices K_1, K_2, ..., $K_n \in M(m, n)$ such that $BK_1 + ABK_2 + \cdots + A^{n-1}BK_n = I$.

For the next construction it is more convenient to use the notion of the Riemann–Liouville derivative. We begin by introducing a notation for compositions of the Riemann–Liouville derivatives with the same order α, $\alpha \in (0,1)$. Let $R_{0+}^{\alpha,0}\psi(t) = \psi(t)$. Then for $j \in \mathbb{N}$, recursively, we put $R_{0+}^{\alpha,j+1}\psi(t) := {}_0D_t^\alpha \left(R_{0+}^{\alpha,j}\psi(t) \right)$.

Theorem 3.16. *Let rank $[A|B] = n$ and $\alpha \in (0,1)$. Let p be such that $S(T - t) \in {}_tI_T^\alpha(L_p)$ and φ be a real function given on $[0,T]$ such that*

(i) $\int_0^T \varphi(t)dt = 1$;

(ii) $R_{0+}^{\alpha,j}\psi(t) \in {}_0I_t^\alpha(L_q)$ for $j = 0, \ldots, n-1$, where

$$\psi(t) = g(t)\left(b - S_0(T)a\right)\varphi(t)$$

and $S(T - t)g(t) = I$, $t \in [0,T]$, for $1/p + 1/q \leq 1 + \alpha$.

Then the control

$$\hat{u}(t) = K_1\psi(t) + K_2\,{}_0D_t^\alpha\psi(t) + \cdots + K_nR_{0+}^{\alpha,n-1}\psi(t),$$

$t \in [0,T]$, transfers a to b at time $T \geq 0$.

Proof. Is similar to that of Theorem 2.24 and can be found in Mozyrska and Torres, 2010. □

3.7 A Noether Theorem for Fractional Optimal Control

The main result of this section is a Noether-type theorem for fractional optimal control problems in the sense of Caputo (Theorem 3.18). As a corollary, we obtain a Noether theorem for the fractional problems of the calculus of variations (Corollary 3.5).

The fractional optimal control problem in the sense of Caputo is introduced, without loss of generality, in Lagrange form:

$$I[q(\cdot), u(\cdot)] = \int_a^b L\left(t, q(t), u(t)\right) dt \longrightarrow \min,$$

$$^C_a D^\alpha_t q(t) = \varphi\left(t, q(t), u(t)\right),$$

(3.66)

where functions $q : [a, b] \to \mathbb{R}^n$ satisfy appropriate boundary conditions. The Lagrangian $L : [a, b] \times \mathbb{R}^n \times \mathbb{R}^m \to \mathbb{R}$ and the velocity vector $\varphi : [a, b] \times \mathbb{R}^n \times \mathbb{R}^m \to \mathbb{R}^n$ are assumed to be functions of class C^1 with respect to all their arguments. We also assume, without loss of generality, that $0 < \alpha \le 1$. In conformity with the calculus of variations, we consider that the control functions $u(\cdot)$ take values on an open set of $\mathbb{R}^m$. Throughout the work we denote by $\partial_i L$, $i = 1, 2, 3$, the partial derivative of function $L(\cdot, \cdot, \cdot)$ with respect to its ith argument.

Definition 3.8 (Process). *An admissible pair* $(q(\cdot), u(\cdot))$ *that satisfies the control system* $^C_a D^\alpha_t q(t) = \varphi\left(t, q(t), u(t)\right)$ *of problem* (3.66) *is said to be a process.*

Remark 3.5. Choosing $\alpha = 1$, Problem (3.66) is reduced to the classical problem of optimal control theory:

$$I[q(\cdot), u(\cdot)] = \int_a^b L\left(t, q(t), u(t)\right) dt \longrightarrow \min,$$

$$\dot{q}(t) = \varphi\left(t, q(t), u(t)\right).$$

(3.67)

Remark 3.6. The fundamental fractional problem of the calculus of variations in the sense of Caputo,

$$I[q(\cdot)] = \int_a^b L\left(t, q(t), {}^C_a D^\alpha_t q(t)\right) \longrightarrow \min,$$

(3.68)

is a particular case of (3.66): we just need to choose $\varphi(t, q, u) = u$.

Using the standard Lagrange multiplier technique, we rewrite problem (3.66) in the following equivalent form:

$$I[q(\cdot), u(\cdot), p(\cdot)] = \int_a^b \left[\mathcal{H}\left(t, q(t), u(t), p(t)\right) - p(t) \cdot {}^C_a D^\alpha_t q(t)\right] dt \longrightarrow \min,$$

(3.69)

where the Hamiltonian $\mathcal{H}$ is defined by

$$\mathcal{H}(t, q, u, p) = L(t, q, u) + p \cdot \varphi(t, q, u) \,. \tag{3.70}$$

Remark 3.7. In the context of classical mechanics, p is interpreted as the generalized momentum. In the optimal control literature, the multiplier p is known as the adjoint variable.

We now proceed with the usual steps for obtaining necessary optimality conditions in the calculus of variations. We begin by computing the variation δI of functional (3.69):

$$\delta I = \int_a^b \left[\partial_2 \mathcal{H} \cdot \delta q + \partial_3 \mathcal{H} \cdot \delta u + \partial_4 \mathcal{H} \cdot \delta p - \delta p \cdot {}_a^C D_t^\alpha q - p \cdot \delta \left({}_a^C D_t^\alpha q \right) \right] dt \,, \tag{3.71}$$

where δq, δu and δp are the variations of q, u, and p respectively. Equation (3.71) is equivalent to

$$\delta I = \int_a^b \left[(\partial_2 \mathcal{H} - {}_t D_b^\alpha p) \cdot \delta q + \partial_3 \mathcal{H} \cdot \delta u + \left(\partial_4 \mathcal{H} - {}_a^C D_t^\alpha q \right) \cdot \delta p \right] dt$$
$$- \left({}_t D_b^{\alpha-1} p \right) \cdot \delta q \Big|_a^b$$

where the fractional derivatives of $p(t)$ are in the sense of Riemann–Liouville (in contrast with the fractional derivative of $q(t)$ which is taken in the sense of Caputo). Standard arguments lead us to the following result.

Theorem 3.17. *If $(q(\cdot), u(\cdot))$ is an optimal process for problem (3.66), then there exists a function $p(\cdot) \in C^1([a, b]; \mathbb{R}^n)$ such that for all $t \in [a, b]$ the tuple $(q(\cdot), u(\cdot), p(\cdot))$ satisfy the following conditions:*

(i) the Hamiltonian system

$$\begin{cases} {}_t D_b^\alpha p(t) = \partial_2 \mathcal{H}(t, q(t), u(t), p(t)) \,, \\ {}_a^C D_t^\alpha q(t) = \partial_4 \mathcal{H}(t, q(t), u(t), p(t)) \,; \end{cases}$$

(ii) the stationary condition

$$\partial_3 \mathcal{H}(t, q(t), u(t), p(t)) = 0 \,;$$

(iii) the transversality condition

$$\left({}_t D_b^{\alpha-1} p \right) \cdot \delta q \Big|_a^b = 0 \,;$$

with $\mathcal{H}$ given by (3.70).

Definition 3.9. *A triple* $(q(\cdot), u(\cdot), p(\cdot))$ *satisfying Theorem 3.17 will be called a fractional Pontryagin extremal.*

Remark 3.8. For the fundamental fractional problem of the calculus of variations in the sense of Caputo (3.68) we have $\mathcal{H} = L + p \cdot u$. It follows from Theorem 3.17 that

$$\begin{aligned} {}^C_a D^\alpha_t q &= u \,, \\ {}_t D^\alpha_b p &= \partial_2 L \,, \\ \partial_3 \mathcal{H} = 0 &\Leftrightarrow p = -\partial_3 L \Rightarrow {}_t D_b{}^\alpha p = -{}_t D^\alpha_b \partial_3 L \,. \end{aligned} \tag{3.72}$$

Comparing both expressions for ${}_t D^\alpha_b p$, we arrive at the fractional Euler–Lagrange equations:

$$\partial_2 L + {}_t D^\alpha_b \partial_3 L = 0 \,. \tag{3.73}$$

In other words, for the problem of the calculus of variations (3.68) the fractional Pontryagin extremals give the fractional Euler–Lagrange extremals.

Remark 3.9. Our optimal control problem (3.66) only involves Caputo fractional derivatives but both Caputo and Riemann–Liouville fractional derivatives appear in the necessary optimality condition given by Theorem 3.17.

The notion of variational invariance for problem (3.66) is defined with the help of the equivalent problem (3.69).

Definition 3.10 (Invariance of (3.66) without time transformation). *We say that functional* (3.66) *is invariant under the one-parameter family of infinitesimal transformations*

$$\begin{cases} \bar{q}(t) = q(t) + \varepsilon \xi(t, q, u, p) + o(\varepsilon) \,, \\ \bar{u}(t) = u(t) + \varepsilon \varsigma(t, q, u, p) + o(\varepsilon) \,, \\ \bar{p}(t) = p(t) + \varepsilon \varrho(t, q, u, p) + o(\varepsilon) \,, \end{cases}$$

if, and only if,

$$\int_{t_a}^{t_b} \left[\mathcal{H}\left(t, q(t), u(t), p(t)\right) - p(t) \cdot {}^C_a D^\alpha_t q(t) \right] dt$$

$$= \int_{t_a}^{t_b} \left[\mathcal{H}\left(t, \bar{q}(t), \bar{u}(t), \bar{p}(t)\right) - \bar{p}(t) \cdot {}^C_a D^\alpha_t \bar{q}(t) \right] dt \quad (3.74)$$

for any subinterval $[t_a, t_b] \subseteq [a, b]$.

Noether's theorem will be proved following similar steps as those for the problems of the calculus of variations in the Riemann–Liouville sense.

Lemma 3.3 (Necessary and sufficient condition of invariance). *If functional* (3.66) *is invariant, in the sense of Definition 3.10, then*

$$\partial_2 \mathcal{H}\left(t, q, u, p\right) \cdot \xi + \partial_3 \mathcal{H}\left(t, q, u, p\right) \cdot \varsigma$$
$$+ \left(\partial_4 \mathcal{H}\left(t, q, u, p\right) - {}_a^C D_t^\alpha q\right) \cdot \varrho - p \cdot {}_a^C D_t^\alpha \xi = 0. \quad (3.75)$$

Proof. Since condition (3.74) is to be valid for any subinterval $[t_a, t_b] \subseteq [a, b]$, we can write (3.74) in the following equivalent form:

$$\mathcal{H}\left(t, q(t), u(t), p(t)\right) - p(t) \cdot {}_a^C D_t^\alpha q(t) = \mathcal{H}\left(t, \bar{q}(t), \bar{u}(t), \bar{p}(t)\right) - \bar{p}(t) \cdot {}_a^C D_t^\alpha \bar{q}(t). \quad (3.76)$$

We differentiate both sides of (3.76) with respect to ε and then substitute ε by zero. The definition and properties of the Caputo fractional derivative permit us to write that

$$0 = \partial_2 \mathcal{H}\left(t, q, u, p\right) \cdot \xi + \partial_3 \mathcal{H}\left(t, q, u, p\right) \cdot \varsigma + \left(\partial_4 \mathcal{H}\left(t, q, u, p\right) - {}_a^C D_t^\alpha q\right) \cdot \varrho$$
$$- p \cdot \frac{d}{d\varepsilon} \left[\frac{1}{\Gamma(n - \alpha)} \int_a^t (t - \theta)^{n-\alpha-1} \left(\frac{d}{d\theta}\right)^n q(\theta) d\theta \right.$$
$$\left. + \frac{\varepsilon}{\Gamma(n - \alpha)} \int_a^t (t - \theta)^{n-\alpha-1} \left(\frac{d}{d\theta}\right)^n \xi \, d\theta \right]_{\varepsilon=0}$$
$$= \partial_2 \mathcal{H}\left(t, q, u, p\right) \cdot \xi(t, q) + \partial_3 \mathcal{H}\left(t, q, u, p\right) \cdot \varsigma + \left(\partial_4 \mathcal{H}\left(t, q, u, p\right) - {}_a^C D_t^\alpha q\right) \cdot \varrho$$
$$- p \cdot \frac{1}{\Gamma(n - \alpha)} \int_a^t (t - \theta)^{n-\alpha-1} \left(\frac{d}{d\theta}\right)^n \xi \, d\theta. \quad (3.77)$$

Expression (3.77) is equivalent to (3.75). □

We propose the following notion of fractional conservation law in the sense of Caputo, which involves the operator $\mathcal{D}_t^\omega$.

Definition 3.11. *Given two functions f and g of class C^1 in the interval $[a, b]$, we introduce the following operator:*

$$\mathcal{D}_t^\omega \left[f, g\right] = -g \, {}_t D_b^\omega f + f \, {}_a^C D_t^\omega g,$$

where $t \in [a, b]$ and $\omega \in \mathbb{R}_0^+$.

Definition 3.12 (Fractional conservation law in the Caputo sense). *A quantity $C_f \left(t, q(t), {}_a^C D_t^\alpha q(t), u(t), p(t)\right)$ is said to be a fractional conservation law in the sense of Caputo if it is possible to write C_f as a sum of* *products,*

$$C_f\left(t, q(t), {}_a^C D_t^\alpha q(t), u(t), p(t)\right)$$

$$= \sum_{i=1}^{r} C_i^1\left(t, q(t), {}_a^C D_t^\alpha q(t), u(t), p(t)\right) \cdot C_i^2\left(t, q(t), {}_a^C D_t^\alpha q(t), u(t), p(t)\right)$$

$$(3.78)$$

for some $r \in \mathbb{N}$, *and for every* $i = 1, \ldots, r$ *the pair* C_i^1 *and* C_i^2 *satisfy one of the following relations:*

$$\mathcal{D}_t^\alpha\left[C_i^1\left(t, q(t), {}_a^C D_t^\alpha q(t), u(t), p(t)\right), C_i^2\left(t, q(t), {}_a^C D_t^\alpha q(t), u(t), p(t)\right)\right] = 0$$

$$(3.79)$$

or

$$\mathcal{D}_t^\alpha\left[C_i^2\left(t, q(t), {}_a^C D_t^\alpha q(t), u(t), p(t)\right), C_i^1\left(t, q(t), {}_a^C D_t^\alpha q(t), u(t), p(t)\right)\right] = 0$$

$$(3.80)$$

along all the fractional Pontryagin extremals (Definition 3.9).

Remark 3.10. If $\alpha = 1$ (3.79) and (3.80) coincide, and C_f (3.78) satisfy the classical definition of conservation law: $\frac{d}{dt}\left[C_f(t, q, \dot{q})\right] = 0$.

Lemma 3.4 (Noether's theorem without transformation of time).
If functional (3.66) *is invariant in the sense of Definition 3.10, then*

$$p(t) \cdot \xi$$

is a fractional conservation law in the sense of Caputo.

Proof. We use the conditions of Theorem 3.17 in the necessary and sufficient condition of invariance (3.75):

$$0 = -\partial_2 \mathcal{H} \cdot \xi - \partial_3 \mathcal{H} \cdot \varsigma - \left(\partial_4 \mathcal{H} - {}_a^C D_t^\alpha q\right) \cdot \varrho + p \cdot {}_a^C D_t^\alpha \xi$$

$$= -\xi \cdot {}_t D_b^\alpha p + p \cdot {}_a^C D_t^\alpha \xi$$

$$= \mathcal{D}_t^\alpha[p, \xi] \ . \qquad \qquad \square$$

Definition 3.13 (Invariance of (3.66)). *Functional* (3.66) *is said to be invariant under the one-parameter infinitesimal transformations*

$$\begin{cases} \bar{t} = t + \varepsilon\tau(t, q, u, p) + o(\varepsilon)\,, \\ \bar{q}(t) = q(t) + \varepsilon\xi(t, q, u, p) + o(\varepsilon)\,, \\ \bar{u}(t) = u(t) + \varepsilon\varsigma(t, q, u, p) + o(\varepsilon)\,, \\ \bar{p}(t) = p(t) + \varepsilon\varrho(t, q, u, p) + o(\varepsilon)\,, \end{cases} \qquad (3.81)$$

if, and only if,

$$\int_{t_a}^{t_b} \left[\mathcal{H}\left(t, q(t), u(t), p(t)\right) - p(t) \cdot {}_a^C D_t^\alpha q(t) \right] dt$$

$$= \int_{\bar{t}(t_a)}^{\bar{t}(t_b)} \left[\mathcal{H}\left(\bar{t}, \bar{q}(\bar{t}), \bar{u}(\bar{t}), \bar{p}(\bar{t})\right) - \bar{p}(\bar{t}) \cdot {}_a^C D_{\bar{t}}^\alpha \bar{q}(\bar{t}) \right] d\bar{t} \quad (3.82)$$

for any subinterval $[t_a, t_b] \subseteq [a, b]$.

The next result provides an extension of Noether's theorem for fractional optimal control problems in the sense of Caputo.

Theorem 3.18 (Fractional Noether's theorem).
If the functional (3.66) is invariant under the one-parameter infinitesimal transformations (3.81), then

$$C_f\left(t, q(t), {}_a^C D_t^\alpha q(t), u(t), p(t)\right)$$
$$= \left[\mathcal{H}(t, q(t), u(t), p(t)) - (1 - \alpha)p(t) \cdot {}_a^C D_t^\alpha q(t) \right] \tau - p(t) \cdot \xi \quad (3.83)$$

is a fractional conservation law in the sense of Caputo (see Definition 3.12).

Proof. Is similar to that of Theorem 3.22 and can be found in Frederico and Torres, 2008b. □

Remark 3.11. If $\alpha = 1$, problem (3.66) takes the classical form (3.67) and Theorem 3.18 gives the conservation law of (Torres, 2002):

$$C(t, q(t), u(t), p(t)) = \left[\mathcal{H}(t, q(t), u(t), p(t)) \right] \tau - p(t) \cdot \xi \,.$$

As a corollary, we obtain the analogous result proved for fractional problems of the calculus of variations in the Riemann–Liouville sense.

Definition 3.14 (Variational invariance of (3.68)).
Functional (3.68) is said to be invariant under the one-parameter family of infinitesimal transformations

$$\begin{cases} \bar{t} = t + \varepsilon \tau(t, q) + o(\varepsilon) \,, \\ \bar{q}(t) = q(t) + \varepsilon \xi(t, q) + o(\varepsilon) \,, \end{cases} \quad (3.84)$$

if, and only if,

$$\int_{t_a}^{t_b} L\left(t, q(t), {}_a^C D_t^\alpha q(t)\right) dt = \int_{\bar{t}(t_a)}^{\bar{t}(t_b)} L\left(\bar{t}, \bar{q}(\bar{t}), {}_a^C D_{\bar{t}}^\alpha \bar{q}(\bar{t}),\right) d\bar{t}$$

for any subinterval $[t_a, t_b] \subseteq [a, b]$.

Corollary 3.5. *If functional* (3.68) *is invariant under the family of transformations* (3.84), *then*

$$C_f\left(t, q, {}_a^C D_t^\alpha q\right) = \partial_3 L\left(t, q, {}_a^C D_t^\alpha q\right) \cdot \xi$$
$$+ \left[L\left(t, q, {}_a^C D_t^\alpha q\right) - \alpha \partial_3 L\left(t, q, {}_a^C D_t^\alpha q\right) \cdot {}_a^C D_t^\alpha q\right] \tau \quad (3.85)$$

is a fractional conservation law in the sense of Caputo.

Proof. The fractional conservation law (3.85) is obtained applying Theorem 3.18 to functional (3.68). □

Remark 3.12. If $\alpha = 1$ problem (3.68) is reduced to the classical problem of the calculus of variations,

$$I[q(\cdot)] = \int_a^b L\left(t, q(t), \dot{q}(t)\right) \longrightarrow \min, \quad (3.86)$$

and one obtains from Corollary 3.5 the standard Noether's theorem proved in 1918:

$$C(t, q, \dot{q}) = \partial_3 L\left(t, q, \dot{q}\right) \cdot \xi(t, q) + \left[L\left(t, q, \dot{q}\right) - \partial_3 L\left(t, q, \dot{q}\right) \cdot \dot{q}\right] \tau(t, q) \quad (3.87)$$

is a conservation law, i.e., (3.87) is constant along all the solutions of the Euler–Lagrange equations

$$\partial_2 L\left(t, q, \dot{q}\right) = \frac{d}{dt} \partial_3 L\left(t, q, \dot{q}\right) \quad (3.88)$$

(these classical equations are obtained from (3.73) putting $\alpha = 1$).

In classical mechanics, when problem (3.86) does not depend explicitly on q, i.e., $L = L(t, \dot{q})$, it follows from (3.72) and (3.88) that the generalized momentum p is a conservation law. This is also an immediate consequence of the classical Noether's theorem: from the invariance with respect to translations on q ($\tau = 0$, $\xi = 1$), it follows from (3.87) that $p = \partial_3 L$ is a conservation law. Another famous example of the application of Noether's theorem in classical mechanics is given by the conservation of energy: when the Lagrangian L in (3.86) is autonomous, i.e., $L = L(q, \dot{q})$, we have invariance under time-translations ($\tau = 1$, $\xi = 0$) and it follows from (3.70), (3.72) and (3.87), that the Hamiltonian $\mathcal{H}$ (which is interpreted as being the energy in classical mechanics) is a conservation law. Surprisingly enough, we show next, as an immediate consequence of our Theorem 3.18, that for the problem (3.66) with a fractional order of differentiation α ($\alpha \neq 1$), the following happens:

(a) similarly to classical mechanics, the generalized momentum p is a fractional conservation law when L and φ do not depend explicitly on q (Example 3.10);

(b) differently from classical mechanics, the Hamiltonian $\mathcal{H}$ is not a fractional conservation law when L and φ are autonomous (Example 3.11).

In situation (b), we obtain from our Theorem 3.18 a new fractional conservation law that involves not only the Hamiltonian $\mathcal{H}$ but also the fractional order of differentiation α, the generalized momentum p, and the Caputo derivative of the state trajectory q (see (3.89) below). This is in agreement with the fact that the fractional calculus of variations provide a very good formalism to model non-conservative mechanics. In the classical case we have $\alpha = 1$ and the new fractional conservation law (3.89) reduces to the expected "conservation of energy" $\mathcal{H}$.

Example 3.10. Let us consider problem (3.66) with $L(t, q, u) = L(t, u)$, $\varphi(t, q, u) = \varphi(t, u)$. Such a problem is invariant under translations on the variable q, i.e., condition (3.82) is verified for $\bar{t} = t$, $\bar{q}(t) = q(t) + \varepsilon$, $\bar{u}(\bar{t}) = u(t)$ and $\bar{p}(\bar{t}) = p(t)$: we have $d\bar{t} = dt$ and condition (3.82) is satisfied since ${}^{C}_{a}D^{\alpha}_{\bar{t}}\bar{q}(\bar{t}) = {}^{C}_{a}D^{\alpha}_{t}q(t)$:

$$
\begin{aligned}
{}^{C}_{a}D^{\alpha}_{\bar{t}}\bar{q}(\bar{t}) &= \frac{1}{\Gamma(n-\alpha)} \int_{\bar{a}}^{\bar{t}} (\bar{t}-\theta)^{n-\alpha-1} \left(\frac{d}{d\theta}\right)^{n} \bar{q}(\theta)d\theta \\
&= \frac{1}{\Gamma(n-\alpha)} \int_{a}^{t} (t-\theta)^{n-\alpha-1} \left(\frac{d}{d\theta}\right)^{n} (q(t) + \varepsilon) \, d\theta \\
&= {}^{C}_{a}D^{\alpha}_{t}q(t) + {}^{C}_{a}D^{\alpha}_{t}\varepsilon \\
&= {}^{C}_{a}D^{\alpha}_{t}q(t).
\end{aligned}
$$

In agreement with (3.81) one has $\xi = 1$ and $\tau = \varsigma = \varrho = 0$. It follows from Theorem 3.18 that $p(t)$ is a fractional conservation law in the sense of Caputo.

Example 3.11. We now consider the autonomous problem (3.66): $L(t, q, u) = L(q, u)$ and $\varphi(t, q, u) = \varphi(q, u)$. This problem is invariant under time translation, i.e., the invariance condition (3.82) is verified for $\bar{t} = t + \varepsilon$, $\bar{q}(\bar{t}) = q(t)$, $\bar{u}(\bar{t}) = u(t)$ and $\bar{p}(\bar{t}) = p(t)$: we have $d\bar{t} = dt$ and (3.82) follows from the fact that $_a^C D_{\bar{t}}^\alpha \bar{q}(\bar{t}) = _a^C D_t^\alpha q(t)$:

$$
\begin{aligned}
_a^C D_{\bar{t}}^\alpha \bar{q}(\bar{t}) &= \frac{1}{\Gamma(n-\alpha)} \int_{\bar{a}}^{\bar{t}} (\bar{t} - \theta)^{n-\alpha-1} \left(\frac{d}{d\theta}\right)^n \bar{q}(\theta)d\theta \\
&= \frac{1}{\Gamma(n-\alpha)} \int_{a+\varepsilon}^{t+\varepsilon} (t + \varepsilon - \theta)^{n-\alpha-1} \left(\frac{d}{d\theta}\right)^n \bar{q}(\theta)d\theta \\
&= \frac{1}{\Gamma(n-\alpha)} \int_a^t (t - s)^{n-\alpha-1} \left(\frac{d}{ds}\right)^n \bar{q}(t + \varepsilon)ds \\
&= _a^C D_t^\alpha \bar{q}(t + \varepsilon) = _a^C D_t^\alpha \bar{q}(\bar{t}) \\
&= _a^C D_t^\alpha q(t) \,.
\end{aligned}
$$

With the notation (3.81) one has $\tau = 1$ and $\xi = \varsigma = \varrho = 0$. We conclude from Theorem 3.18 that

$$
\mathcal{H}(t,q,u,p) - (1 - \alpha)p \cdot _a^C D_t^\alpha q \tag{3.89}
$$

is a fractional conservation law in the sense of Caputo. For $\alpha = 1$ (3.89) represents the "conservation of the total energy":

$$
\mathcal{H}(t,q(t),u(t),p(t)) = constant\,, \quad t \in [a,b]\,,
$$

for any Pontryagin extremal $(q(\cdot), u(\cdot), p(\cdot))$ of the problem.

3.8   Fractional Riesz–Caputo Derivatives

In this section we prove Noether's theorem for fractional variational problems with Riesz–Caputo derivatives. Both Lagrangian and Hamiltonian formulations are obtained.

3.8.1   *Riesz–Caputo Conservation of Momentum*

We begin by defining the fractional functional under consideration.

Problem 3.1. *The fractional problem of the calculus of variations in the sense of Riesz–Caputo is to find the stationary functions of the functional*

$$
I[q(\cdot)] = \int_a^b L\left(t, q(t), _a^{RC} D_b^\alpha q(t)\right) dt\,, \tag{3.90}
$$

where $[a,b] \subset \mathbb{R}$, $a < b$, $0 < \alpha \le 1$, and the admissible functions $q : t \mapsto q(t)$ and the Lagrangian $L : (t, q, v_l) \mapsto L(t, q, v_l)$ are assumed to be functions of class C^2:

$$
q(\cdot) \in C^2\left([a,b]; \mathbb{R}^n\right);
$$
$$
L(\cdot, \cdot, \cdot) \in C^2\left([a,b] \times \mathbb{R}^n \times \mathbb{R}^n; \mathbb{R}\right).
$$

Remark 3.13. When $\alpha = 1$ the functional (3.90) is reduced to the classical functional of the calculus of variations:

$$I[q(\cdot)] = \int_a^b L\left(t, q(t), \dot{q}(t)\right) dt. \tag{3.91}$$

Theorem 3.19. *If $q(\cdot)$ is an extremizer of* (3.90), *then it satisfies the following fractional Euler–Lagrange equation in the sense of Riesz–Caputo:*

$$\partial_2 L\left(t, q(t), {}^{RC}_{a}D_b^\alpha q(t)\right) - {}^{R}_{a}D_b^\alpha \partial_3 L\left(t, q(t), {}^{RC}_{a}D_b^\alpha q(t)\right) = 0 \tag{3.92}$$

for all $t \in [a, b]$.

Remark 3.14. The functional (3.90) involves Riesz–Caputo fractional derivatives only. However, both Riesz–Caputo and Riesz fractional derivatives appear in the fractional Euler–Lagrange equation (3.92).

Remark 3.15. Let $\alpha = 1$. Then the fractional Euler–Lagrange equation in the sense of Riesz–Caputo (3.92) is reduced to the classical Euler–Lagrange equation:

$$\partial_2 L\left(t, q(t), \dot{q}(t)\right) - \frac{d}{dt} \partial_3 L\left(t, q(t), \dot{q}(t)\right) = 0.$$

Theorem 3.19 leads to the concept of fractional extremal in the sense of Riesz–Caputo.

Definition 3.15. *A function $q(\cdot)$ that is a solution of* (3.92) *is said to be a fractional Riesz–Caputo extremal for functional* (3.90).

In order to prove a fractional Noether's theorem, we begin by introducing the notion of variational invariance and by formulating a necessary condition of invariance without transformation of the independent variable t.

Definition 3.16 (Invariance of (3.90)). *Functional* (3.90) *is said to be invariant under an ε-parameter group of infinitesimal transformations $\bar{q}(t) = q(t) + \varepsilon\xi(t, q(t)) + o(\varepsilon)$ if*

$$\int_{t_a}^{t_b} L\left(t, q(t), {}^{RC}_{a}D_b^\alpha q(t)\right) dt = \int_{t_a}^{t_b} L\left(t, \bar{q}(t), {}^{RC}_{a}D_b^\alpha \bar{q}(t)\right) dt \tag{3.93}$$

for any subinterval $[t_a, t_b] \subseteq [a, b]$.

The next theorem establishes a necessary condition of invariance.

Theorem 3.20 (Necessary condition of invariance).
If functional (3.90) *is invariant in the sense of Definition 3.16, then*

$$\partial_2 L\left(t, q(t), {}_a^{RC}D_b^\alpha q(t)\right) \cdot \xi(t, q(t))$$
$$+ \partial_3 L\left(t, q(t), {}_a^{RC}D_b^\alpha q(t)\right) \cdot {}_a^{RC}D_b^\alpha \xi(t, q(t)) = 0\,. \quad (3.94)$$

Proof. Having in mind that condition (3.93) is valid for any subinterval $[t_a, t_b] \subseteq [a, b]$, we can remove the integral signs in (3.93). Differentiating this condition with respect to ε, then substituting $\varepsilon = 0$, and using the definitions and properties of the fractional derivatives, we arrive at the intended conclusion:

$$0 = \partial_2 L\left(t, q(t), {}_a^{RC}D_b^\alpha q(t)\right) \cdot \xi(t, q) + \partial_3 L\left(t, q(t), {}_a^{RC}D_b^\alpha q(t)\right)$$

$$\cdot \frac{d}{d\varepsilon}\left[\frac{1}{\Gamma(n-\alpha)}\int_a^b |t-\theta|^{n-\alpha-1}\left(\frac{d}{d\theta}\right)^n \bar{q}(\theta)d\theta\right]_{\varepsilon=0}$$

$$= \partial_2 L\left(t, q, {}_a^{RC}D_b^\alpha q\right) \cdot \xi(t, q)$$

$$+ \partial_3 L\left(t, q, {}_a^{RC}D_b^\alpha q\right) \cdot \frac{d}{d\varepsilon}\left[\frac{1}{\Gamma(n-\alpha)}\int_a^b |t-\theta|^{n-\alpha-1}\left(\frac{d}{d\theta}\right)^n q(\theta)d\theta\right.$$

$$\left. + \frac{\varepsilon}{\Gamma(n-\alpha)}\int_a^b |t-\theta|^{n-\alpha-1}\left(\frac{d}{d\theta}\right)^n \xi(\theta, q)d\theta\right]_{\varepsilon=0}$$

$$= \partial_2 L\left(t, q, {}_a^{RC}D_b^\alpha q\right) \cdot \xi(t, q)$$

$$+ \partial_3 L\left(t, q, {}_a^{RC}D_b^\alpha q\right) \cdot \frac{1}{\Gamma(n-\alpha)}\int_a^b |t-\theta|^{n-\alpha-1}\left(\frac{d}{d\theta}\right)^n \xi(\theta, q)d\theta$$

$$= \partial_2 L\left(t, q, {}_a^{RC}D_b^\alpha q\right) \cdot \xi(t, q) + \partial_3 L\left(t, q, {}_a^{RC}D_b^\alpha q\right) \cdot {}_a^{RC}D_b^\alpha \xi(t, q)\,. \qquad \square$$

Remark 3.16. Let $\alpha = 1$. From (3.94) we obtain the classical condition of invariance of the calculus of variations without transformation of the independent variable t:

$$\partial_2 L\left(t, q, \dot{q}\right) \cdot \xi(t, q) + \partial_3 L\left(t, q, \dot{q}\right) \cdot \dot{\xi}(t, q) = 0\,.$$

The following definition is useful in order to introduce an appropriate concept of fractional conserved quantity in the sense of Riesz–Caputo.

Definition 3.17. *Given two functions f and g of class C^1 in the interval $[a, b]$, we introduce the following operator:*

$$\mathcal{D}_t^\gamma\left(f, g\right) = g \cdot {}_a^R D_b^\gamma f + f \cdot {}_a^{RC}D_b^\gamma g\,,$$

where $t \in [a, b]$ and $\gamma \in \mathbb{R}_0^+$.

Remark 3.17. We note that the operator $\mathcal{D}_t^\gamma$ involves both Riesz and Riesz–Caputo fractional derivatives.

Remark 3.18. In the classical context one has $\gamma = 1$ and

$$\mathcal{D}_t^1 (f,g) = f' \cdot g + f \cdot g' = \frac{d}{dt}(f \cdot g) = \mathcal{D}_t^1 (g,f) \,.$$

Roughly speaking, $\mathcal{D}_t^\gamma (f,g)$ is a fractional version of the derivative of the product of f with g. Differently from the classical context, in the fractional case one has, in general, $\mathcal{D}_t^\gamma (f,g) \neq \mathcal{D}_t^\gamma (g,f)$.

We now prove the fractional Noether's theorem in the sense of Riesz–Caputo without transformation of the independent variable t.

Theorem 3.21. *If functional* (3.90) *is invariant in the sense of Definition 3.16, then*

$$\mathcal{D}_t^\alpha \left[\partial_3 L \left(t, q(t), {}_a^{RC} D_b^\alpha q(t) \right), \xi(t, q(t)) \right] = 0 \qquad (3.95)$$

along any fractional Riesz–Caputo extremal $q(t)$, $t \in [a,b]$ (Definition 3.15).

Proof. Using the fractional Euler–Lagrange equation (3.92), we have:

$$\partial_2 L \left(t, q, {}_a^{RC} D_b^\alpha q \right) = {}_a^R D_b^\alpha \partial_3 L \left(t, q, {}_a^{RC} D_b^\alpha q \right) \,. \qquad (3.96)$$

Replacing (3.96) in the necessary condition of invariance (3.94), we get:

$${}_a^R D_b^\alpha \partial_3 L \left(t, q, {}_a^{RC} D_b^\alpha q \right) \cdot \xi(t,q) + \partial_3 L \left(t, q, {}_a^{RC} D_b^\alpha q \right) \cdot {}_a^{RC} D_t^\alpha \xi(t,q) = 0 \,. \qquad (3.97)$$

By definition of the operator $\mathcal{D}_t^\gamma (f,g)$ it follows from (3.97) that

$$\mathcal{D}_t^\alpha \left[\partial_3 L \left(t, q, {}_a^{RC} D_b^\alpha q \right), \xi(t,q) \right] = 0 \,.$$

$\square$

Remark 3.19. In the particular case when $\alpha = 1$ we get from the fractional conservation law in the sense of Riesz–Caputo (3.95) the classical Noether's conservation law of momentum:

$$\frac{d}{dt} \left[\partial_3 L \left(t, q(t), \dot{q}(t) \right) \cdot \xi(t, q(t)) \right] = 0$$

along any Euler–Lagrange extremal $q(\cdot)$ of (3.91). For this reason, we call the fractional law (3.95) the fractional Riesz–Caputo conservation of momentum.

3.8.2   The Noether Theorem in the Riesz–Caputo Sense

The next definition gives a more general notion of invariance for the integral functional (3.90). The main result of this section, Theorem 3.22, is formulated with the help of this definition.

Definition 3.18 (Invariance of (3.90)). *The integral functional* (3.90) *is said to be invariant under the one-parameter group of infinitesimal transformations*

$$\begin{cases} \bar{t} = t + \varepsilon\tau(t, q(t)) + o(\varepsilon)\,, \\ \bar{q}(t) = q(t) + \varepsilon\xi(t, q(t)) + o(\varepsilon)\,, \end{cases}$$

if

$$\int_{t_a}^{t_b} L\left(t, q(t), {}_{a}^{RC}D_b^\alpha q(t)\right) dt = \int_{\bar{t}(t_a)}^{\bar{t}(t_b)} L\left(\bar{t}, \bar{q}(\bar{t}), {}_{a}^{RC}D_b^\alpha \bar{q}(\bar{t})\right) d\bar{t}$$

for any subinterval $[t_a, t_b] \subseteq [a, b]$.

Our next theorem gives a generalization of Noether's theorem for fractional problems of the calculus of variations in the sense of Riesz–Caputo.

Theorem 3.22 (Noether's fractional theorem). *If the integral functional* (3.90) *is invariant in the sense of Definition 3.18, then*

$$\mathcal{D}_t^\alpha\left[\partial_3 L\left(t, q, {}_{a}^{RC}D_t^\alpha q\right), \xi(t, q)\right]$$
$$+ \mathcal{D}_t^\alpha\left[L\left(t, q, {}_{a}^{RC}D_t^\alpha q\right) - \alpha\partial_3 L\left(t, q, {}_{a}^{RC}D_t^\alpha q\right) \cdot {}_{a}^{RC}D_b^\alpha q, \tau(t, q)\right] = 0 \quad (3.98)$$

along any fractional Riesz–Caputo extremal $q(\cdot)$.

Proof. We reparameterize the time (the independent variable t) with a Lipschitzian transformation

$$[a, b] \ni t \longmapsto \sigma f(\lambda) \in [\sigma_a, \sigma_b]$$

that satisfies

$$t'_\sigma = \frac{dt(\sigma)}{d\sigma} = f(\lambda) = 1 \; if \; \lambda = 0\,. \tag{3.99}$$

In this way one reduces (3.90) to an autonomous integral functional

$$\bar{I}[t(\cdot), q(t(\cdot))] = \int_{\sigma_a}^{\sigma_b} L\left(t(\sigma), q(t(\sigma)), {}_{\sigma_a}^{RC}D_{\sigma_b}^\alpha q(t(\sigma))\right) t'_\sigma d\sigma, \tag{3.100}$$

where $t(\sigma_a) = a$ and $t(\sigma_b) = b$. Using the definitions and properties of fractional derivatives, we find that

$$
{}^{RC}_{\sigma_a}D^\alpha_{\sigma_b}q(t(\sigma))
$$

$$
= \frac{1}{\Gamma(n-\alpha)} \int_{\frac{a}{f(\lambda)}}^{\frac{b}{f(\lambda)}} |\sigma f(\lambda) - \theta|^{n-\alpha-1} \left(\frac{d}{d\theta(\sigma)}\right)^n q\left(\theta f^{-1}(\lambda)\right) d\theta
$$

$$
= \frac{(t'_\sigma)^{-\alpha}}{\Gamma(n-\alpha)} \int_{\frac{a}{(t'_\sigma)^2}}^{\frac{b}{(t'_\sigma)^2}} |\sigma - s|^{n-\alpha-1} \left(\frac{d}{ds}\right)^n q(s)ds
$$

$$
= (t'_\sigma)^{-\alpha} \, {}^{RC}_{\chi}D^\alpha_\omega q(\sigma) \quad \left(\chi = \frac{a}{(t'_\sigma)^2}, \; \omega = \frac{b}{(t'_\sigma)^2}\right).
$$

We then have

$$
\bar{I}[t(\cdot), q(t(\cdot))] = \int_{\sigma_a}^{\sigma_b} L\left(t(\sigma), q(t(\sigma)), (t'_\sigma)^{-\alpha} \, {}^{RC}_{\chi}D^\alpha_\omega q(\sigma)\, q(\sigma)\right) t'_\sigma d\sigma
$$

$$
\doteq \int_{\sigma_a}^{\sigma_b} \bar{L}_f\left(t(\sigma), q(t(\sigma)), t'_\sigma, \, {}^{RC}_{\chi}D^\alpha_\omega q(\sigma)\right) d\sigma
$$

$$
= \int_a^b L\left(t, q(t), \, {}^{RC}_a D^\alpha_b q(t)\right) dt
$$

$$
= I[q(\cdot)].
$$

If the integral functional (3.90) is invariant in the sense of Definition 3.18, then the integral functional (3.100) is invariant in the sense of Definition 3.16. It follows from Theorem 3.21 that

$$
\mathcal{D}^\alpha_t \left[\partial_4 \bar{L}_f, \xi\right] + \mathcal{D}^\alpha_t \left[\frac{\partial}{\partial t'_\sigma} \bar{L}_f, \tau\right] = 0 \tag{3.101}
$$

is a fractional conserved law in the sense of Riesz–Caputo. For $\lambda = 0$, the condition (3.99) allows us to write that

$$
{}^{RC}_{\chi}D^\alpha_\omega q(\sigma) = {}^{RC}_a D^\alpha_b q(t),
$$

and therefore we find that

$$
\partial_4 \bar{L}_f = \partial_3 L \tag{3.102}
$$

and

$$
\frac{\partial}{\partial t'_\sigma} \bar{L}_f = \partial_4 \bar{L}_f \cdot \frac{\partial}{\partial t'_\sigma} \left[\frac{(t'_\sigma)^{-\alpha}}{\Gamma(n-\alpha)} \int_\chi^\omega |\sigma - s|^{n-\alpha-1} \left(\frac{d}{ds}\right)^n q(s)\,ds\right] t'_\sigma + L
$$

$$
= \partial_4 \bar{L}_f \cdot \left[\frac{-\alpha(t'_\sigma)^{-\alpha-1}}{\Gamma(n-\alpha)} \int_\chi^\omega |\sigma - s|^{n-\alpha-1} \left(\frac{d}{ds}\right)^n q(s)\,ds\right] t'_\sigma + L
$$

$$
= -\alpha \partial_3 L \cdot {}^{RC}_a D^\alpha_b q + L. \tag{3.103}
$$

Substituting the quantities (3.102) and (3.103) into (3.101), we obtain the fractional conservation law in the sense of Riesz–Caputo (3.98).   □

Remark 3.20. In the particular case $\alpha = 1$ we obtain from (3.98) the classical Noether's conservation law:

$$\frac{d}{dt}\left[\partial_3 L\left(t, q, \dot{q}\right) \cdot \xi(t, q) + \left(L(t, q, \dot{q}) - \partial_3 L\left(t, q, \dot{q}\right) \cdot \dot{q}\right) \tau(t, q)\right] = 0$$

along any Euler–Lagrange extremal $q(\cdot)$ of (3.91).

3.8.3 *Optimal Control of Riesz–Caputo Systems*

We now adopt the Hamiltonian formalism. The fractional optimal control problem in the sense of Riesz–Caputo is introduced, without loss of generality, in Lagrange form:

$$I[q(\cdot), u(\cdot)] = \int_a^b L\left(t, q(t), u(t)\right) dt \longrightarrow \min, \qquad (3.104)$$

subject to the fractional differential system

$$_a^{RC} D_a^\alpha q(t) = \varphi\left(t, q(t), u(t)\right) \qquad (3.105)$$

and initial condition

$$q(a) = q_a. \qquad (3.106)$$

The Lagrangian $L : [a, b] \times \mathbb{R}^n \times \mathbb{R}^m \to \mathbb{R}$ and the fractional velocity vector $\varphi : [a, b] \times \mathbb{R}^n \times \mathbb{R}^m \to \mathbb{R}^n$ are assumed to be functions of class C^1 with respect to all their arguments. We also assume, without loss of generality, that $0 < \alpha \leq 1$. In conformity with the calculus of variations, we consider that the control functions $u(\cdot)$ take values on an open set of $\mathbb{R}^m$.

Definition 3.19. *The fractional differential system* (3.105) *is said to be a fractional control system in the sense of Riesz–Caputo.*

Remark 3.21. In the particular case $\alpha = 1$ the problem (3.104)–(3.106) is reduced to the classical optimal control problem

$$I[q(\cdot), u(\cdot)] = \int_a^b L\left(t, q(t), u(t)\right) dt \longrightarrow \min, \qquad (3.107)$$

$$\dot{q}(t) = \varphi\left(t, q(t), u(t)\right), \quad q(a) = q_a. \qquad (3.108)$$

Remark 3.22. The fractional functional of the calculus of variations in the sense of Riesz–Caputo (3.90) is obtained from (3.104)–(3.105) choosing $\varphi(t, q, u) = u$.

Definition 3.20 (Fractional process in the Riesz–Caputo sense).
An admissible pair $(q(\cdot), u(\cdot))$ that satisfies the fractional control system (3.105) of the fractional optimal control problem (3.104)–(3.106), $t \in [a, b]$, is said to be a fractional process in the sense of Riesz–Caputo.

Theorem 3.23. *If $(q(\cdot), u(\cdot))$ is a fractional process of problem (3.104)–(3.106) in the sense of Riesz–Caputo, then there exists a co-vector function $p(\cdot) \in PC^1([a, b]; \mathbb{R}^n)$ such that for all $t \in [a, b]$ the triple $(q(\cdot), u(\cdot), p(\cdot))$ satisfy the following conditions:*

(i) the Hamiltonian system

$$\begin{cases} {}_a^{RC}D_b^\alpha q(t) = \partial_4 \mathcal{H}(t, q(t), u(t), p(t)) \,, \\ {}_a^R D_b^\alpha p(t) = -\partial_2 \mathcal{H}(t, q(t), u(t), p(t)) \,; \end{cases}$$

(ii) the stationary condition

$$\partial_3 \mathcal{H}(t, q(t), u(t), p(t)) = 0 \,;$$

where the Hamiltonian $\mathcal{H}$ is given by

$$\mathcal{H}(t, q, u, p) = L(t, q, u) + p \cdot \varphi(t, q, u) \,. \tag{3.109}$$

Definition 3.21. *A triple $(q(\cdot), u(\cdot), p(\cdot))$ satisfying Theorem 3.23 will be called a fractional Pontryagin extremal in the sense of Riesz–Caputo.*

Remark 3.23. In the case of the fractional calculus of variations in the sense of Riesz–Caputo one has $\varphi(t, q, u) = u$ (Remark 3.22) and $\mathcal{H} = L + p \cdot u$. From Theorem 3.23 we get ${}_a^{RC}D_a^\alpha q = u$ and ${}_a^R D_b^\alpha p = -\partial_2 L$ from the Hamiltonian system, and from the stationary condition $\partial_3 \mathcal{H} = 0$ it follows that $p = -\partial_3 L$, thus ${}_a^R D_b^\alpha p = -{}_a^R D_b^\alpha \partial_3 L$. Comparing both expressions for ${}_a^R D_b^\alpha p$, we arrive at the fractional Euler–Lagrange equations (3.92):

$$\partial_2 L = {}_a^R D_b^\alpha \partial_3 L.$$

Minimizing (3.104) subject to (3.105) is equivalent, by the Lagrange multiplier rule, to minimizing

$$J[q(\cdot), u(\cdot), p(\cdot)] = \int_a^b \left[\mathcal{H}(t, q(t), u(t), p(t)) - p(t) \cdot {}_a^{RC}D_a^\alpha q(t) \right] dt \tag{3.110}$$

with $\mathcal{H}$ given by (3.109).

Remark 3.24. Theorem 3.23 is easily proved applying the optimality condition (3.92) to the equivalent functional (3.110).

The notion of variational invariance for (3.104)–(3.105) is defined with the help of the augmented functional (3.110).

Definition 3.22 (Variational invariance of (3.104)–(3.105)). *We say that the integral functional* (3.110) *is invariant under the one-parameter family of infinitesimal transformations*

$$\begin{cases} \bar{t} = t + \varepsilon\tau(t, q(t), u(t), p(t)) + o(\varepsilon)\,, \\ \bar{q}(t) = q(t) + \varepsilon\xi(t, q(t), u(t), p(t)) + o(\varepsilon)\,, \\ \bar{u}(t) = u(t) + \varepsilon\varrho(t, q(t), u(t), p(t)) + o(\varepsilon)\,, \\ \bar{p}(t) = p(t) + \varepsilon\varsigma(t, q(t), u(t), p(t)) + o(\varepsilon)\,, \end{cases} \tag{3.111}$$

if

$$\left[\mathcal{H}(\bar{t}, \bar{q}(\bar{t}), \bar{u}(\bar{t}), \bar{p}(\bar{t})) - \bar{p}(\bar{t}) \cdot {}^{RC}_{\bar{a}}D_{\bar{b}}{}^{\alpha}\bar{q}(\bar{t}) \right] d\bar{t}$$
$$= \left[\mathcal{H}(t, q(t), u(t), p(t)) - p(t) \cdot {}^{RC}_{a}D_{b}^{\alpha}q(t) \right] dt\,. \tag{3.112}$$

The next theorem provides us with an extension of Noether's theorem to the wider fractional context of optimal control in the sense of Riesz–Caputo.

Theorem 3.24. *If* (3.104)–(3.105) *is variationally invariant, in the sense of Definition 3.22, then*

$$\mathcal{D}_t^{\alpha} \left[\mathcal{H}(t, q(t), u(t), p(t)) - (1 - \alpha) p(t) \cdot {}^{RC}_{a}D_{b}^{\alpha}q(t), \tau(t, q(t)) \right]$$
$$- \mathcal{D}_t^{\alpha} \left[p(t), \xi(t, q(t)) \right] = 0 \tag{3.113}$$

along any fractional Pontryagin extremal $(q(\cdot), u(\cdot), p(\cdot))$ *of problem* (3.104)–(3.106).

Proof. The fractional conservation law (3.113) in the sense of Riesz–Caputo is obtained by applying Theorem 3.22 to the equivalent functional (3.110). □

Remark 3.25. In the particular case $\alpha = 1$, the fractional optimal control problem (3.104)–(3.106) is reduced to the standard optimal control problem (3.107)–(3.108). In this situation one gets from Theorem 3.24 the Noether-type theorem associated with the classical optimal control problem: invariance under a one-parameter family of infinitesimal transformations (3.111) implies that

$$\mathcal{H}(t, q(t), u(t), p(t))\tau(t, q(t)) - p(t) \cdot \xi(t, q(t)) = constant \tag{3.114}$$

along all the Pontryagin extremals (we obtain the conservation law (3.114) by choosing $\alpha = 1$ in (3.113)).

Theorem 3.24 gives a new and interesting result for autonomous fractional variational problems. Let us consider an autonomous fractional optimal control problem, i.e., (3.104) and (3.105), with the Lagrangian L and the fractional velocity vector φ not depending explicitly on the independent variable t:

$$I[q(\cdot), u(\cdot)] = \int_a^b L\left(q(t), u(t)\right) dt \longrightarrow \min, \qquad (3.115)$$

$$_a^{RC}D_b^\alpha q(t) = \varphi\left(q(t), u(t)\right). \qquad (3.116)$$

Corollary 3.6. *For the autonomous fractional problem* (3.115)–(3.116)

$$_a^R D_b^\alpha \left[\mathcal{H}(t, q(t), u(t), p(t)) + (\alpha - 1) p(t) \cdot {}_a^{RC}D_b^\alpha q(t)\right] = 0$$

along any fractional Pontryagin extremal $(q(\cdot), u(\cdot), p(\cdot))$.

Proof. As the Hamiltonian $\mathcal{H}$ does not depend explicitly on the independent variable t, we can easily see that (3.115)–(3.116) is invariant under translation of the time variable: the condition of invariance (3.112) is satisfied with $\bar{t}(t) = t + \varepsilon$, $\bar{q}(t) = q(t)$, $\bar{u}(t) = u(t)$, and $\bar{p}(t) = p(t)$. Indeed, given that $d\bar{t} = dt$, the invariance condition (3.112) is verified if $_{\bar{a}}^{RC}D_{\bar{b}}^\alpha \bar{q}(\bar{t}) = {}_a^{RC}D_b^\alpha q(t)$. This is true because

$$_{\bar{a}}^{RC}D_{\bar{b}}^\alpha \bar{q}(\bar{t}) = \frac{1}{\Gamma(n-\alpha)} \int_{\bar{a}}^{\bar{b}} |\bar{t} - \theta|^{n-\alpha-1} \left(\frac{d}{d\theta}\right)^n \bar{q}(\theta) d\theta$$

$$= \frac{1}{\Gamma(n-\alpha)} \int_{a+\varepsilon}^{b+\varepsilon} |t + \varepsilon - \theta|^{n-\alpha-1} \left(\frac{d}{d\theta}\right)^n \bar{q}(\theta) d\theta$$

$$= \frac{1}{\Gamma(n-\alpha)} \int_a^b |t - s|^{n-\alpha-1} \left(\frac{d}{ds}\right)^n \bar{q}(t + \varepsilon) ds$$

$$= {}_a^{RC}D_b^\alpha \bar{q}(t + \varepsilon) = {}_a^{RC}D_b^\alpha \bar{q}(\bar{t})$$

$$= {}_a^{RC}D_b^\alpha q(t).$$

Using the notation in (3.111) we have $\tau = 1$ and $\xi = \varrho = \varsigma = 0$. From Theorem 3.24 we arrive at the intended conclusion. $\qquad \square$

Corollary 3.6 asserts that, unlike the classical autonomous problem of optimal control, for (3.115)–(3.116) the fractional Hamiltonian $\mathcal{H}$ is not conserved. Instead of $\frac{d}{dt}(H) = 0$ we have

$$_a^R D_b^\alpha \left[\mathcal{H} + (\alpha - 1) p(t) \cdot {}_a^{RC}D_b^\alpha q(t)\right] = 0, \qquad (3.117)$$

i.e., fractional conservation of the Hamiltonian $\mathcal{H}$ plus a quantity that depends on the fractional order α of differentiation. This seems to be explained by violation of the homogeneity of space-time caused by the fractional derivatives, $\alpha \neq 1$. In the particular $\alpha = 1$ we obtain from (3.117) the classical result: the Hamiltonian $\mathcal{H}$ is preserved along all the Pontryagin extremals of (3.107)–(3.108).

To illustrate our results, we consider in this section two examples where the fractional Lagrangian does not depend explicitly on the independent variable t (autonomous case). In both examples we use our Corollary 3.6 to establish the fractional conservation laws.

Example 3.12. Let us consider the following fractional problem of the calculus of variations:

$$I[q(\cdot)] = \frac{1}{2} \int_0^1 \left({}^{RC}_0 D_1^\alpha q(t) \right)^2 dt \longrightarrow \min .$$

The Hamiltonian (3.109) takes the form $\mathcal{H} = -\frac{1}{2} p^2$. From Corollary 3.6 we conclude that

$$ {}^R_0 D_1^\alpha \left[\frac{p^2(t)}{2} (1 - 2\alpha) \right] = 0 . \tag{3.118}$$

Example 3.13. Consider now the fractional optimal control problem

$$I[q(\cdot)] = \frac{1}{2} \int_0^1 \left[q^2(t) + u^2(t) \right] dt \longrightarrow \min ,$$
$$ {}^{RC}_0 D_1^\alpha q(t) = -q(t) + u(t) ,$$

under the initial condition $q(0) = 1$. The Hamiltonian $\mathcal{H}$ defined by (3.109) takes the following form:

$$\mathcal{H} = \frac{1}{2} \left(q^2 + u^2 \right) + p(-q + u).$$

It follows from our Corollary 3.6 that

$$ {}^R_0 D_1^\alpha \left[\frac{1}{2} \left(q^2 + u^2 \right) + \alpha p(-q + u) \right] = 0 \tag{3.119}$$

along any fractional Pontryagin extremal $(q(\cdot), u(\cdot), p(\cdot))$ of the problem.

For $\alpha = 1$ the conservation laws (3.118) and (3.119) give the well-known results of conservation of energy.

3.9 Multidimensional Lagrangians

This section presents the necessary and sufficient optimality conditions for fractional Caputo variational problems with multiple integrals. These conditions are then applied to physical problems. The fractional Noether-type theorem for conservative and non-conservative generalized mechanical systems is proved.

Partial Riemann–Liouville integrals and derivatives were defined in Section 2.6. Here, we define the partial Caputo derivatives by

$$
{}^{C}_{a_k}D^{\alpha_k}_{x_k}f(x_1,\ldots,x_n)
$$
$$
= \frac{1}{\Gamma(1-\alpha_k)}\int_{a_k}^{x_k}(x_k-t_k)^{-\alpha_k}\frac{\partial}{\partial t_k}f(x_1,\ldots,x_{k-1},t_k,x_{k+1},\ldots,x_n)dt_k,
$$

$$
{}^{C}_{x_k}D^{\alpha_k}_{b_k}f(x_1,\ldots,x_n)
$$
$$
= -\frac{1}{\Gamma(1-\alpha_k)}\int_{x_k}^{b_k}(t_k-x_k)^{-\alpha_k}\frac{\partial}{\partial t_k}f(x_1,\ldots,x_{k-1},t_k,x_{k+1},\ldots,x_n)dt_k.
$$

3.9.1 Multidimensional Euler–Lagrange Equations

We discuss double integrals. The treatment for $n > 2$ dimensions is similar.

Problem 1: Given $F(x_1,x_2,u,p,q,r,s) \in C^1$ and the rectangle $R = [a_1,b_1] \times [a_2,b_2]$. Then, among functions $u(x_1,x_2) : \mathbb{R}^2 \to \mathbb{R}$, such that $u|_{\partial R} = \varphi$, for some given function φ defined on ∂R, to find $\tilde{u}(x_1,x_2)$ giving a minimum for

$$
\mathcal{J}(u) = \int\int_R F[u](x_1,x_2)dx_1dx_2, \tag{3.120}
$$

where the operator $[\cdot]$ is defined in the following way:

$$
[u](x_1,x_2) := (x_1,x_2,u(x_1,x_2),{}^{C}_{a_1}D^{\alpha_1}_{x_1}u(x_1,x_2),{}^{C}_{a_2}D^{\alpha_2}_{x_2}u(x_1,x_2),
$$
$$
{}^{C}_{x_1}D^{\beta_1}_{b_1}u(x_1,x_2),{}^{C}_{x_2}D^{\beta_2}_{b_2}u(x_1,x_2)).
$$

We assume that:

(a) The set of admissible functions for *Problem 1* consists of all functions $u(x_1,x_2) : \mathbb{R}^2 \to \mathbb{R}$, such that ${}^{C}_{a_i}D^{\alpha_i}_{x_i}u,{}^{C}_{x_i}D^{\beta_i}_{b_i}u \in C(R,\mathbb{R})$, $i = 1,2$, and $u|_{\partial R} = \varphi$;

(b) $F \in C^1(R \times \mathbb{R}^5;\mathbb{R})$;

(c) ${}_{x_i}D^{\alpha_i}_{b_i}\partial_{i+3}F$ and ${}_{a_i}D^{\beta_i}_{x_i}\partial_{i+5}F$, $i = 1,2$, are continuous on R for all admissible functions.

Theorem 3.25. *Let $\tilde{u}$ be a solution to Problem 1. Then $\tilde{u}$ satisfies the two-dimensional fractional Euler–Lagrange differential equation:*

$$
\partial_3 F[u](x_1,x_2) + \sum_{i=1}^{2}{}_{x_i}D^{\alpha_i}_{b_i}\partial_{i+3}F[u](x_1,x_2) + \sum_{i=1}^{2}{}_{a_i}D^{\beta_i}_{x_i}\partial_{i+5}F[u](x_1,x_2) = 0.
$$
$$
\tag{3.121}
$$

Proof. Assume that $\tilde{u}$ is a solution to *Problem 1*. Let $h(x_1, x_2)$ be any fixed function on R such that $h|_{\partial R} = 0$ and $^C_{a_i}D^{\alpha_i}_{x_i}h, ^C_{x_i}D^{\beta_i}_{b_i}h \in C(R, \mathbb{R})$, $i = 1, 2$. We embed $\tilde{u}$ in the one-parameter family $\{u = \tilde{u} + \varepsilon h : |\varepsilon| < \varepsilon_0, \varepsilon_0 > 0\}$. Then $\varphi(\varepsilon) := \mathcal{J}(\tilde{u} + \varepsilon h)$ must have a minimum at $\varepsilon = 0$. Hence, $\varphi'(0) = 0$ is necessary for $\tilde{u}$ to provide a minimum for $\mathcal{J}(u)$. We have

$$0 = \varphi'(0)$$

$$= \int\int_R \left(\partial_3 F[\tilde{u}](x_1, x_2)h(x_1, x_2) + \sum_{i=1}^{2} \partial_{i+3}F[\tilde{u}](x_1, x_2)^C_{a_i}D^{\alpha_i}_{x_i}h(x_1, x_2) \right.$$

$$\left. + \sum_{i=1}^{2} \partial_{i+5}F[\tilde{u}](x_1, x_2)^C_{x_i}D^{\beta_i}_{b_i}h(x_1, x_2) \right) dx_1 dx_2.$$

$$(3.122)$$

Under our assumptions we can apply the Fubini theorem and the integration by parts formulas (3.3) and (3.4). Hence,

$$\int\int_R \left(\sum_{i=1}^{2} \partial_{i+3}F[\tilde{u}](x_1, x_2)^C_{a_i}D^{\alpha_i}_{x_i}h(x_1, x_2) \right.$$

$$\left. + \sum_{i=1}^{2} \partial_{i+5}F[\tilde{u}](x_1, x_2)^C_{x_i}D^{\beta_i}_{b_i}h(x_1, x_2) \right) dx_1 dx_2$$

$$= \int\int_R \left(\sum_{i=1}^{2} {}_{x_i}D^{\alpha_i}_{b_i}\partial_{i+3}F[\tilde{u}](x_1, x_2) + \sum_{i=1}^{2} {}_{a_i}D^{\beta_i}_{x_i}\partial_{i+5}F[\tilde{u}](x_1, x_2) \right)$$

$$\times h(x_1, x_2)dx_1 dx_2$$

$$+ \int_{a_2}^{b_2} \left({}_{x_1}I^{1-\alpha_1}_{b_1}\partial_4 F[\tilde{u}](x_1, x_2) - {}_{a_1}I^{1-\beta_1}_{x_1}\partial_5 F[\tilde{u}](x_1, x_2) \right) h(x_1, x_2)\Big|^{x_1=b_1}_{x_1=a_1} dx_2$$

$$+ \int_{a_1}^{b_1} \left({}_{x_2}I^{1-\alpha_2}_{b_2}\partial_6 F[\tilde{u}](x_1, x_2) - {}_{a_2}I^{1-\beta_2}_{x_2}\partial_7 F[\tilde{u}](x_1, x_2) \right) h(x_1, x_2)\Big|^{x_2=b_2}_{x_2=a_2} dx_1.$$

Combining this equality with (3.122), and remembering that $h|_{\partial R} = 0$, we find

$$\int\int_R \left(\partial_3 F[\tilde{u}](x_1, x_2) + \sum_{i=1}^{2} {}_{x_i}D^{\alpha_i}_{b_i}\partial_{i+3}F[\tilde{u}](x_1, x_2) \right.$$

$$\left. + \sum_{i=1}^{2} {}_{a_i}D^{\beta_i}_{x_i}\partial_{i+5}F[\tilde{u}](x_1, x_2) \right) h(x_1, x_2)dx_1 dx_2 = 0.$$

Since h is an arbitrary function, by the fundamental lemma of the calculus of variations (see Section 2.2 in Giaquinta and Hildebrandt, 1996), $\tilde{u}$ satisfies (3.121). $\qquad\square$

Definition 3.23. *A function $u = u(x_1, x_2)$ that is a solution of (3.121) is said to be a fractional extremal surface of $\mathcal{J}(u)$.*

3.9.2   Natural Boundary Conditions

Problem 2: Let $F(x_1, x_2, u, p, q, r, s) \in C^1$ and R be the rectangle $R = [a_1, b_1] \times [a_2, b_2]$. Among functions $u(x_1, x_2) : \mathbb{R}^2 \to \mathbb{R}$, find $\tilde{u}(x_1, x_2)$ giving a minimum to (3.120).

Theorem 3.26. *Let $\tilde{u}$ be a solution to Problem 2. Then $\tilde{u}$ satisfies the two-dimensional fractional Euler–Lagrange differential equation (3.121) together with the following boundary conditions:*

(i) $\left. {}_{x_1}I_{b_1}^{1-\alpha_1}\partial_4 F[\tilde{u}](x_1, x_2) - {}_{a_1}I_{x_1}^{1-\beta_1}\partial_5 F[\tilde{u}](x_1, x_2)\right|_{x_1=a_1} = 0$ *for all $x_2 \in (a_2, b_2)$;*

(ii) $\left. {}_{x_1}I_{b_1}^{1-\alpha_1}\partial_4 F[\tilde{u}](x_1, x_2) - {}_{a_1}I_{x_1}^{1-\beta_1}\partial_5 F[\tilde{u}](x_1, x_2)\right|_{x_1=b_1} = 0$ *for all $x_2 \in (a_2, b_2)$;*

(iii) $\left. {}_{x_2}I_{b_2}^{1-\alpha_2}\partial_6 F[\tilde{u}](x_1, x_2) - {}_{a_2}I_{x_2}^{1-\beta_2}\partial_7 F[\tilde{u}](x_1, x_2)\right|_{x_2=a_2} = 0$ *for all $x_1 \in (a_1, b_1)$;*

(iv) $\left. {}_{x_2}I_{b_2}^{1-\alpha_2}\partial_6 F[\tilde{u}](x_1, x_2) - {}_{a_2}I_{x_2}^{1-\beta_2}\partial_7 F[\tilde{u}](x_1, x_2)\right|_{x_2=b_2} = 0$ *for all $x_1 \in (a_1, b_1)$.*

Proof.   Proceeding as in the proof of Theorem 3.25, we obtain

$$\int\int_R \left(\partial_3 F[\tilde{u}](x_1, x_2) + \sum_{i=1}^{2} {}_{x_i}D_{b_i}^{\alpha_i}\partial_{i+3} F[\tilde{u}](x_1, x_2) \right.$$

$$\left. + \sum_{i=1}^{2} {}_{a_i}D_{x_i}^{\beta_i}\partial_{i+5} F[\tilde{u}](x_1, x_2) \right) h(x_1, x_2)dx_1 dx_2$$

$$+ \int_{a_2}^{b_2} \left({}_{x_1}I_{b_1}^{1-\alpha_1}\partial_4 F[\tilde{u}](x_1, x_2) - {}_{a_1}I_{x_1}^{1-\beta_1}\partial_5 F[\tilde{u}](x_1, x_2) \right) h(x_1, x_2) \Bigg|_{\substack{x_1=b_1 \\ x_1=a_1}} dx_2$$

$$+ \int_{a_1}^{b_1} \left({}_{x_2}I_{b_2}^{1-\alpha_2}\partial_6 F[\tilde{u}](x_1, x_2) - {}_{a_2}I_{x_2}^{1-\beta_2}\partial_7 F[\tilde{u}](x_1, x_2) \right) h(x_1, x_2) \Bigg|_{\substack{x_2=b_2 \\ x_2=a_2}} dx_1$$

$$= 0 \quad (3.123)$$

where h is an arbitrary continuous function. In particular, the above equation holds for $h|_{\partial R} = 0$. For such h the second and the third addends of (3.123) vanish and $\tilde{u}$ satisfies (3.121). With this result, equation (3.123) takes the form

$$
\int_{a_2}^{b_2} \left({}_{x_1}I_{b_1}^{1-\alpha_1}\partial_4 F[\tilde{u}](x_1,x_2) - {}_{a_1}I_{x_1}^{1-\beta_1}\partial_5 F[\tilde{u}](x_1,x_2) \right) h(x_1,x_2) \Bigg|_{x_1=b_1} dx_2
$$

$$
- \int_{a_2}^{b_2} \left({}_{x_1}I_{b_1}^{1-\alpha_1}\partial_4 F[\tilde{u}](x_1,x_2) - {}_{a_1}I_{x_1}^{1-\beta_1}\partial_5 F[\tilde{u}](x_1,x_2) \right) h(x_1,x_2) \Bigg|_{x_1=a_1} dx_2
$$

$$
+ \int_{a_1}^{b_1} \left({}_{x_2}I_{b_2}^{1-\alpha_2}\partial_6 F[\tilde{u}](x_1,x_2) - {}_{a_2}I_{x_2}^{1-\beta_2}\partial_7 F[\tilde{u}](x_1,x_2) \right) h(x_1,x_2) \Bigg|_{x_2=b_2} dx_1
$$

$$
- \int_{a_1}^{b_1} \left({}_{x_2}I_{b_2}^{1-\alpha_2}\partial_6 F[\tilde{u}](x_1,x_2) - {}_{a_2}I_{x_2}^{1-\beta_2}\partial_7 F[\tilde{u}](x_1,x_2) \right) h(x_1,x_2) \Bigg|_{x_2=a_2} dx_1
$$

$$
= 0. \quad (3.124)
$$

Since h is an arbitrary function, we can consider the subclass of functions for which $h \equiv 0$ on $[a_1,b_1] \times \{a_2\} \cup [a_1,b_1] \times \{b_2\} \cup \{b_1\} \times [a_2,b_2]$. For such h, equation (3.124) reduces to

$$
\int_{a_2}^{b_2} \left({}_{x_1}I_{b_1}^{1-\alpha_1}\partial_4 F[\tilde{u}](x_1,x_2) - {}_{a_1}I_{x_1}^{1-\beta_1}\partial_5 F[\tilde{u}](x_1,x_2) \right) h(x_1,x_2) \Bigg|_{x_1=a_1} dx_2
$$

$$
= 0.
$$

By the fundamental lemma of the calculus of variations (see Section 2.2 in Giaquinta and Hildebrandt, 1996), we obtain

$$
{}_{x_1}I_{b_1}^{1-\alpha_1}\partial_4 F[\tilde{u}](x_1,x_2) - {}_{a_1}I_{x_1}^{1-\beta_1}\partial_5 F[\tilde{u}](x_1,x_2) \Big|_{x_1=a_1} = 0.
$$

The other natural boundary conditions are proved similarly, by appropriate choices of h. □

3.9.3 Sufficient Condition for Optimality

The following sufficient condition for an extremal to be a global minimizer extends the well-known convexity condition in the classical calculus of variations.

Definition 3.24. *Given a function L, we say that $L(x, y_1, \ldots, y_n)$ is jointly convex in $(y_1, \ldots, y_n)$, if $\partial_i L$, $i = 2, \ldots, n+1$, exist, are continuous and* *verify the following condition:*

$$L(x, y_1 + h_1, \ldots, y_n + h_n) - L(x, y_1, \ldots, y_n) \geq \sum_{i=2}^{n+1} \partial_i L(x, y_1, \ldots, y_n) h_i$$

for all $(x, y_1, \ldots, y_n), (x, y_1 + h_1, \ldots, y_n + h_n) \in \mathbb{R}^{n+1}$.

Theorem 3.27. *Let* $F(x_1, x_2, u, p, q, r, s)$ *be a jointly convex with respect* (u, p, q, r, s). *If* $\tilde{u}$ *satisfies the two-dimensional fractional Euler–Lagrange differential equation* (3.121) *and conditions* (i)–(iv) *of Theorem 3.26, then* $\tilde{u}$ *is a global minimizer to Problem 2.*

Proof. Assume that $\tilde{u}$ is a solution to *Problem 2*. Since $F(x_1, x_2, u, p, q, r, s)$ is jointly convex with respect to (u, p, q, r, s), for any h such that ${}_{a_i}^C D_{x_i}^{\alpha_i} h, {}_{x_i}^C D_{b_i}^{\beta_i} h \in C(R, \mathbb{R})$, $i = 1, 2$, we have

$$\mathcal{J}(\tilde{u} + h) - \mathcal{J}(\tilde{u}) = \int \int_R (F[\tilde{u} + h](x_1, x_2) - F[\tilde{u}](x_1, x_2)) \, dx_1 dx_2$$

$$\geq \int \int_R \left(\partial_3 F[\tilde{u}](x_1, x_2) h(x_1, x_2) + \sum_{i=1}^2 \partial_{i+3} F[\tilde{u}](x_1, x_2) {}_{a_i}^C D_{x_i}^{\alpha_i} h(x_1, x_2) \right.$$

$$\left. + \sum_{i=1}^2 \partial_{i+5} F[\tilde{u}](x_1, x_2) {}_{x_i}^C D_{b_i}^{\beta_i} h(x_1, x_2) \right) dx_1 dx_2.$$

Proceeding as in the proof of Theorem 3.25, we obtain

$$\mathcal{J}(\tilde{u} + h) - \mathcal{J}(\tilde{u}) \geq \int \int_R \left(\partial_3 F[\tilde{u}](x_1, x_2) + \sum_{i=1}^2 {}_{x_i} D_{b_i}^{\alpha_i} \partial_{i+3} F[\tilde{u}](x_1, x_2) \right.$$

$$\left. + \sum_{i=1}^2 {}_{a_i} D_{x_i}^{\beta_i} \partial_{i+5} F[\tilde{u}](x_1, x_2) \right) h(x_1, x_2) dx_1 dx_2$$

$$+ \int_{a_2}^{b_2} \left({}_{x_1} I_{b_1}^{1-\alpha_1} \partial_4 F[\tilde{u}](x_1, x_2) - {}_{a_1} I_{x_1}^{1-\beta_1} \partial_5 F[\tilde{u}](x_1, x_2) \right) h(x_1, x_2) \Big|_{x_1=a_1}^{x_1=b_1} dx_2$$

$$+ \int_{a_1}^{b_1} \left({}_{x_2} I_{b_2}^{1-\alpha_2} \partial_6 F[\tilde{u}](x_1, x_2) - {}_{a_2} I_{x_2}^{1-\beta_2} \partial_7 F[\tilde{u}](x_1, x_2) \right) h(x_1, x_2) \Big|_{x_2=a_2}^{x_2=b_2} dx_1.$$

As $\tilde{u}$ satisfies equation (3.121) and conditions (i)–(iv) of Theorem 3.26 we have $\mathcal{J}(\tilde{u} + h) - \mathcal{J}(\tilde{u}) \geq 0$. $\qquad \square$

3.9.4   Examples

In the following we present physical examples. The main aim is to illustrate the Euler–Lagrange equations when we have a Lagrangian depending on fractional derivatives.

Example 3.14. Consider a minimizer of

$$J(u) = \frac{1}{2} \int \int_R \sum_{i=1}^{2} \left({}_{a_i}^C D_{x_i}^{\alpha_i} u(x_1, x_2) \right)^2 dx_1 dx_2 \qquad (3.125)$$

in a set of functions that satisfy condition $u = f$ on ∂R, where f is a given function defined on the boundary ∂R of $R = [a_1, b_1] \times [a_2, b_2]$. By Theorem 3.25 a minimizer of (3.125) satisfies the following equation:

$$\sum_{i=1}^{2} {}_{x_i} D_{b_i}^{\alpha_i} {}_{a_i}^C D_{x_i}^{\alpha_i} u(x_1, x_2) = 0. \qquad (3.126)$$

Observe that if α_i goes to 1, then the operator ${}_{a_i}^C D_{x_i}^{\alpha_i}$, $i = 1, 2$, can be replaced with $\frac{\partial}{\partial x_i}$, and the operator ${}_{x_i} D_{b_i}^{\alpha_i}$, $i = 1, 2$, can be replaced with $-\frac{\partial}{\partial x_i}$ (Podlubny, 1999). Thus, for $\alpha \to 1$, problem (3.125) becomes the Dirichlet integral of u over R and equation (3.126) becomes the Laplace equation that arises in the potential theory of electrostatic fields.

Example 3.15. Consider a motion in a medium whose displacement may be described by a scalar function $u(t, x)$, where $x = (x_1, x_2)$. For example, this function might represent the transverse displacement of a membrane. Suppose that the kinetic energy $\mathcal{K}$ and the potential energy $\mathcal{P}$ of the medium are given by

$$\mathcal{K}\left(\frac{\partial u}{\partial t} \right) = \frac{1}{2} \int \rho \left(\frac{\partial u}{\partial t} \right)^2 dx, \quad \mathcal{P}(u) = \frac{1}{2} \int k \|\nabla u\|^2 dx,$$

where $\rho(x)$ is the mass density and $k(x)$ is the stiffness, both assumed positive. Then, the Lagrangian is given by $\mathcal{L} = \mathcal{K} - \mathcal{P}$ and the classical action functional is

$$J(u) = \frac{1}{2} \int \int_R \left(\rho \left(\frac{\partial u}{\partial t} \right)^2 - k \|\nabla u\|^2 \right) dx dt.$$

When we have the Lagrangian with the kinetic term depending on a fractional derivative, then the fractional action functional has the form

$$J(u) = \frac{1}{2} \int \int_R \left(\rho \left({}_0^C D_t^\alpha u \right)^2 - k \|\nabla u\|^2 \right) dx dt. \qquad (3.127)$$

The fractional Euler–Lagrange equation satisfied by a stationary point of (3.127) is

$$\rho_t D_{t_1 0}^{\alpha}{}^{C} D_t^{\alpha} u - \nabla(k\nabla u) = 0.$$

If ρ and k are constants, then equation $_t D_{t_1 0}^{\alpha}{}^{C} D_t^{\alpha} u - c^2 \Delta u = 0$, where $c^2 = k/\rho$ can be called the time-fractional wave equation. In the case when the kinetic and potential energy depend on fractional derivatives, the action functional for the system has the form

$$\mathcal{J}(u) = \frac{1}{2} \int \int_R \left[\rho \left(_0^C D_t^{\alpha} u \right)^2 - k((_{a_1}^C D_{x_1}^{\alpha_1} u)^2 + (_{a_2}^C D_{x_2}^{\alpha_2} u)^2) \right] dx dt. \quad (3.128)$$

The fractional Euler–Lagrange equation satisfied by a stationary point of (3.128) is

$$\rho_t D_{t_1 0}^{\alpha}{}^{C} D_t^{\alpha} u - k(_{x_1} D_{b_1}^{\alpha_1}{}^{C}{}_{a_1} D_{x_1}^{\alpha_1} u + _{x_2} D_{b_2}^{\alpha_2}{}^{C}{}_{a_2} D_{x_2}^{\alpha_2} u) = 0.$$

If ρ and k are constants, then

$$_t D_{t_1 0}^{\alpha}{}^{C} D_t^{\alpha} u - c^2(_{x_1} D_{b_1}^{\alpha_1}{}^{C}{}_{a_1} D_{x_1}^{\alpha_1} u + _{x_2} D_{b_2}^{\alpha_2}{}^{C}{}_{a_2} D_{x_2}^{\alpha_2} u) = 0$$

can be called the space- and time-fractional wave equation. Observe that in the limit $\alpha, \alpha_1, \alpha_2 \to 1$, one obtains the classical wave equation

$$\frac{\partial^2 u}{\partial t^2} - c^2 \Delta u = 0$$

with wave-speed c.

3.9.5 *Fractional Noether-Type Theorem*

In order to prove a fractional Noether's theorem, we begin by introducing the notion of variational invariance.

Definition 3.25. *Functional* (3.120) *is said to be invariant under an ε-parameter group of infinitesimal transformations* $\bar{u}(x_1, x_2) = u(x_1, x_2) + \varepsilon \xi(x_1, x_2, u(x_1, x_2)) + o(\varepsilon)$ *if*

$$\int \int_{R^*} F[u](x_1, x_2) dx_1 dx_2 = \int \int_{R^*} F[\bar{u}](x_1, x_2) dx_1 dx_2 \quad (3.129)$$

for any $R^* \subseteq R$.

The next theorem establishes a necessary condition of invariance.

Theorem 3.28. *If functional* (3.120) *is invariant in the sense of Definition 3.25, then*

$$\partial_3 F[u](x_1, x_2)\xi(x_1, x_2, u(x_1, x_2))$$

$$+ \sum_{i=1}^{2} \Big[\partial_{i+3} F[u](x_1, x_2) {}_{a_i}^{C} D_{x_i}^{\alpha_i} \xi(x_1, x_2, u(x_1, x_2))$$

$$+ \partial_{i+5} F[u](x_1, x_2) {}_{x_i}^{C} D_{b_i}^{\beta_i} \xi(x_1, x_2, u(x_1, x_2))\Big] = 0. \quad (3.130)$$

Proof. Similar to the proof of Theorem 2.21. □

The following definition is similar to Definition 3.11 and is useful in order to formulate a fractional Noether-type theorem.

Definition 3.26. *For a given ordered pair of functions f and g, we introduce the following operators:*

$$\mathcal{D}^\gamma (f, g) = -g \cdot {}_{x_i} D_{b_i}^\gamma f + f \cdot {}_{a_i}^{C} D_{x_i}^\gamma g,$$

and

$$\mathcal{D}_*^\gamma (f, g) = -g \cdot {}_{a_i} D_{x_i}^\gamma f + f \cdot {}_{x_i}^{C} D_{b_i}^\gamma g,$$

where $x_i \in [a_i, b_i]$, $i = 1, 2$, and $\gamma \in \mathbb{R}_0^+$. We assume that the above partial fractional derivatives are continuous on R.

We shall prove the fractional Noether theorem without transformation of the independent variables.

Theorem 3.29. *If functional* (3.120) *is invariant in the sense of Definition 3.25, then*

$$\sum_{i=1}^{2} \{ \mathcal{D}^{\alpha_i} (\partial_{i+3} F[u](x_1, x_2), \xi(x_1, x_2, u(x_1, x_2)))$$

$$+ \mathcal{D}_*^{\beta_i} (\partial_{i+5} F[u](x_1, x_2), \xi(x_1, x_2, u(x_1, x_2))) \} = 0 \quad (3.131)$$

along any extremal surface $u = u(x_1, x_2)$ of $\mathcal{J}(u)$.

Proof. Using the fractional Euler–Lagrange equation (3.121), we have

$$\partial_3 F[u](x_1, x_2) = - \sum_{i=1}^{2} {}_{x_i} D_{b_i}^{\alpha_i} \partial_{i+3} F[u](x_1, x_2) - \sum_{i=1}^{2} {}_{a_i} D_{x_i}^{\beta_i} \partial_{i+5} F[u](x_1, x_2).$$

$$(3.132)$$

Substituting (3.132) into the necessary condition of invariance (3.130), we find

$$-\sum_{i=1}^{2} \xi(x_1, x_2, u(x_1, x_2))_{x_i} D_{b_i}^{\alpha_i} \partial_{i+3} F[u](x_1, x_2)$$

$$-\sum_{i=1}^{2} \xi(x_1, x_2, u(x_1, x_2))_{a_i} D_{x_i}^{\beta_i} \partial_{i+5} F[u](x_1, x_2)$$

$$+\sum_{i=1}^{2} \partial_{i+3} F[u](x_1, x_2)_{a_i}^{C} D_{x_i}^{\alpha_i} \xi(x_1, x_2, u(x_1, x_2))$$

$$+\sum_{i=1}^{2} \partial_{i+5} F[u](x_1, x_2)_{x_i}^{C} D_{b_i}^{\beta_i} \xi(x_1, x_2, u(x_1, x_2)) = 0.$$

By definition of operators $\mathcal{D}^{\gamma}$ and $\mathcal{D}_{*}^{\gamma}$, we obtain equation (3.131). $\qquad\square$

Remark 3.26. In the particular case when α_i goes to 1 and $\beta_i = 0$, $i = 1, 2$, we obtain, from the fractional conservation law (3.131) the classical Noether conservation law without transformation of the independent variables (conservation of current):

$$\sum_{i=1}^{2} \frac{\partial}{\partial x_i} \left(\xi(x_1, x_2, u(x_1, x_2)) \partial_{i+3} F[u](x_1, x_2) \right) = 0$$

along any extremal surface of

$$\mathcal{J}(u) = \int\int_{R} F(x_1, x_2, u(x_1, x_2), \partial_1 u(x_1, x_2), \partial_2 u(x_1, x_2)) dx_1 dx_2.$$

Example 3.16. Let

$$\mathcal{J}(u) = \int\int_{R} L(x_1, x_2, {}_{a_1}^{C} D_{x_1}^{\alpha_1} u(x_1, x_2), {}_{a_2}^{C} D_{x_2}^{\alpha_2} u(x_1, x_2), {}_{x_1}^{C} D_{b_1}^{\beta_1} u(x_1, x_2),$$

$$ {}_{x_2}^{C} D_{b_2}^{\beta_2} u(x_1, x_2)) dx_1 dx_2. \quad (3.133)$$

An easy computation shows that functional (3.133) is invariant under the transformations $\bar{u}(x_1, x_2) = u(x_1, x_2) + \varepsilon r$, where r is a constant. In this case, by Theorem 3.29, we obtain

$$\sum_{i=1}^{2} \Big\{ {}_{x_i} D_{b_i}^{\alpha_i} \partial_{i+3} L(x_1, x_2, {}_{a_1}^{C} D_{x_1}^{\alpha_1} u(x_1, x_2), {}_{a_2}^{C} D_{x_2}^{\alpha_2} u(x_1, x_2), {}_{x_1}^{C} D_{b_1}^{\beta_1} u(x_1, x_2)$$

$$+ {}_{a_i} D_{x_i}^{\beta_i} \partial_{i+5} L(x_1, x_2, {}_{a_1}^{C} D_{x_1}^{\alpha_1} u(x_1, x_2), {}_{a_2}^{C} D_{x_2}^{\alpha_2} u(x_1, x_2), {}_{x_1}^{C} D_{b_1}^{\beta_1} u(x_1, x_2) \Big\} = 0,$$

which is consistent with the Euler–Lagrange equation (3.121).

3.10   On the Second Noether Theorem

In 1918 Emmy Noether published a paper that is now the famous Noether's theorem. But, in fact, Noether proved two theorems in her 1918 paper (Noether, 1918). The first theorem explains the correspondence between conserved quantities and continuous symmetry transformations that depend on constant parameters. Such transformations are global transformations. Familiar examples from classical mechanics include the connections between spatial translations and conservation of linear momentum; spatial rotations and conservation of angular momentum; and time translations and conservation of energy (Gelfand and Fomin, 2000; Logan, 1977). The second theorem, less well known, guarantees syzygies between the Euler–Lagrange equations for a variational problem which is invariant under transformations that depend on arbitrary functions and their derivatives. Such transformations are local transformations. The statement of the theorem for this case is very general and, aside from its application to general relativity, it applies in a wide variety of other cases, for example, quantum chromodynamics and other gauge field theories. From the second theorem, one has identities between Lagrange expressions and their derivatives. Noether called these identities "dependencies". For example, the Bianchi identities in the general theory of relativity are examples of such "dependencies". In electrodynamics, if the Lagrangian represents a charged particle interacting with an electromagnetic field, one finds that it is invariant under the combined action of the so-called gauge transformation of the first kind on the charged particle field, and a gauge transformation of the second kind on the electromagnetic field. The conservation of charge is a result of this invariance, following from Noether's second theorem. For a complete history of Noether's two theorems on variational symmetries see Brading, 2002; Kosmann-Schwarzbach, 2010 and also Carinena *et al.*, 2005; Hydon and Mansfield, 2011; Logan, 1974; Torres, 2003 for some other generalizations.

3.10.1   *Single Integral Case*

Consider a system characterized by a set of functions

$$x^i(t), \quad i = 1, \ldots, n, \qquad (3.134)$$

depending on time t. We can simplify the notation by interpreting (3.134) as a vector function $x = (x^1, \ldots, x^n)$. Define the action functional in the form

$$J(x) = \int_a^b L(t, x(t), {}_a^C D_t^\alpha x(t)) dt, \tag{3.135}$$

where:

(a) ${}_a^C D_t^\alpha x(t) := \left({}_a^C D_t^{\alpha_1} x^1(t), \ldots, {}_a^C D_t^{\alpha_n} x^n(t)\right)$, $0 < \alpha_i \le 1$, $i = 1, \ldots, n$;

(b) $x \in C^1([a, b], \mathbb{R}^n)$;

(c) $L \in C^1([a, b] \times \mathbb{R}^{2n}, \mathbb{R})$;

(d) $t \to \frac{\partial L}{\partial_a^C D_t^{\alpha_k} x^k} \in AC([a, b])$ for every $x \in C^1([a, b], \mathbb{R}^n)$, $k = 1, \ldots, n$.

We define the admissible set of functions $A([a, b])$ by

$$A([a, b]) := \{x \in C^1([a, b], \mathbb{R}^n) : x(a) = x_a, x(b) = x_b, x_a, x_b \in \mathbb{R}^n\}.$$

Theorem 3.30 (cf. Theorem 3.2). *A necessary condition for the function $x \in A([a, b])$ to provide an extremum for the functional (3.135) is that its components satisfy the n fractional equations*

$$\frac{\partial L}{\partial x^k} + {}_t D_b^{\alpha_k} \frac{\partial L}{\partial_a^C D_t^{\alpha_k} x^k} = 0, \quad k = 1, \ldots, n$$

for $t \in [a, b]$.

Define

$$E_k^f(L) := \frac{\partial L}{\partial x^k} + {}_t D_b^{\alpha_k} \frac{\partial L}{\partial_a^C D_t^{\alpha_k} x^k}.$$

We shall call $E_k^f(L)$ the fractional Lagrange expressions. The invariance transformations that we consider are infinitesimal transformations that depend upon arbitrary functions and their fractional derivatives in the sense of Caputo. Let

$$\begin{cases} \bar{t} = t, \\ \bar{x}^k(t) = x^k(t) + T^{k1}(p_1(t)) + \cdots + T^{kr}(p_r(t)), \quad k = 1, \ldots, n, \end{cases} \tag{3.136}$$

where T^{ks} are linear fractional differential operators and p_s, $s = 1, \ldots, r$, are r arbitrary, independent C^1 functions defined on $[a, b]$. Then, we consider four types of fractional differential operators:

(a) Operator of the first kind

$$T^{ks} = T_1^{ks} := a_0^{ks}(t) + a_1^{ks}(t) {}_a^C D_t^{\beta_{ks1}} + \cdots + a_l^{ks}(t) {}_a^C D_t^{\beta_{ksl}}, 0 < \beta_{ksi} \le 1,$$

and ${}_a^C D_t^{\beta_{ksi}} p_s \in C^1([a, b], \mathbb{R})$, $a_i^{ks} \in C^1([a, b], \mathbb{R})$, $s = 1, \ldots, r$, $i = 0, \ldots, l$.

(b) Operator of the second kind

$$T^{ks} = T_2^{ks} := a_0^{ks}(t) + a_1^{ks}(t)_t^C D_b^{\beta_{ks1}} + \cdots + a_l^{ks}(t)_t^C D_b^{\beta_{ksl}}, \quad 0 < \beta_{ksi} \le 1,$$

and $_t^C D_b^{\beta_{ksi}} p_s \in C^1([a,b],\mathbb{R})$, $a_i^{ks} \in C^1([a,b],\mathbb{R})$, $s = 1,\ldots,r$, $i = 0,\ldots,l$.

(c) Operator of the third kind

$$T^{ks} = T_3^{ks} := a_0^{ks}(t) + a_1^{ks}(t)_a^C D_t^{\beta_{ks}} + a_2^{ks}(t)_a^C D_t^{1+\beta_{ks}}$$

$$+ \cdots + a_l^{ks}(t)_a^C D_t^{l-1+\beta_{ks}},$$

$0 < \beta_{ks} \le 1$, $p_s \in C^l([a,b],\mathbb{R})$, and $_a^C D_t^{i-1+\beta_{ks}} p_s \in C^1([a,b],\mathbb{R})$, $a_0^{ks}, a_i^{ks} \in C^1([a,b],\mathbb{R})$, $s = 1,\ldots,r$, $i = 1,\ldots,l$.

(d) Operator of the fourth kind

$$T^{ks} = T_4^{ks} := a_0^{ks}(t) + a_1^{ks}(t)_t^C D_b^{\beta_{ks}} + a_1^{ks}(t)_t^C D_b^{1+\beta_{ks}}$$

$$+ \cdots + a_l^{ks}(t)_t^C D_b^{l-1+\beta_{ks}},$$

$0 < \beta_{ks} \le 1$, $p_s \in C^l([a,b],\mathbb{R})$, and $_t^C D_b^{i-1+\beta_{ks}} p_s \in C^1([a,b],\mathbb{R})$, $a_0^{ks}, a_i^{ks} \in C^1([a,b],\mathbb{R})$, $s = 1,\ldots,r$, $i = 1,\ldots,l$.

Now we define the formal adjoint operator $\tilde{T}^{ks}$ of a fractional differential operator T^{ks} similar in spirit to the classical case, by integration by parts:

$$\int_a^b q T^{ks}(p_s) dt = \int_a^b p_s \tilde{T}^{ks}(q) dt + [\cdot]_{t=a}^{t=b}, \tag{3.137}$$

where $[\cdot]_{t=a}^{t=b}$ represents the boundary terms. Therefore, the adjoints of T_i^{ks}, $i = 1,\ldots,4$, are given by expressions

$$\int_a^b q T_1^{ks}(p_s) dt = \int_a^b p_s \tilde{T}_1^{ks}(q)$$

$$= \int_a^b p_s \left(a_0^{ks} q + \sum_{i=1}^l {}_t D_b^{\beta_{ksi}}(a_i^{ks} q) \right) dt + [\cdot]_{t=a}^{t=b}, \tag{3.138}$$

$$\int_a^b q T_2^{ks}(p_s) dt = \int_a^b p_s \tilde{T}_2^{ks}(q)$$

$$= \int_a^b p_s \left(a_0^{ks} q + \sum_{i=1}^l {}_a D_t^{\beta_{ksi}}(a_i^{ks} q) \right) dt + [\cdot]_{t=a}^{t=b}, \tag{3.139}$$

$$\int_a^b q T_3^{ks}(p_s) dt = \int_a^b p_s \tilde{T}_3^{ks}(q)$$

$$= \int_a^b p_s \left(a_0^{ks} q + \sum_{i=0}^{l-1} {}_t D_b^{i+\beta_{ks}}(a_{i+1}^{ks} q) \right) dt + [\cdot]_{t=a}^{t=b}, \tag{3.140}$$

$$\int_a^b q T_4^{ks}(p_s) dt = \int_a^b p_s \tilde{T}_4^{ks}(q)$$

$$= \int_a^b p_s \left(a_0^{ks} q + \sum_{i=0}^{l-1} {}_a D_t^{i+\beta_{ks}}(a_{i+1}^{ks} q) \right) dt + [\cdot]_{t=a}^{t=b}. \tag{3.141}$$

Note that, if $p_s(a) = p_s(b) = 0$, then boundary terms in (3.138) and (3.139) vanish. Similarly, if $p_s(a) = p_s^{(1)}(a) = \ldots = p_s^{(l-1)}(a) = 0$ and $p_s(b) = p_s^{(1)}(b) = \ldots = p_s^{(l-1)}(b) = 0$, then the boundary terms in (3.140) and (3.141) vanish. Now we define invariance.

Definition 3.27. *The functional* (3.135) *is invariant under transformations* (3.136) *if, and only if, for all $x \in C^1([a,b], \mathbb{R}^n)$ we have*

$$\int_a^b L(t, \bar{x}(t), {}_a^C D_t^\alpha \bar{x}(t)) dt = \int_a^b L(t, x(t), {}_a^C D_t^\alpha x(t)) dt.$$

Theorem 3.31. *If functional* (3.135) *is invariant under transformations* (3.136), *then there exist r identities of the form*

$$\sum_{k=1}^n \tilde{T}^{ks} \left(E_k^f(L) \right) = 0, \quad s = 1, \ldots, r, \tag{3.142}$$

where $\tilde{T}^{ks}$ is the adjoint of T^{ks}.

Proof. We give the proof only for the case $T^{ks} = T_1^{ks}$; other cases can be proved similarly. By Definition 3.27,

$$0 = \int_a^b L(t, \bar{x}(t), {}_a^C D_t^\alpha \bar{x}(t)) dt - \int_a^b L(t, x(t), {}_a^C D_t^\alpha x(t)) dt$$

$$= \int_a^b \left(L(t, \bar{x}(t), {}_a^C D_t^\alpha \bar{x}(t)) - L(t, x(t), {}_a^C D_t^\alpha x(t)) \right) dt.$$

Then, by the Taylor formula,

$$0 = \sum_{k=1}^n \int_a^b \left[\frac{\partial L}{\partial x^k} T_1^{ks}(p_s) + \frac{\partial L}{\partial {}_a^C D_t^{\alpha_k} x^k} {}_a^C D_t^{\alpha_k} T_1^{ks}(p_s) \right] dt, \tag{3.143}$$

where $T_1^{ks}(p_s) = \sum_{s=1}^r T_1^{ks}(p_s)$. The second term in the integrand may be integrated by parts (see formula (3.3)) to obtain

$$\int_a^b \frac{\partial L}{\partial_a^C D_t^{\alpha_k} x^k} {}_a^C D_t^{\alpha_k} T_1^{ks}(p_s)$$

$$= T_1^{ks}(p_s)_t I_b^{1-\alpha_k} \frac{\partial L}{\partial_a^C D_t^{\alpha_k} x^k}\Big|_{x=a}^{x=b} + \int_a^b T_1^{ks}(p_s)_t D_b^{\alpha_k} \frac{\partial L}{\partial_a^C D_t^{\alpha_k} x^k} dt. \quad (3.144)$$

Since p_s are arbitrary, we may choose p_s such that $p_s(a) = p_s(b) = 0$ and ${}_a^C D_t^{\beta_{ksi}} p_s(t)|_{t=b} = 0$, $s = 1, \ldots, r$, $i = 1, \ldots, l$; and if $\beta_{ksi} = 1$, then also ${}_a^C D_t^{\beta_{ksi}} p_s(t)|_{t=a} = 0$. Therefore, the boundary term in (3.144) vanishes and substituting (3.144) into (3.143) we get

$$0 = \sum_{k=1}^n \int_a^b \left[\frac{\partial L}{\partial x^k} + {}_t D_b^{\alpha_k} \frac{\partial L}{\partial_a^C D_t^{\alpha_k} x^k} \right] T_1^{ks}(p_s) dt.$$

Using the definition of the adjoint operator $\tilde{T}_1^{ks}$ of a fractional differential operator T_1^{ks}, that is, equation (3.138), we get

$$0 = \sum_{k=1}^n \int_a^b \sum_{s=1}^r \tilde{T}_1^{ks} \left(E_k^f(L) \right) p_s dt + [\cdot]_{t=a}^{t=b}.$$

Again, appealing to the arbitrariness of p_s, we can force the boundary term to vanish, and finally, by the fundamental lemma of calculus of variations, we conclude that

$$\sum_{k=1}^n \tilde{T}_1^{ks} \left(E_k^f(L) \right) = 0, \quad s = 1, \ldots, r.$$

$\square$

Remark 3.27. Notice that if we put $\beta_{ks} = 1$ in transformations of the third or the fourth kind, then we obtain infinitesimal transformations:

$$\begin{cases} \bar{t} = t \\ \bar{x}^k(t) = x^k + B^{ks}(p_s) + \ldots, \end{cases}$$

where

$$B^{ks} = b_0^{ks}(t) + b_1^{ks}(t)\frac{d}{dt} + b_2^{ks}(t)\frac{d^2}{dt^2} + \cdots + b_l^{ks}(t)\frac{d^l}{dt^l}, \quad k = 1, \ldots, n.$$

In this case the adjoint operator $\tilde{B}^{ks}$ of the differential operator B^{ks} is given by

$$\tilde{B}^{ks}(q) = b_0^{ks} q + \sum_{i=1}^l (-1)^i \frac{d^i}{dt^i}(b_i^{ks} q), \quad k = 1, \ldots, n$$

and the identities (3.142) take the form

$$\sum_{k=1}^{n} b_0^{ks}(E_k(L)) + \sum_{k=1}^{n}\sum_{i=1}^{l}(-1)^i \frac{d^i}{dt^i}\left(b_i^{ks}E_k(L)\right) = 0, \quad s = 1,\ldots,r,$$

which are exactly the Noether identities (see Brading, 2002; Logan, 1974).

Remark 3.28. The fractional differential operators T_1^{ks}, T_2^{ks}, T_3^{ks} and T_4^{ks} can of course be combined, that is, we can consider infinitesimal transformations that depend upon arbitrary functions and their fractional derivatives in the sense of Caputo: left and right with various orders.

3.10.2 *Multiple Integral Case*

Consider a system characterized by a set of functions

$$u^j(t, x_1,\ldots,x_m), \quad j = 1,\ldots,n, \tag{3.145}$$

depending on time t and the space coordinates $x_1,\ldots,x_m$. We can simplify the notation by interpreting (3.145) as a vector function $u = (u^1,\ldots,u^n)$ and writing $t = x_0$, $x = (x_0, x_1,\ldots,x_m)$, $dx = dx_0 dx_1 \cdots dx_m$. Then (3.145) becomes simply $u(x)$ and is called a vector field. Define the action functional in the form

$$\mathcal{J}(u) = \int_{\Omega} \mathcal{L}(x, u, {}^C\nabla^\alpha u) dx, \tag{3.146}$$

where $\Omega = R \times [a_0, b_0]$, $R = [a_1, b_1] \times \ldots \times [a_m, b_m]$, and ${}^C\nabla^\alpha$ is the operator

$$\left({}^C_{a_0}D^{\alpha_0}_{x_0}, {}^C_{a_1}D^{\alpha_1}_{x_1}, \cdots, {}^C_{a_m}D^{\alpha_m}_{x_m}\right),$$

where $\alpha = (\alpha_0, \alpha_1,\ldots,\alpha_m)$, $0 < \alpha_i \leq 1$, $i = 0,\ldots,m$. The function $\mathcal{L}(x, u, {}^C\nabla^\alpha u)$ is called the fractional Lagrangian density of the field. We assume that:

(a) $u^j \in C^1(\Omega, \mathbb{R})$, $j = 1,\ldots,n$;
(b) $\mathcal{L} \in C^1(\mathbb{R}^{m+1} \times \mathbb{R}^n \times \mathbb{R}^{n(m+1)}; \mathbb{R})$;
(c) $x \to \frac{\partial \mathcal{L}}{\partial {}^C_{a_i}D^{\alpha_i}_{x_i}u^j}$ for every $u^j \in C^1(\Omega, \mathbb{R})$ are C^1 functions, $j = 1,\ldots,n$, $i = 0,\ldots,m$.

Define the admissible set of functions $A(\Omega)$ by

$$A(\Omega) := \{u : \Omega \to \mathbb{R}^n : u(x) = \varphi(x) \quad \text{for} \quad x \in \partial\Omega\},$$

where $\varphi : \partial\Omega \to \mathbb{R}^n$ is a given function.

Applying the principle of stationary action to (3.146) we obtain the multidimensional fractional Euler–Lagrange equations for the field.

Theorem 3.32 (cf. Theorem 3.25). *A necessary condition for the function* $u \in A(\Omega)$ *to provide an extremum for the action functional* (3.146) *it that its components satisfy the* n *multidimensional fractional Euler–Lagrange equations:*

$$\frac{\partial \mathcal{L}}{\partial u^j} + \sum_{i=0}^{m} x_i D_{b_i}^{\alpha_i} \frac{\partial \mathcal{L}}{\partial_{a_i}^C D_{x_i}^{\alpha_i} u^j} = 0, \quad j = 1, \ldots, n.$$

As before, we define

$$E_j^f(\mathcal{L}) := \frac{\partial \mathcal{L}}{\partial u^j} + \sum_{i=0}^{n} x_i D_{b_i}^{\alpha_i} \frac{\partial \mathcal{L}}{\partial_{a_i}^C D_{x_i}^{\alpha_i} u^j},$$

which are called the fractional Lagrange expressions. We shall study infinitesimal transformations that depend upon arbitrary functions of independent variables and their partial fractional derivatives in the sense of Caputo. Let

$$\begin{cases} \bar{x} = x, \\ \bar{u}^j(t) = u^j(x) + T^{j1}(p_1(x)) + \cdots + T^{jr}(p_r(x)), \quad j = 1, \ldots, n, \end{cases}$$

(3.147)

where T^{js} are linear fractional differential operators and p_s, $s = 1, \ldots, r$ are the r arbitrary, independent C^1 functions defined on Ω. Then, we consider two types of fractional differential operators:

(a) Operator of the first kind

$$T^{js} = T_1^{js} := c^{js}(x) + \sum_{i=0}^{m} c_i^{js}(x)_{a_i}^C D_{x_i}^{\beta_{jsi}}, \quad 0 < \beta_{jsi} \leq 1,$$

and $_{a_i}^C D_{x_i}^{\beta_{jsi}} p_s$, c^{js}, c_i^{js} are C^1 functions defined on Ω, $s = 1, \ldots, r$, $i = 1, \ldots, m$.

(b) Operator of the second kind

$$T^{js} = T_1^{js} := c^{js}(x) + \sum_{i=0}^{m} c_i^{js}(x)_{x_i}^C D_{b_i}^{\beta_{jsi}}, \quad 0 < \beta_{jsi} \leq 1,$$

and $_{x_i}^C D_{b_i}^{\beta_{jsi}} p$, c^{js}, c_i^{js} are C^1 functions defined on Ω, $s = 1, \ldots, r$, $i = 1, \ldots, m$.

We define invariance similarly to the one-dimensional case.

Definition 3.28. *The functional* (3.146) *is invariant under transformations* (3.147) *if, and only if, for all* $u \in C^1(\Omega, \mathbb{R}^n)$ *we have*

$$\int_\Omega \mathcal{L}(x, \bar{u}, {}^C \nabla^\alpha \bar{u}) dx = \int_\Omega \mathcal{L}(x, u, {}^C \nabla^\alpha u) dx.$$

Theorem 3.33. *If functional* (3.146) *is invariant under transformations* (3.147), *then there exist r identities of the form*

$$\sum_{j=1}^{n} \tilde{T}^{js} \left(E_j^f(\mathcal{L}) \right) = 0, \quad s = 1, \ldots, r,$$

where $\tilde{T}^{js}$ is the adjoint of T^{js}.

Proof. We give the proof only for the case $T^{js} = T_1^{js}$; the other case can be proved similarly. By Definition 3.28, we have

$$0 = \int_\Omega \mathcal{L}(x, \tilde{u}, {}^C\nabla^\alpha \bar{u}) dx - \int_\Omega \mathcal{L}(x, u, {}^C\nabla^\alpha u) dx$$

$$= \int_\Omega \left(\mathcal{L}(x, \tilde{u}, {}^C\nabla^\alpha \bar{u}) - \mathcal{L}(x, u, {}^C\nabla^\alpha u) \right) dx.$$

Then, by the Taylor formula

$$0 = \sum_{j=1}^{n} \int_\Omega \left(\frac{\partial \mathcal{L}}{\partial u^j} T_1^{js}(p_s) + \sum_{i=0}^{m} \frac{\partial \mathcal{L}}{\partial_{a_i}^C D_{x_i}^{\alpha_i} u^j} {}_{a_i}^C D_{x_i}^{\alpha_i} T_1^{js}(p_s) \right) dx, \quad (3.148)$$

where $T_1^{js}(p_s) = \sum_{s=1}^{r} T_1^{js}(p_s)$. The Fubini theorem allows us to rewrite integrals as iterated integrals so that we can use the integration by parts formula (3.4):

$$\int_\Omega \sum_{i=0}^{m} \frac{\partial \mathcal{L}}{\partial_{a_i}^C D_{x_i}^{\alpha_i} u^j} {}_{a_i}^C D_{x_i}^{\alpha_i} T_1^{js}(p_s) dx$$

$$= \int_\Omega \sum_{i=0}^{m} {}_{x_i} D_{b_i}^{\alpha_i} \frac{\partial \mathcal{L}}{\partial_{a_i}^C D_{x_i}^{\alpha_i} u^j} T_1^{js}(p_s) dx + [\cdot]|_{\partial\Omega}, \quad j = 1, \ldots, n, \quad (3.149)$$

where $[\cdot]|_{\partial\Omega}$ represent the boundary terms – $m + 1$-volumes integrals. Since p_s are arbitrary, we may choose p_s such that: $p_s(x)|_{\partial\Omega} = 0$ and ${}_{a_i}^C D_{x_i}^{\beta_{jsi}} p_s(t)|_{\partial\Omega} = 0$, $s = 1, \ldots, r$, $i = 1, \ldots, l$. Therefore, the boundary term in (3.149) vanishes and substituting (3.149) into (3.148) we get

$$0 = \sum_{j=1}^{n} \int_\Omega \left(\frac{\partial \mathcal{L}}{\partial u^j} + \sum_{i=0}^{m} {}_{x_i} D_{b_i}^{\alpha_i} \frac{\partial \mathcal{L}}{\partial_{a_i}^C D_{x_i}^{\alpha_i} u^j} \right) T_1^{js}(p_s) dx.$$

Now we proceed as in the one-dimensional case and define the adjoint operator $\tilde{T}_1^{js}$ of a fractional differential operator T_1^{js} by

$$\int_\Omega q(x) T_1^{js}(p_s(x)) dx = \int_\Omega p_s(x) \tilde{T}_1^{js}(q(x)) dx + [\cdot]|_{\partial\Omega},$$

$j = 1, \ldots, n$, $s = 1, \ldots, r$, where we use the Fubini theorem. Again, appealing to the arbitrariness of p_s, we can force the boundary term to vanish (by putting $p_s(x)|_{\partial\Omega} = 0$). Therefore,

$$0 = \sum_{j=1}^{n} \int_{\Omega} \sum_{s=1}^{r} \tilde{T}_1^{js} \left(\frac{\partial \mathcal{L}}{\partial u^j} + \sum_{i=0}^{m} x_i D_{b_i}^{\alpha_i} \frac{\partial \mathcal{L}}{\partial_{a_i}^{C} D_{x_i}^{\alpha_i} u^j} \right) p_s dx.$$

Finally, by the fundamental lemma of the calculus of variations, we conclude that

$$\sum_{j=1}^{n} \tilde{T}_1^{js} \left(E_j^f(\mathcal{L}) \right) = 0, \quad s = 1, \ldots, r.$$

$\square$

Remark 3.29. The adjoints of T_i^{js}, $i = 1, 2$, are given by the expressions

$$\tilde{T}_1^{js}(q) = c^{js} q + \sum_{i=0}^{m} x_i D_{b_i}^{\beta_{jsi}} (c_i^{js} q), \quad j = 1, \ldots, n,$$

$$\tilde{T}_2^{js}(q) = c^{js} q + \sum_{i=0}^{m} a_i D_{x_i}^{\beta_{jsi}} (c_i^{js} q), \quad j = 1, \ldots, n.$$

Remark 3.30. The fractional differential operators T_1^{js} and T_2^{js} can of course be combined, that is, we can consider infinitesimal transformations that depend upon arbitrary functions and their partial fractional derivatives in the sense of Caputo: left and right with various orders.

3.10.3 *Example*

In order to illustrate our result we will use the Lagrangian density for the electromagnetic field (see Gelfand and Fomin, 2000):

$$\mathcal{L} = \frac{1}{8\pi} (\mathbf{E}^2 - \mathbf{H}^2), \tag{3.150}$$

where $\mathbf{E}$ and $\mathbf{H}$ are the electric field vector and the magnetic field vector, respectively. Following Baleanu *et al.*, 2009, we shall generalize (3.150) to the fractional Lagrangian density by changing the classical partial derivatives into fractional. Let $x = (x_0, x_1, x_2, x_3) \in \Omega$, $\mathbf{A}(x) = (A_1(x), A_2(x), A_3(x))$, and $A_0(x)$ be a vector potential and a scalar potential, respectively. They are defined by setting

$$\mathbf{E} = grad^{(\alpha_1, \alpha_2, \alpha_3)} A_0 - {}_{a_0}^{C} D_{x_0}^{\alpha_0} \mathbf{A}, \quad \mathbf{H} = curl \mathbf{A}, \quad 0 < \alpha_i \le 1, \quad i = 0, \ldots, 3,$$

$$\tag{3.151}$$

where

$$grad^{(\alpha_1,\alpha_2,\alpha_3)} A_0 = \mathbf{i}_{a_1}^C D_{x_1}^{\alpha_1} A_0 + \mathbf{j}_{a_2}^C D_{x_2}^{\alpha_2} A_0 + \mathbf{k}_{a_3}^C D_{x_3}^{\alpha_3} A_0,$$

$$_{a_0}^C D_{x_0}^{\alpha_0} \mathbf{A} = \mathbf{i}_{a_0}^C D_{x_0}^{\alpha_0} A_1 + \mathbf{j}_{a_0}^C D_{x_0}^{\alpha_0} A_2 + \mathbf{k}_{a_0}^C D_{x_0}^{\alpha_0} A_3,$$

$$curl \mathbf{A} = \mathbf{i}(_{a_2}^C D_{x_2}^{\alpha_2} A_3 - {}_{a_3}^C D_{x_3}^{\alpha_3} A_2) + \mathbf{j}(_{a_3}^C D_{x_3}^{\alpha_3} A_1 - {}_{a_1}^C D_{x_1}^{\alpha_1} A_3)$$
$$+ \mathbf{k}(_{a_1}^C D_{x_1}^{\alpha_1} A_2 - {}_{a_2}^C D_{x_2}^{\alpha_2} A_1).$$

Replacing $\mathbf{E}$ and $\mathbf{H}$ in (3.150) by their expressions (3.151), we obtain the fractional Lagrangian density

$$\mathcal{L} = \frac{1}{8\pi} \left[(grad^{(\alpha_1,\alpha_2,\alpha_3)} A_0 - {}_{a_0}^C D_{x_0}^{\alpha_0} \mathbf{A})^2 - (curl \mathbf{A})^2 \right]. \tag{3.152}$$

Note that, similarly to the integer case, the potential $(A_0, \mathbf{A})$ is not uniquely determined by the vectors $\mathbf{E}$ and $\mathbf{H}$. Namely, $\mathbf{E}$ and $\mathbf{H}$ do not change if we make a gauge transformation:

$$\tilde{A}_j(x) = A_j(x) + {}_{a_j}^C D_{x_j}^{\alpha_j} f(x), \quad j = 0, \ldots, 3, \tag{3.153}$$

where $f : \Omega \to \mathbb{R}$ is an arbitrary function of class C^2 in all of its arguments. Therefore, the Lagrangian density (3.152), and hence the action functional, is invariant under transformation (3.153). By Theorem (3.33), we conclude that

$$\sum_{j=0}^{3} x_j D_{b_j}^{\alpha_j} \left(E_j^f(\mathcal{L}) \right) = 0,$$

where $E_j^f(\mathcal{L})$ are Lagrange expressions corresponding to (3.152). Equations $E_j^f(\mathcal{L}) = 0$ do not uniquely determine the potential $(A_0, \mathbf{A})$ and to avoid this lack of uniqueness, the fractional Lorentz condition can be imposed on $(A_0, \mathbf{A})$.

3.11 Further Generalizations

In this section we study fractional variational problems in terms of a combined fractional Caputo derivative. This idea goes back at least as far as Klimek, 2001, where, based on the Riemann–Liouville fractional derivatives, the symmetric fractional derivative was introduced. Klimek's approach (Klimek, 2001) is obtained in our framework as a particular case, by choosing parameter γ to be $1/2$. Necessary optimality conditions of Euler–Lagrange type for the basic, isoperimetric, and Lagrange variational problems are proved, as well as transversality and sufficient optimality conditions. This allows us to obtain the necessary and sufficient Pareto optimality conditions for multiobjective fractional variational problems.

Let $\alpha, \beta \in (0,1]$ and $\gamma \in [0,1]$. We define the fractional derivative operator $^{C}D_{\gamma}^{\alpha,\beta}$ by

$$^{C}D_{\gamma}^{\alpha,\beta} := \gamma\, {_{a}^{C}D_{x}^{\alpha}} + (1-\gamma)\, {_{x}^{C}D_{b}^{\beta}}, \qquad (3.154)$$

which acts on $f \in AC([a,b])$ in the expected way:

$$^{C}D_{\gamma}^{\alpha,\beta} f(x) = \gamma\, {_{a}^{C}D_{x}^{\alpha}} f(x) + (1-\gamma)\, {_{x}^{C}D_{b}^{\beta}} f(x).$$

Note that $^{C}D_{0}^{\alpha,\beta} f(x) = {_{x}^{C}D_{b}^{\beta}} f(x)$ and $^{C}D_{1}^{\alpha,\beta} f(x) = {_{a}^{C}D_{x}^{\alpha}} f(x)$. The operator (3.154) is obviously linear. Using equations (3.3) and (3.4) it is easy to derive the following rule of fractional integration by parts for $^{C}D_{\gamma}^{\alpha,\beta}$:

$$\int_{a}^{b} g(x)\, {^{C}D_{\gamma}^{\alpha,\beta}} f(x)dx = \gamma\left[f(x)\, {_{x}I_{b}^{1-\alpha}} g(x)\right]_{x=a}^{x=b}$$

$$+ (1-\gamma)\left[-f(x)\, {_{a}I_{x}^{1-\beta}} g(x)\right]_{x=a}^{x=b} + \int_{a}^{b} f(x) D_{1-\gamma}^{\beta,\alpha} g(x)dx, \quad (3.155)$$

where $D_{1-\gamma}^{\beta,\alpha} := (1-\gamma)\, {_{a}D_{x}^{\beta}} + \gamma\, {_{x}D_{b}^{\alpha}}$. Let $N \in \mathbb{N}$ and $f = [f_1, \dots, f_N]$: $[a,b] \to \mathbb{R}^N$ with $f_i \in AC([a,b])$, $i = 1, \dots, N$; $\alpha, \beta, \gamma \in \mathbb{R}^N$ with $\alpha_i, \beta_i \in (0,1]$ and $\gamma_i \in [0,1]$, $i = 1, \dots, N$. Then,

$$^{C}D_{\gamma}^{\alpha,\beta} f(x) := \left[{^{C}D_{\gamma_1}^{\alpha_1,\beta_1}} f_1(x), \dots, {^{C}D_{\gamma_N}^{\alpha_N,\beta_N}} f_N(x)\right].$$

Let $\mathbf{D}$ denote the set of all functions $y : [a,b] \to \mathbb{R}^N$ such that $^{C}D_{\gamma}^{\alpha,\beta} y$ exists and is continuous on the interval $[a,b]$. We endow $\mathbf{D}$ with the following norm:

$$\|y\|_{1,\infty} := \max_{a \leq x \leq b} \|y(x)\| + \max_{a \leq x \leq b} \|{^{C}D_{\gamma}^{\alpha,\beta}} y(x)\|,$$

where $\|\cdot\|$ is a norm in $\mathbb{R}^N$. We denote by $\partial_i K$, $i = 1, \dots, M$ ($M \in \mathbb{N}$), the partial derivative of function $K : \mathbb{R}^M \to \mathbb{R}$ with respect to its ith argument. Let $\lambda \in \mathbb{R}^r$. For simplicity of notation we introduce the operators $[\cdot]_{\gamma}^{\alpha,\beta}$ and $\lambda\{\cdot\}_{\gamma}^{\alpha,\beta}$ by

$$[y]_{\gamma}^{\alpha,\beta}(x) := \left(x, y(x), {^{C}D_{\gamma}^{\alpha,\beta}} y(x)\right),$$

$$\lambda\{y\}_{\gamma}^{\alpha,\beta}(x) := \left(x, y(x), {^{C}D_{\gamma}^{\alpha,\beta}} y(x), \lambda_1, \dots, \lambda_r\right).$$

3.11.1   Calculus of Variations via $^C\mathbf{D}_\gamma^{\alpha,\beta}$

Let us begin with the following fundamental problem:

$$\mathcal{J}(y) = \int_a^b L[y]_\gamma^{\alpha,\beta}(x)\,dx \longrightarrow \min \qquad (3.156)$$

where $L \in C^1([a,b] \times \mathbb{R}^{2N}; \mathbb{R})$, over all $y \in \mathbf{D}$ satisfying the boundary conditions

$$y(a) = y^a, \quad y(b) = y^b, \qquad (3.157)$$

where $y^a, y^b \in \mathbb{R}^N$ are given. The set of admissible trajectories is defined as $\mathcal{D} := \{y \in \mathbf{D} : y \text{ satisfies } (3.157)\}$.

The next theorem gives the fractional Euler–Lagrange equation for the problem (3.156)–(3.157).

Theorem 3.34 (Fractional Euler–Lagrange equations). *Let y, $y = (y_1, \ldots, y_N)$, be a local minimizer to problem (3.156)–(3.157). Then, y satisfies the following system of N fractional Euler–Lagrange equations:*

$$\partial_i L[y]_\gamma^{\alpha,\beta}(x) + D_{1-\gamma_{i-1}}^{\beta_{i-1},\alpha_{i-1}}\partial_{N+i}L[y]_\gamma^{\alpha,\beta}(x) = 0, \quad i = 2, \ldots N+1, \quad (3.158)$$

for all $x \in [a,b]$.

Proof. Suppose that y is a local minimizer for $\mathcal{J}$. Let h be an arbitrary admissible variation for problem (3.156)–(3.157), i.e., $h_i(a) = h_i(b) = 0$, $i = 1, \ldots, N$. Based on the differentiability properties of L and Theorem 1.1, a necessary condition for y to be a local minimizer is given by

$$\frac{\partial}{\partial \varepsilon}\mathcal{J}(y + \varepsilon h)\bigg|_{\varepsilon=0} = 0\,,$$

that is,

$$\int_a^b \left[\sum_{i=2}^{N+1} \partial_i L[y]_\gamma^{\alpha,\beta}(x)h_{i-1}(x) \right.$$

$$\left. + \sum_{i=2}^{N+1} \partial_{N+i}L[y]_\gamma^{\alpha,\beta}(x)^C D_{\gamma_{i-1}}^{\alpha_{i-1},\beta_{i-1}}h_{i-1}(x)\right]dx = 0. \quad (3.159)$$

Using formulae (3.155) of integration by parts in the second term of the integrand function, we get

$$\int_a^b \left[\sum_{i=2}^{N+1} \partial_i L[y]_\gamma^{\alpha,\beta}(x) + D_{1-\gamma_{i-1}}^{\beta_{i-1},\alpha_{i-1}} \partial_{N+i} L[y]_\gamma^{\alpha,\beta}(x) \right] h_{i-1}(x)dx$$

$$+ \gamma \left[\sum_{i=2}^{N+1} h_{i-1}(x) \left({}_x I_b^{1-\alpha_{i-1}} \partial_{N+i} L[y]_\gamma^{\alpha,\beta}(x) \right) \right] \Bigg|_{x=a}^{x=b}$$

$$- (1-\gamma) \left[\sum_{i=2}^{N+1} h_{i-1}(x) \left({}_a I_x^{1-\beta_{i-1}} \partial_{N+i} L[y]_\gamma^{\alpha,\beta}(x) \right) \right] \Bigg|_{x=a}^{x=b} = 0. \quad (3.160)$$

Since $h_i(a) = h_i(b) = 0$, $i = 1, \ldots, N$, by the fundamental lemma of the calculus of variations we deduce that

$$\partial_i L[y]_\gamma^{\alpha,\beta}(x) + D_{1-\gamma_{i-1}}^{\beta_{i-1},\alpha_{i-1}} \partial_{N+i} L[y]_\gamma^{\alpha,\beta}(x) = 0, \quad i = 2, \ldots, N+1,$$

for all $x \in [a,b]$.   □

Recall that if α and β go to 1, then ${}_a^C D_x^\alpha$ can be replaced with $\frac{d}{dx}$ and ${}_x^C D_b^\beta$ with $-\frac{d}{dx}$. Thus, if $\gamma = 1$ or $\gamma = 0$, then for $\alpha, \beta \to 1$ we obtain a corresponding result in the classical context of the calculus of variations.

Let $l \in \{1, \ldots, N\}$. Assume that $y(a) = y^a$, $y_i(b) = y_i^b$, $i = 1, \ldots, N$, $i \neq l$, but $y_l(b)$ is free. Then, $h_l(b)$ is free and by equations (3.158) and (3.160) we obtain

$$\left[\gamma {}_x I_b^{1-\alpha_l} \partial_{N+1+l} L[y]_\gamma^{\alpha,\beta}(x) - (1-\gamma) {}_a I_x^{1-\beta_l} \partial_{N+1+l} L[y]_\gamma^{\alpha,\beta}(x) \right] \Bigg|_{x=b} = 0.$$

Let us consider now the case when $y(a) = y^a$, $y_i(b) = y_i^b$, $i = 1, \ldots, N$, $i \neq l$, and $y_l(b)$ is free but restricted by a terminal condition $y_l(b) \leq y_l^b$. Then, in the optimal solution y, we have two possible types of outcome: $y_l(b) < y_l^b$ or $y_l(b) = y_l^b$. If $y(b) < y_l^b$, then there are admissible neighboring paths with terminal values both above and below $y_l(b)$, so that $h_l(b)$ can take either sign. Therefore, the transversality condition is

$$\left[\gamma {}_x I_b^{1-\alpha_l} \partial_{N+1+l} L[y]_\gamma^{\alpha,\beta}(x) - (1-\gamma) {}_a I_x^{1-\beta_l} \partial_{N+1+l} L[y]_\gamma^{\alpha,\beta}(x) \right] \Bigg|_{x=b} = 0$$
$$(3.161)$$

for $y_l(b) < y_l^b$. The other outcome $y_l(b) = y_l^b$ only admits the neighboring paths with terminal value $\tilde{y}_l(b) \leq y_l(b)$. Assuming, without loss of generality, that $h_l(b) \geq 0$, this means that $\varepsilon \leq 0$. Hence, the transversality condition, which has it root in the first order condition (3.159), must be changed to an inequality. For a minimization problem, the $\leq$ type of inequality is called for, and we obtain

$$\left[\gamma_x I_b^{1-\alpha_l} \partial_{N+1+l} L[y]_\gamma^{\alpha,\beta}(x) - (1-\gamma)_a I_x^{1-\beta_l} \partial_{N+1+l} L[y]_\gamma^{\alpha,\beta}(x) \right]\Big|_{x=b} \leq 0$$
(3.162)

for $y_l(b) = y_l^b$. Combining (3.161) and (3.162), we may write the following transversality condition for a minimization problem:

$$\left[\gamma_x I_b^{1-\alpha_l} \partial_{N+1+l} L[y]_\gamma^{\alpha,\beta}(x) - (1-\gamma)_a I_x^{1-\beta_l} \partial_{N+1+l} L[y]_\gamma^{\alpha,\beta}(x) \right]\Big|_{x=b} \leq 0,$$

for $y_l(b) \leq y_l^b$, and

$$\left[\gamma_x I_b^{1-\alpha_l} \partial_{N+1+l} L[y]_\gamma^{\alpha,\beta}(x) - (1-\gamma)_a I_x^{1-\beta_l} \partial_{N+1+l} L[y]_\gamma^{\alpha,\beta}(x) \right]\Big|_{x=b}$$
$$\times (y_l(b) - y_l^b) = 0.$$

The $^{\mathbf{C}}\mathbf{D}_\gamma^{\alpha,\beta}$ fractional isoperimetric problem

Let us now consider the isoperimetric problem that consists of minimizing (3.156) over all $y \in \mathbf{D}$ satisfying r isoperimetric constraints

$$\mathcal{G}^j(y) = \int_a^b G^j[y]_\gamma^{\alpha,\beta}(x)dx = l_j, \quad j = 1, \ldots, r,$$
(3.163)

where $G^j \in C^1([a,b] \times \mathbb{R}^{2N}; \mathbb{R})$, $j = 1, \ldots, r$, and boundary conditions (3.157).

Necessary optimality conditions for isoperimetric problems can be obtained by the following general theorem (cf. Theorem 1.5).

Theorem 3.35. *Let $\mathcal{J}, \mathcal{G}^1, \ldots, \mathcal{G}^r$ be functionals defined in a neighborhood of y and having continuous first variations in this neighborhood. Suppose that y is a local minimizer to (3.156) subject to the boundary conditions (3.157) and the isoperimetric constrains (3.163). Assume that there are functions $h^1, \ldots, h^r \in \mathbf{D}$ such that the matrix $A = (a_{kl})$, $a_{kl} := \delta\mathcal{G}^k(y; h^l)$, has maximal rank r. Then there exist constants $\lambda_1, \ldots, \lambda_r \in \mathbb{R}$ such that the functional*

$$\mathcal{F} := \mathcal{J} - \sum_{j=1}^r \lambda_j \mathcal{G}^j$$

satisfies

$$\delta\mathcal{F}(y; h) = 0$$
(3.164)

for all $h \in \mathbf{D}$.

Suppose now that assumptions of Theorem 3.35 hold. Then, equation (3.164) is fulfilled for every $h \in \mathbf{D}$. Let us consider function h such that $h(a) = h(b) = 0$. Then, we have

$$
0 = \delta \mathcal{F}(y; h) = \frac{\partial}{\partial \varepsilon} \mathcal{F}(y + \varepsilon h)|_{\varepsilon=0}
$$

$$
= \int_a^b \left[\sum_{i=2}^{N+1} \partial_i F_\lambda \{y\}_\gamma^{\alpha,\beta}(x) h_{i-1}(x) \right.
$$

$$
\left. + \sum_{i=2}^{N+1} \partial_{N+i} F_\lambda \{y\}_\gamma^{\alpha,\beta}(x)^C D_{\gamma_{i-1}}^{\alpha_{i-1},\beta_{i-1}} h_{i-1}(x) \right] dx,
$$

where the function $F : [a,b] \times \mathbb{R}^{2N} \times \mathbb{R}^r \to \mathbb{R}$ is defined by

$$
F_\lambda \{y\}_\gamma^{\alpha,\beta}(x) := L[y]_\gamma^{\alpha,\beta}(x) - \sum_{j=1}^r \lambda_j G^j [y]_\gamma^{\alpha,\beta}(x).
$$

On account of the above, and in a similar way to the proof of Theorem 3.34, we obtain

$$
\partial_i F_\lambda \{y\}_\gamma^{\alpha,\beta}(x) + D_{1-\gamma_{i-1}}^{\beta_{i-1},\alpha_{i-1}} \partial_{N+i} F_\lambda \{y\}_\gamma^{\alpha,\beta}(x) = 0, \quad i = 2, \ldots N+1. \tag{3.165}
$$

Therefore, we have the following necessary optimality condition for the fractional isoperimetric problems:

Theorem 3.36. *Let assumptions of Theorem 3.35 hold. If y is a local minimizer to the isoperimetric problem given by (3.156), (3.157) and (3.163), then y satisfies the system of N fractional Euler–Lagrange equations (3.165) for all $x \in [a,b]$.*

Suppose now that constraints (3.163) are characterized by inequalities

$$
\mathcal{G}^j(y) = \int_a^b G^j [y]_\gamma^{\alpha,\beta}(x) dx \le l_j, \quad j = 1, \ldots, r.
$$

In this case we can set

$$
\int_a^b \left(G^j [y]_\gamma^{\alpha,\beta}(x) - \frac{l_j}{b-a} \right) dx + \int_a^b (\phi_j(x))^2 dx = 0,
$$

$j = 1, \ldots, r$, where ϕ_j have the same continuity properties as y_i. Therefore, we obtain the following problem:

$$
\hat{\mathcal{J}}(y) = \int_a^b \hat{L}(x, y(x),^C D_{\gamma_{i-1}}^{\alpha_{i-1},\beta_{i-1}} y(x), \phi_1(x), \ldots, \phi_r(x)) \, dx \longrightarrow \min
$$

subject to r isoperimetric constraints

$$\int_a^b \left[G^j[y]_\gamma^{\alpha,\beta}(x) - \frac{l_j}{b-a} + (\phi_j(x))^2 \right] dx = 0, \quad j = 1, \ldots, r,$$

and boundary conditions (3.157). Under assumptions of Theorem 3.36, we conclude that there exist constants $\lambda_j \in \mathbb{R}$, $j = 1, \ldots, r$, for which the system of equations

$$D_{1-\gamma_{i-1}}^{\beta_{i-1},\alpha_{i-1}} \partial_{N+i} \hat{F}(x, y(x), {}^C D_{\gamma_{i-1}}^{\alpha_{i-1},\beta_{i-1}} y(x), \lambda_1, \ldots, \lambda_r, \phi_1(x), \ldots, \phi_r(x))$$

$$+ \partial_i \hat{F}(x, y(x), {}^C D_{\gamma_{i-1}}^{\alpha_{i-1},\beta_{i-1}} y(x), \lambda_1, \ldots, \lambda_r, \phi_1(x), \ldots, \phi_r(x)) = 0, \quad (3.166)$$

$i = 2, \ldots, N+1$, $\hat{F} = \hat{L} + \sum_{j=1}^r \lambda_j (G^j - \frac{l_j}{b-a} + \phi_j^2)$ and

$$\lambda_j \phi_j(x) = 0, \quad j = 1, \ldots, r, \tag{3.167}$$

hold for all $x \in [a, b]$. Note that it is enough to assume that the regularity condition holds for the constraints which are active at the local minimizer y (constraint $\mathcal{G}^k$ is active at y if $\mathcal{G}^k(y) = l_k$). Indeed, suppose that $l < r$ constraints, say $\mathcal{G}^1, \ldots, \mathcal{G}^l$ for simplicity, are active at the local minimizer y, and there are functions $h^1, \ldots, h^l \in \mathbf{D}$ such that the matrix

$$B = (b_{kj}), \quad b_{kj} := \delta \mathcal{G}^k(y; h^j), \quad k, j = 1, \ldots, l < r$$

has maximal rank l. Since the inequality constraints $\mathcal{G}^{l+1}, \ldots, \mathcal{G}^r$ are inactive, the condition (3.167) is trivially satisfied by taking $\lambda_{l+1} = \cdots = \lambda_r = 0$. On the other hand, since the inequality constraints $\mathcal{G}^1, \ldots, \mathcal{G}^l$ are active and satisfy a regularity condition at y, the conclusion that there exist constants $\lambda_j \in \mathbb{R}$, $j = 1, \ldots, r$, such that (3.166) holds, follow from Theorem 3.36. Moreover, (3.167) is trivially satisfied for $j = 1, \ldots, l$.

The ${}^C\mathbf{D}_\gamma^{\alpha,\beta}$ fractional Lagrange problem

Let us consider the following Lagrange problem, which consists of minimizing (3.156) over all $y \in \mathbf{D}$ satisfying r independent constraints $(r < N)$

$$G^j[y]_\gamma^{\alpha,\beta}(x) = 0, \quad j = 1, \ldots, r, \tag{3.168}$$

and boundary conditions (3.157). In mechanics, constraints of type (3.168) are called nonholonomic. The independence of the r constraints $G^j \in C^1([a, b] \times \mathbb{R}^{2N}; \mathbb{R})$ means that there should exist a nonvanishing Jacobian determinant of order r, such as $\left| \frac{\partial(G^1, \ldots, G^r)}{\partial(p_{N+2}, \ldots, p_{N+2+r})} \right| \neq 0$. Of course, any r of p_j, $j = N+2, \ldots, 2N+1$, can be used, not necessarily the first r.

Theorem 3.37. *A function y that is a solution to problem* (3.156)–(3.157) *subject to r independent constraints ($r < N$)* (3.168) *satisfies, for suitably chosen functions λ_j, $j = 1, \ldots, r$, the system of N fractional Euler–Lagrange equations*

$$\partial_i F[y, \lambda]_\gamma^{\alpha,\beta}(x) + D_{1-\gamma_{i-1}}^{\beta_{i-1},\alpha_{i-1}} \partial_{N+i} F[y, \lambda]_\gamma^{\alpha,\beta}(x) = 0, \quad x \in [a, b],$$

$i = 2, \ldots, N + 1$, *where*

$$F[y, \lambda]_\gamma^{\alpha,\beta}(x) = L[y]_\gamma^{\alpha,\beta}(x) + \sum_{j=1}^{r} \lambda_j(x) G^j[y]_\gamma^{\alpha,\beta}(x).$$

Proof. Suppose that $y = (y_1, \ldots, y_N)$ is the solution to the problem defined by (3.156), (3.157), and (3.168). Let $h = (h_1, \ldots, h_N)$ be an arbitrary admissible variation, i.e., $h_i(a) = h_i(b) = 0$, $i = 1, \ldots, N$, and $G^j[y + \varepsilon h]_\gamma^{\alpha,\beta}(x) = 0$, $j = 1, \ldots, r$, where $\varepsilon \in \mathbb{R}$ is a small parameter. Because $y = (y_1, \ldots, y_N)$ is a solution to the problem defined by (3.156), (3.157), and (3.168), it follows that

$$\frac{\partial}{\partial \varepsilon} \mathcal{J}(y + \varepsilon h) \bigg|_{\varepsilon = 0} = 0,$$

that is,

$$\int_a^b \left[\sum_{i=2}^{N+1} \partial_i L[y]_\gamma^{\alpha,\beta}(x) h_{i-1}(x) \right.$$

$$\left. + \sum_{i=2}^{N+1} \partial_{N+i} L[y]_\gamma^{\alpha,\beta}(x) {}^C D_{\gamma_{i-1}}^{\alpha_{i-1},\beta_{i-1}} h_{i-1}(x) \right] dx = 0, \quad (3.169)$$

and, for $j = 1, \ldots, r$,

$$\sum_{i=2}^{N+1} \partial_i G^j[y]_\gamma^{\alpha,\beta}(x) h_{i-1}(x) + \sum_{i=2}^{N+1} \partial_{N+i} G^j[y]_\gamma^{\alpha,\beta}(x) {}^C D_{\gamma_{i-1}}^{\alpha_{i-1},\beta_{i-1}} h_{i-1}(x) = 0.$$

$$(3.170)$$

Multiplying the jth equation of the system (3.170) by the unspecified function $\lambda_j(x)$, for all $j = 1, \ldots, r$, integrating with respect to x, and adding the left-hand sides (all equal to zero for any choice of the λ_j) to the integrand of (3.169), we obtain

$$\int_a^b \left[\sum_{i=2}^{N+1} \left(\partial_i L[y]_\gamma^{\alpha,\beta}(x) + \sum_{j=1}^r \lambda_j(x)\partial_i G^j[y]_\gamma^{\alpha,\beta}(x) \right) h_{i-1}(x) \right.$$

$$+ \sum_{i=2}^{N+1} \left(\partial_{N+i} L[y]_\gamma^{\alpha,\beta}(x) + \sum_{j=1}^r \lambda_j(x)\partial_{N+i} G^j[y]_\gamma^{\alpha,\beta}(x) \right)$$

$$\times \left({}^C D_{\gamma_{i-1}}^{\alpha_{i-1},\beta_{i-1}} h_{i-1}(x) \right) \bigg] dx$$

$$= \int_a^b \left[\sum_{i=2}^{N+1} \partial_i F[y,\lambda]_\gamma^{\alpha,\beta}(x) h_i(x) + \sum_{i=2}^{N+1} \partial_{N+i} F[y,\lambda]_\gamma^{\alpha,\beta}(x) \right.$$

$$\times \left({}^C D_{\gamma_{i-1}}^{\alpha_{i-1},\beta_{i-1}} h_{i-1}(x) \right) \bigg] dx$$

$$= 0,$$

where $F[y,\lambda]_\gamma^{\alpha,\beta}(x) = L[y]_\gamma^{\alpha,\beta}(x) + \sum_{j=1}^r \lambda(x)G^j[y]_\gamma^{\alpha,\beta}(x)$. Integrating by parts,

$$\int_a^b \left[\sum_{i=2}^{N+1} \partial_i F[y,\lambda]_\gamma^{\alpha,\beta}(x) + D_{1-\gamma_{i-1}}^{\beta_{i-1},\alpha_{i-1}} \partial_{N+i} F[y,\lambda]_\gamma^{\alpha,\beta}(x) \right] h_{i-1}(x) dx = 0.$$

$$(3.171)$$

Because of (3.170), we cannot regard the N functions $h_1,\ldots,h_N$ as being free for arbitrary choice. There is a subset of r of these functions whose assignment is restricted by the assignment of the remaining $(N-r)$. We can assume, without loss of generality, that $h_1,\ldots,h_r$ are the functions of the set whose dependence upon the choice of the arbitrary $h_{r+1},\ldots,h_N$ is governed by (3.170). We now assign the functions $\lambda_1,\ldots,\lambda_r$ to be the set of r functions such that, for all x between a and b, the coefficients of $h_1,\ldots,h_r$ in the integrand of (3.171) vanish. That is, $\lambda_1,\ldots,\lambda_r$ are chosen so as to satisfy

$$\partial_i F[y,\lambda]_\gamma^{\alpha,\beta}(x) + D_{1-\gamma_{i-1}}^{\beta_{i-1},\alpha_{i-1}} \partial_{N+i} F[y,\lambda]_\gamma^{\alpha,\beta}(x) = 0, \qquad (3.172)$$

$i = 2,\ldots,r+1$, $x \in [a,b]$. With this choice, (3.171) gives

$$\int_a^b \left[\sum_{i=r+2}^{N+1} \partial_i F[y,\lambda]_\gamma^{\alpha,\beta}(x) + D_{1-\gamma_{i-1}}^{\beta_{i-1},\alpha_{i-1}} \partial_{N+i} F[y,\lambda]_\gamma^{\alpha,\beta}(x) \right] h_{i-1}(x) dx = 0.$$

Since the functions $h_{r+1},\ldots,h_N$ are arbitrary, we may employ the fundamental lemma of the calculus of variations to conclude that

$$\partial_i F[y,\lambda]_\gamma^{\alpha,\beta}(x) + D_{1-\gamma_{i-1}}^{\beta_{i-1},\alpha_{i-1}} \partial_{N+i} F[y,\lambda]_\gamma^{\alpha,\beta}(x) = 0, \qquad (3.173)$$

$i = r+2,\ldots,N+1$, for all $x \in [a,b]$. $\qquad\square$

Remark 3.31. In order to determine the $(N + r)$ unknown functions $y_1, \ldots, y_n, \lambda_1, \ldots, \lambda_r$, we must consider the system of $(N + r)$ equations, consisting of (3.168), (3.172), and (3.173), together with the $2N$ boundary conditions (3.157).

Assume now that the constraints, instead of (3.168), are characterized by inequalities:

$$G^j[y]_\gamma^{\alpha,\beta}(x) \leq 0, \quad j = 1, \ldots, r.$$

In this case we can set

$$G^j[y]_\gamma^{\alpha,\beta}(x) + (\phi_j(x))^2 = 0, \quad j = 1, \ldots, r,$$

where ϕ_j have the some continuity properties as y_i. Therefore, we obtain the following problem:

$$\hat{\mathcal{J}}(y) = \int_a^b \hat{L}(x, y(x), {}^C D_\gamma^{\alpha,\beta} y(x), \phi_1(x), \ldots, \phi_r(x)) \, dx \longrightarrow \min \quad (3.174)$$

subject to r independent constraints $(r < N)$

$$G^j[y]_\gamma^{\alpha,\beta}(x) + (\phi_j(x))^2 = 0, \quad j = 1, \ldots, r, \quad\quad\quad (3.175)$$

and boundary conditions (3.157). Applying Theorem 3.37 we get the following result.

Theorem 3.38. *A set of functions* $y_1, \ldots, y_N$, $\phi_1, \ldots, \phi_r$, *which is a solution to problem* (3.174)–(3.175), *satisfies, for suitably chosen* λ_j, $j = 1, \ldots, r$, *the following system of equations:*

$$D_{1-\gamma_{i-1}}^{\beta_{i-1}, \alpha_{i-1}} \partial_{N+i} \hat{F}(x, y(x), {}^C D_\gamma^{\alpha,\beta} y(x), \lambda_1(x), \ldots, \lambda_r(x), \phi_1(x), \ldots, \phi_r(x))$$

$$+ \partial_i \hat{F}(x, y(x), {}^C D_\gamma^{\alpha,\beta} y(x), \lambda_1(x), \ldots, \lambda_r(x), \phi_1(x), \ldots, \phi_r(x)) = 0,$$

$i = 2, \ldots, N + 1$, *for all* $x \in [a, b]$, *where* $\hat{F} = \hat{L} + \sum_{j=1}^r \lambda_j(G^j + \phi_j^2)$ *and* $\lambda_j(x)\phi_j(x) = 0$, $j = 1, \ldots, r$.

Sufficient condition of optimality

In this section we provide sufficient optimality conditions for the elementary and the isoperimetric problem of the ${}^C D_\gamma^{\alpha,\beta}$ fractional calculus of variations. Similarly to what happens in the classical calculus of variations, some conditions of convexity are in order.

Theorem 3.39. *Let* $L(\underline{x}, y, v)$ *be jointly convex in* (y, v). *If* y *satisfies the system of* N *fractional Euler–Lagrange equations* (3.158), *then* y *is a global minimizer to problem* (3.156)–(3.157).

Proof.   Is similar to that of Theorem 2.14.   □

Theorem 3.40. *Let* $F(\underline{x}, y, v, \bar{\lambda}) = L(\underline{x}, y, v) - \sum_{j=1}^{r} \bar{\lambda}_j G^j(\underline{x}, y, v)$ *be jointly convex in* (y, v), *for some constants* $\bar{\lambda}_j \in \mathbb{R}$, $j = 1, \ldots, r$. *If* y^0 *satisfies the system of* N *fractional Euler–Lagrange equations* (3.165), *then* y^0 *is a minimizer to the isoperimetric problem defined by* (3.156), (3.157) *and* (3.163).

Proof.   Is similar to that of Theorem 3.12 and can be found in Malinowska and Torres, 2012a.   □

Problems with classical and combined Caputo derivatives

We now consider variational functionals with a Lagrangian depending on a combined Caputo fractional derivative and the classical derivative. Euler–Lagrange equations are proved for the basic and isoperimetric problems, as well as the transversality conditions.

For $f = [f_1, \ldots, f_N] : [a, b] \to \mathbb{R}^N$, $N \in \mathbb{N}$, and $f_i \in AC([a,b])$, $i = 1, \ldots, N$, we put

$$^C D_\gamma^{\alpha,\beta} f(x) := \left[{}^C D_\gamma^{\alpha,\beta} f_1(x), \ldots, {}^C D_\gamma^{\alpha,\beta} f_N(x) \right].$$

Consider the following functional:

$$\mathcal{J}(y) = \int_a^b L\left(x, y(x), y'(x), {}^C D_\gamma^{\alpha,\beta} y(x)\right) dx, \qquad (3.176)$$

where $x \in [a, b]$ is the independent variable; $y(x) \in \mathbb{R}^N$ is a real vector variable; $y'(x) \in \mathbb{R}^N$ with y' the first derivative of y; ${}^C D_\gamma^{\alpha,\beta} y(x) \in \mathbb{R}^N$ stands for the combined fractional derivative of y evaluated in x; and $L \in C^1\left([a, b] \times \mathbb{R}^{3N}; \mathbb{R}\right)$. Let $\mathbf{D}$ denote the set of all functions $y : [a, b] \to \mathbb{R}^N$ such that y' and ${}^C D_\gamma^{\alpha,\beta} y$ exist and are continuous on the interval $[a, b]$. We endow $\mathbf{D}$ with the norm

$$\|y\|_{1,\infty} := \max_{a \le x \le b} \|y(x)\| + \max_{a \le x \le b} \|y'(x)\| + \max_{a \le x \le b} \left\| {}^C D_\gamma^{\alpha,\beta} y(x) \right\|,$$

where $\|\cdot\|$ is a norm in $\mathbb{R}^N$. Let $\lambda \in \mathbb{R}^r$. For simplicity of notation we introduce the operators $[\cdot]_\gamma^{\alpha,\beta}$ and $\lambda \{\cdot\}_\gamma^{\alpha,\beta}$ defined by

$$[y]_\gamma^{\alpha,\beta}(x) := \left(x, y(x), y'(x), {}^C D_\gamma^{\alpha,\beta} y(x)\right),$$

$$\lambda \{y\}_\gamma^{\alpha,\beta}(x) := \left(x, y(x), y'(x), {}^C D_\gamma^{\alpha,\beta} y(x), \lambda_1, \ldots, \lambda_r\right).$$

The Euler–Lagrange equation

We begin with the following problem of the fractional calculus of variations.

Problem 3.2. *Find a function $y \in \mathbf{D}$ for which the functional* (3.176), *that is,*

$$\mathcal{J}(y) = \int_a^b L\,[y]_\gamma^{\alpha,\beta}\,(x)dx, \qquad (3.177)$$

subject to given boundary conditions

$$y(a) = y^a, \quad y(b) = y^b, \qquad (3.178)$$

$y^a, y^b \in \mathbb{R}^N$, *achieves a minimum.*

We state the Euler–Lagrange equations for Problem 3.2.

Theorem 3.41. *If $y = (y_1, \ldots, y_N)$ is a local minimizer to Problem 3.2, then y satisfies the system of N Euler–Lagrange equations*

$$\partial_i L\,[y]_\gamma^{\alpha,\beta}\,(x) - \frac{d}{dx}\partial_{N+i}L\,[y]_\gamma^{\alpha,\beta}\,(x) + D_{1-\gamma}^{\beta,\alpha}\partial_{2N+i}L\,[y]_\gamma^{\alpha,\beta}\,(x) = 0, \quad (3.179)$$

$i = 2, \ldots, N+1$, *for all $x \in [a, b]$.*

Proof. Is adapted from proofs of Theorem 2.1 and Theorem 3.34, and can be found in Odzijewicz, Malinowska and Torres, 2012b. □

When the Lagrangian L does not depend on fractional derivatives, then Theorem 3.41 reduces to the classical result (see, e.g., Troutman, 1996). The fractional Euler–Lagrange equations via Caputo derivatives can also be obtained as corollaries of Theorem 3.41. Theorem 3.34 is obtained by choosing a Lagrangian that does not depend on the classical derivatives.

Corollary 3.7. *Let $y = (y_1, \ldots, y_N)$ be a local minimizer to problem*

$$\mathcal{J}(y) = \int_a^b L\left(x, y(x), {}^C D_\gamma^{\alpha,\beta} y(x)\right) dx \longrightarrow \min$$

$$y(a) = y^a, \quad y(b) = y^b,$$

$y^a, y^b \in \mathbb{R}^N$. *Then, y satisfies the system of N fractional Euler–Lagrange equations*

$$\partial_i L[y](x) + D_{1-\gamma}^{\beta,\alpha}\partial_{N+i}L[y](x) = 0, \qquad (3.180)$$

$i = 2, \ldots N+1$, *for all $x \in [a, b]$.*

Transversality conditions

Let $l \in \{1, \ldots, N\}$. Assume now that in Problem 3.2 the boundary conditions (3.178) are substituted by

$$y(a) = y^a, \quad y_i(b) = y_i^b, \ i = 1, \ldots, N \text{ for } i \neq l, \text{ and } y_l(b) \text{ is free} \quad (3.181)$$

or

$$y(a) = y^a, \quad y_i(b) = y_i^b, \ i = 1, \ldots, N \text{ for } i \neq l, \text{ and } y_l(b) \leq y_l^b. \quad (3.182)$$

Theorem 3.42. *If $y = (y_1, \ldots, y_N)$ is a solution to Problem 3.2 with either (3.181) or (3.182) as boundary conditions instead of (3.178), then y satisfies the system of Euler–Lagrange equations (3.179). Moreover, under the boundary conditions (3.181) the extra transversality condition*

$$\left[\partial_{N+l+1} L \left[y\right]_\gamma^{\alpha,\beta} (x) + \gamma_x I_b^{1-\alpha} \partial_{2N+l+1} L \left[y\right]_\gamma^{\alpha,\beta} (x) \right.$$

$$\left. -(1-\gamma)_a I_x^{1-\beta} \partial_{2N+l+1} L \left[y\right]_\gamma^{\alpha,\beta} (x) \right]_{x=b} = 0 \quad (3.183)$$

holds; under the boundary conditions (3.182) the extra transversality condition

$$\left[\partial_{N+l+1} L \left[y\right]_\gamma^{\alpha,\beta} (x) + \gamma_x I_b^{1-\alpha} \partial_{2N+l+1} L \left[y\right]_\gamma^{\alpha,\beta} (x) \right.$$

$$\left. -(1-\gamma)_a I_x^{1-\beta} \partial_{2N+l+1} L \left[y\right]_\gamma^{\alpha,\beta} (x) \right]_{x=b} \leq 0 \quad (3.184)$$

holds, with (3.183) taking place if $y_l(b) < y_l^b$.

When the Lagrangian does not depend on fractional derivatives, then the left-hand sides of (3.183) and (3.184) reduce to the classical expression $\partial_{N+l+1} L (x, y(x), y'(x))$ (for instance, when $N = 1$ and $y(a)$ is fixed with $y(b)$ free, then we get the well-known natural boundary condition $\partial_3 L (b, y(b), y'(b)) = 0$). In the particular case when the Lagrangian does not depend on the classical derivatives, $\gamma = 0$, $N = 1$, and we have boundary conditions (3.181), then one obtains from Theorem 3.42 the following corollary.

Corollary 3.8. *If y is a local minimizer to problem*

$$\mathcal{J}(y) = \int_a^b L \left(x, y(x), {}_x^C D_b^\alpha y(x)\right) dx \longrightarrow \min$$

$$y(a) = y_a \quad (y(b) \text{ is free}),$$

then y satisfies the fractional Euler–Lagrange equation

$$\partial_2 L \left(x, y(x), {}_x^C D_b^\alpha y(x)\right) + {}_a D_x^\alpha \partial_3 L \left(x, y(x), {}_x^C D_b^\alpha y(x)\right) = 0.$$

Moreover,

$$\left[{}_a I_x^{1-\alpha} \partial_3 L(x, y(x), {}_x^C D_b^\alpha y(x)) \right]_{x=b} = 0.$$

The isoperimetric problem

Now we consider the following problem of the calculus of variations.

Problem 3.3. *Minimize functional* (3.177) *subject to given boundary conditions* (3.178) *and* r *isoperimetric constraints*

$$\mathcal{G}^j(y) = \int_a^b G^j\,[y]_\gamma^{\alpha,\beta}\,(x)dx = \xi_j, \quad j = 1,\ldots,r, \tag{3.185}$$

where $G^j \in C^1\left([a,b] \times \mathbb{R}^{3N};\mathbb{R}\right)$ *and* $\xi_j \in \mathbb{R}$ *for* $j = 1,\ldots,r$.

In order to obtain the necessary optimality conditions for the combined fractional isoperimetric problem (Problem 3.3), we make use of Theorem 3.35.

Theorem 3.43. *Let the assumptions of Theorem 3.35 hold. If* y *is a local minimizer to Problem 3.3, then* y *satisfies the system of* N *fractional Euler–Lagrange equations*

$$\partial_i F_\lambda\,\{y\}_\gamma^{\alpha,\beta}\,(x) - \frac{d}{dx}\partial_{N+i} F_\lambda\,\{y\}_\gamma^{\alpha,\beta}\,(x) + D_{1-\gamma}^{\beta,\alpha}\partial_{2N+i} F_\lambda\,\{y\}_\gamma^{\alpha,\beta}\,(x) = 0,$$
$$\tag{3.186}$$

$i = 2,\ldots,N+1$, *for all* $x \in [a,b]$, *where function* $F : [a,b] \times \mathbb{R}^{3N} \times \mathbb{R}^r \to \mathbb{R}$ *is defined by*

$$F_\lambda\,\{y\}_\gamma^{\alpha,\beta}\,(x) := L\,[y]_\gamma^{\alpha,\beta}\,(x) - \sum_{j=1}^r \lambda_j G^j\,[y]_\gamma^{\alpha,\beta}\,(x).$$

Proof.   Is similar to that of Theorem 3.36 and can be found in Odzijewicz, Malinowska and Torres, 2012b.   □

An illustrative example

Let $\alpha \in (0,1)$, $N = 1$, $\gamma = 1$, and $\xi \in \mathbb{R}$. Consider the following fractional isoperimetric problem:

$$\mathcal{J}(y) = \int_0^1 \left(y'(x) + {}_0^C D_x^\alpha y(x)\right)^2 dx \longrightarrow \min$$

$$\mathcal{G}(y) = \int_0^1 \left(y'(x) + {}_0^C D_x^\alpha y(x)\right) dx = \xi \tag{3.187}$$

$$y(0) = 0\,,\ y(1) = \int_0^1 E_{1-\alpha}\left(-(1-t)^{1-\alpha}\right)\xi dt.$$

In this problem we make use of the Mittag–Leffler function $E_\alpha(z)$. We recall that the Mittag–Leffler function is defined by

$$E_\alpha(z) = \sum_{k=0}^{\infty} \frac{z^k}{\Gamma(\alpha k + 1)}.$$

This function appears naturally in the solution of fractional differential equations, as a generalization of the exponential function (Camargo *et al.*, 2009). Indeed, while a linear second order ordinary differential equation with constant coefficients presents an exponential function as a solution, in the fractional case the Mittag–Leffler functions emerge (Kilbas, Srivastava and Trujillo, 2006).

In our example (3.187) the function F of Theorem 3.43 is given by

$$F(x, y, y', {}_{0}^{C}D_x^\alpha y, \lambda) = \left(y' + {}_{0}^{C}D_x^\alpha y\right)^2 - \lambda \left(y' + {}_{0}^{C}D_x^\alpha y\right).$$

One can easily check that y, such that

$$y(x) = \int_0^x E_{1-\alpha}\left(-(x-t)^{1-\alpha}\right)\xi dt \tag{3.188}$$

(a) is not an extremal for $\mathcal{G}$;
(b) satisfies $y' + {}_{0}^{C}D_x^\alpha y = \xi$ (see, e.g., p. 328, Example 5.24, of Kilbas, Srivastava and Trujillo, 2006).

Moreover, (3.188) satisfies the Euler–Lagrange equations (3.186) for $\lambda = 2\xi$, i.e.,

$$-\frac{d}{dx}\left(2\left(y' + {}_{0}^{C}D_x^\alpha y\right) - 2\xi\right) + {}_{x}D_1^\alpha\left(2\left(y' + {}_{0}^{C}D_x^\alpha y\right) - 2\xi\right) = 0.$$

We conclude that (3.188) is an extremal for the isoperimetric problem (3.187).

Remark 3.32. When $\alpha \to 1$ the isoperimetric constraint is redundant with the boundary conditions, and the fractional isoperimetric problem (3.187) simplifies to the classical variational problem

$$\mathcal{J}(y) = 4 \int_0^1 (y'(x))^2 dx \longrightarrow \min$$

$$y(0) = 0, \quad y(1) = \frac{\xi}{2}. \tag{3.189}$$

Our fractional extremal (3.188) gives $y(x) = \frac{\xi}{2}x$, which is exactly the minimizer of (3.189).

Remark 3.33. Choose $\xi = 1$. When $\alpha \to 0$ one gets from (3.187) the classical isoperimetric problem

$$\mathcal{J}(y) = \int_0^1 (y'(x) + y(x))^2 \, dx \longrightarrow \min$$

$$\mathcal{G}(y) = \int_0^1 y(x) dx = \frac{1}{e} \qquad (3.190)$$

$$y(0) = 0 \quad y(1) = 1 - \frac{1}{e}.$$

Our extremal (3.188) is then reduced to the classical extremal $y(x) = 1 - e^{-x}$ of the isoperimetric problem (3.190).

Remark 3.34. Let $\alpha = \frac{1}{2}$. Then (3.187) gives the following fractional isoperimetric problem:

$$\mathcal{J}(y) = \int_0^1 \left(y'(x) + {}_0^C D_x^{\frac{1}{2}} y(x) \right)^2 \, dx \longrightarrow \min$$

$$\mathcal{G}(y) = \int_0^1 \left(y'(x) + {}_0^C D_x^{\frac{1}{2}} y(x) \right) \, dx = \xi \qquad (3.191)$$

$$y(0) = 0, \quad y(1) = \xi \left(\mathrm{erfc}(1) e - \frac{2}{\sqrt{\pi}} - 1 \right),$$

where erfc is the complementary error function defined by

$$\mathrm{erfc}(z) = \frac{2}{\sqrt{\pi}} \int_z^\infty exp(-t^2) dt.$$

The extremal (3.188) for the particular fractional isoperimetric problem (3.191) is

$$y(x) = \xi \left(e^x \mathrm{erfc}(\sqrt{x}) - \frac{2\sqrt{x}}{\sqrt{\pi}} - 1 \right).$$

3.11.2 *Multiobjective Fractional Optimization*

Multiobjective optimization is a natural extension of the traditional optimization of a single-objective function. If the objective functions are commensurate, minimizing a one-objective function minimizes all criteria and the problem can be solved using traditional optimization techniques. However, if the objective functions are incommensurate, or competing, then the minimization of one objective function requires a compromise in another objective. Here we consider multiobjective fractional variational problems with a finite number $d \geq 1$ of objective (cost) functionals

$$\left(\mathcal{J}^1(y), \ldots, \mathcal{J}^d(y)\right) = \left(\int_a^b L^1[y]_\gamma^{\alpha,\beta}(x)\,dx, \ldots, \int_a^b L^d[y]_\gamma^{\alpha,\beta}(x)\,dx\right) \longrightarrow \min \tag{3.192}$$

subject to the boundary conditions

$$y(a) = y^a, \quad y(b) = y^b, \tag{3.193}$$

$y^a, y^b \in \mathbb{R}^N$, and r $(r < N)$ independent constraints

$$G^j[y]_\gamma^{\alpha,\beta}(x) \leq 0, \quad j = 1, \ldots, r, \tag{3.194}$$

where $L^i, G^j \in C^1([a, b] \times \mathbb{R}^{2N}; \mathbb{R})$, $i = 1, \ldots, d$, $j = 1, \ldots, r$. We would like to find a function $y \in \mathbf{D}$, satisfying constraints (3.193) and (3.194), that renders the minimum value to each functional $\mathcal{J}^i$, $i = 1, \ldots, d$, simultaneously. The competition between objectives gives rise to the necessity to distinguish between the difference of multiobjective optimization and traditional single-objective optimization. Competition causes the lack of complete order for multiobjective optimization problems. The concept of Pareto optimality is therefore used to characterize a solution to the multi-objective optimization problem. We define

$$\mathcal{E} := \{y \in \mathbf{D} : y \text{ satisfies conditions (3.193) and (3.194)}\}.$$

Definition 3.29. *A function* $\bar{y} \in \mathcal{E}$ *is called a Pareto optimal solution to problem* (3.192)–(3.194) *if* $y \in \mathcal{E}$ *does not exist with*

$$\forall i \in \{1, \ldots, d\} : \mathcal{J}^i(y) \leq \mathcal{J}^i(\bar{y}) \quad \wedge \quad \exists i \in \{1, \ldots, d\} : \mathcal{J}^i(y) < \mathcal{J}^i(\bar{y}).$$

Definition 3.29 introduces the notion of global Pareto optimality. Another important concept is the one of local Pareto optimality.

Definition 3.30. *A function* $\bar{y} \in \mathcal{E}$ *is called a local Pareto optimal solution to problem* (3.192)–(3.194) *if there exists* $\delta > 0$ *for which there is no* $y \in \mathcal{E}$ *with* $\|y - \bar{y}\|_{1,\infty} < \delta$ *and*

$$\forall i \in \{1, \ldots, d\} : \mathcal{J}^i(y) \leq \mathcal{J}^i(\bar{y}) \quad \wedge \quad \exists i \in \{1, \ldots, d\} : \mathcal{J}^i(y) < \mathcal{J}^i(\bar{y}).$$

Remark 3.35. Any global Pareto optimal solution is locally Pareto optimal.

We obtain a sufficient condition for Pareto optimality by modifying the original multiobjective fractional problem (3.192)–(3.194) into the following weighting problem:

$$\sum_{i=1}^{d} w_i \int_a^b L^i[y]_\gamma^{\alpha,\beta}(x)\,dx \longrightarrow \min \qquad (3.195)$$

subject to $y \in \mathcal{E}$, where $w_i \geq 0$ for all $i = 1, \ldots, d$, and $\sum_{i=1}^{d} w_i = 1$.

Theorem 3.44. *The solution of the weighting problem* (3.195) *is Pareto optimal if the weighting coefficients are positive, that is, $w_i > 0$ for all $i = 1, \ldots, d$. Moreover, the unique solution of the weighting problem* (3.195) *is Pareto optimal.*

Proof. Let $\bar{y} \in \mathcal{E}$ be a solution to problem (3.195) with $w_i > 0$ for all $i = 1, \ldots, d$. Suppose that $\bar{y}$ is not Pareto optimal. Then, there exists y such that $\mathcal{J}^i(y) \leq \mathcal{J}^i(\bar{y})$ for all $i = 1, \ldots, d$ and $\mathcal{J}^j(y) < \mathcal{J}^j(\bar{y})$ for at least one j. Since $w_i > 0$ for all $i = 1, \ldots, d$, we have $\sum_{i=1}^{d} w_i \mathcal{J}^i(y) < \sum_{i=1}^{d} w_i \mathcal{J}^i(\bar{y})$. This contradicts the minimality of $\bar{y}$. Now, let $\bar{y}$ be the unique solution to (3.195). If $\bar{y}$ is not Pareto optimal, then $\sum_{i=1}^{d} w_i \mathcal{J}^i(y) \leq \sum_{i=1}^{d} w_i \mathcal{J}^i(\bar{y})$. This contradicts the uniqueness of $\bar{y}$. $\square$

Therefore, by varying the weights over the unit simplex

$$\left\{ w = (w_1, \ldots, w_d) : w_i \geq 0, \ \sum_{i=1}^{d} w_i = 1 \right\}$$

ones obtains, in principle, different Pareto optimal solutions. The next theorem provides a necessary and sufficient condition for Pareto optimality.

Theorem 3.45. *A function $\bar{y} \in \mathcal{E}$ is Pareto optimal to problem* (3.192)–(3.194) *if, and only if, it is a solution to the scalar fractional variational problem*

$$\int_a^b L^i[y]_\gamma^{\alpha,\beta}(x)\,dx \longrightarrow \min$$

subject to $y \in \mathcal{E}$ and

$$\int_a^b L^j[y]_\gamma^{\alpha,\beta}(x)\,dx \leq \int_a^b L^j[\bar{y}]_\gamma^{\alpha,\beta}(x)\,dx, \quad j = 1, \ldots, d, \quad j \neq i,$$

for each $i = 1, \ldots, d$.

Proof. Suppose that $\bar{y}$ is Pareto optimal. Then $\bar{y} \in \mathcal{C}_k = \{y \in \mathcal{E} : \mathcal{J}^j(y) \le \mathcal{J}^j(\bar{y}), j = 1, \ldots, d, j \ne k\}$ for all k, so $\mathcal{C}_k \ne \emptyset$. If $\bar{y}$ does not minimize $\mathcal{J}^k(y)$ on the constrained set $\mathcal{C}_k$ for some k, then there exists $y \in \mathcal{E}$ such that $\mathcal{J}^k(y) < \mathcal{J}^k(\bar{y})$ and $\mathcal{J}^j(y) \le \mathcal{J}^j(\bar{y})$ for all $j \ne k$. This contradicts the Pareto optimality of $\bar{y}$. Now, suppose that $\bar{y}$ minimize each $\mathcal{J}^k(y)$ on the constrained set $\mathcal{C}_k$. If $\bar{y}$ is not Pareto optimal, then there exists y such that $\mathcal{J}^i(y) \le \mathcal{J}^i(\bar{y})$ for all $i = 1, \ldots, d$ and $\mathcal{J}^j(y) < \mathcal{J}^j(\bar{y})$ for at least one j. This contradicts the minimality of y for $\mathcal{J}^j(y)$ on $\mathcal{C}_j$. $\square$

Remark 3.36. For a function $y \in \mathcal{E}$ to be Pareto optimal to problem (3.192)–(3.194), it is necessary for it to be a solution to the fractional isoperimetric problems

$$\int_a^b L^i[y]_\gamma^{\alpha,\beta}(x)\, dx \longrightarrow \min$$

subject to $y \in \mathcal{E}$ and

$$\int_a^b L^j[y]_\gamma^{\alpha,\beta}(x)\, dx = \int_a^b L^j[\bar{y}]_\gamma^{\alpha,\beta}(x)\, dx, \quad j = 1, \ldots, d, \quad j \ne i,$$

for all $i = 1, \ldots, d$. Therefore, the necessary optimality conditions for the fractional isoperimetric problems (see Theorem 3.36) are also necessary for fractional Pareto optimality.

We illustrate our results with two multiobjective fractional variational problems.

Example 3.17. Let $\bar{y}(x) = E_\alpha(x^\alpha)$, $x \in [0, 1]$, where E_α is the Mittag–Leffler function. Consider the following multiobjective fractional variational problem ($N = 1$, $\gamma = 1$, and $d = 2$):

$$\left(\mathcal{J}^1(y), \mathcal{J}^2(y)\right) = \left(\int_0^1 ({}_0^C D_x^\alpha y(x))^2\, dx, \int_0^1 \bar{y}(x){}_0^C D_x^\alpha y(x)\, dx\right) \longrightarrow \min \tag{3.196}$$

subject to

$$y(0) = 0, \quad y(1) = E_\alpha(1). \tag{3.197}$$

Observe that $\bar{y}$ satisfies the necessary Pareto optimality conditions (see Remark 3.36). Indeed, as shown in Example 3.4 and Example 3.5, $\bar{y}$ is a solution to the isoperimetric problem

$$\mathcal{J}^1(y) = \int_0^1 ({}_0^C D_x^\alpha y(x))^2\, dx \longrightarrow \min$$

subject to

$$\int_0^1 \bar{y}(x) {}_0^C D_x^\alpha y(x)\, dx = \int_0^1 (\bar{y}(x))^2\, dx.$$

Consider now the following fractional isoperimetric problem:

$$\mathcal{J}^2(y) = \int_0^1 \bar{y}(x) {}_0^C D_x^\alpha y(x)\, dx \longrightarrow \min$$

subject to

$$\int_0^1 ({}_0^C D_x^\alpha y(x))^2\, dx = \int_0^1 ({}_0^C D_x^\alpha \bar{y}(x))^2\, dx.$$

Let us apply Theorem 3.35. The equality ${}_x D_1^\alpha y(x) = 0$ holds if, and only if, $y(x) = d(1-x)^{\alpha-1}$ with $d \in \mathbb{R}$ (see Corollary 2.1 of Kilbas, Srivastava and Trujillo, 2006). Hence, $\bar{y}$ does not satisfy equation ${}_x D_1^\alpha({}_0^C D_x^\alpha y) = 0$. The augmented function is

$$F_\lambda \{y\}_\gamma^{\alpha,\beta}(x) = \bar{y}(x) {}_0^C D_x^\alpha y(x) - \lambda ({}_0^C D_x^\alpha y(x))^2,$$

and the corresponding fractional Euler–Lagrange equation gives

$${}_x D_1^\alpha (\bar{y}(x) - 2\lambda {}_0^C D_x^\alpha y(x)) = 0.$$

A solution to this equation is $\lambda = \frac{1}{2}$ and $y = \bar{y}$. Therefore, by Remark 3.36, $y = \bar{y}$ is a candidate Pareto optimal solution to problem (3.196)–(3.197).

Example 3.18. Consider the following multiobjective fractional variational problem:

$$\left(\mathcal{J}^1(y), \mathcal{J}^2(y)\right)$$
$$= \left(\int_0^1 \frac{1}{2} ({}_0^C D_x^\alpha y(x) - f(x))^2\, dx, \int_0^1 \frac{1}{2} ({}_0^C D_x^\alpha y(x))^2\, dx \right) \longrightarrow \min \quad (3.198)$$

subject to

$$y(0) = 0, \quad y(1) = \chi, \quad \chi \in \mathbb{R}, \qquad (3.199)$$

where f is a fixed function. In this case we have $N = 1$, $\gamma = 1$, and $d = 2$. By Theorem 3.44, Pareto optimal solutions to problem (3.198)–(3.199) can be found by considering the family of problems

$$w \int_0^1 \frac{1}{2} ({}_0^C D_x^\alpha y(x) - f(x))^2\, dx + (1-w) \int_0^1 \frac{1}{2} {}_0^C D_x^\alpha y(x)\, dx \longrightarrow \min \quad (3.200)$$

subject to

$$y(0) = 0, \quad y(1) = \chi, \quad \chi \in \mathbb{R}, \qquad (3.201)$$

where $w \in [0, 1]$. Let us now fix w. By Theorem 3.34, a solution to problem (3.200)–(3.201) satisfies the fractional Euler–Lagrange equation

$$_x D_1^\alpha \left(_0^C D_x^\alpha y(x) - wf(x) \right) = 0. \tag{3.202}$$

Moreover, by Theorem 3.39, a solution to (3.202) is a global minimizer to problem (3.200)–(3.201). Therefore, solving equation (3.202) for $w \in [0, 1]$, we are able to obtain Pareto optimal solutions to problem (3.198)–(3.199). In order to solve equation (3.202), firstly we use Corollary 2.1 of Kilbas, Srivastava and Trujillo, 2006, to get the following equation:

$$_0^C D_x^\alpha y(x) - wf(x) = d(1 - x)^{\alpha - 1}, \quad d \in \mathbb{R}. \tag{3.203}$$

Equation (3.203) needs to be solved numerically. We did numerical simu-

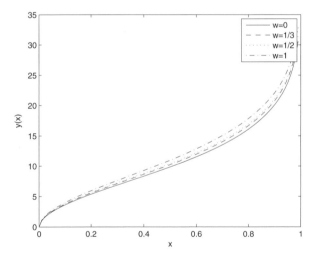

Fig. 3.2 The minimizer of Example 3.18 for $f(x) = e^x$ and $\alpha = \frac{1}{2}$.

lations using the MATLAB solver `fode` for linear Fractional-Order Differential Equations (FODE) with constant coefficients, developed by Merrikh-Bayat, 2007. The results for $\alpha = 1/2$, $f(x) = e^x$, and different values of the parameter w can be seen in Fig. 3.2. Numerical results for different values of α show that when $\alpha \to 1$ the fractional solution converges to the solution of the classical problem of the calculus of variations.

3.11.3   *Problems Depending on Indefinite Integrals*

We obtain necessary optimality conditions for variational problems with a Lagrangian depending on a Caputo fractional derivative, a fractional and an indefinite integral. The main results give fractional Euler–Lagrange type equations and natural boundary conditions, which provide a generalization of the previous results. Isoperimetric problems, problems with holonomic constraints and those dependent on higher-order Caputo derivatives, as well as fractional Lagrange problems, are considered.

The fundamental problem

Let $\alpha \in (0, 1]$ and $\beta > 0$. The problem that we address is stated in the following way. Minimize the cost functional

$$J(y) = \int_a^b L(x, y(x), {}_a^C D_x^\alpha y(x), {}_a I_x^\beta y(x), z(x))dx, \qquad (3.204)$$

where the variable z is defined by

$$z(x) = \int_a^x l(t, y(t), {}_a^C D_t^\alpha y(t), {}_a I_t^\beta y(t))dt,$$

subject to the boundary conditions

$$y(a) = y_a \quad \text{and} \quad y(b) = y_b. \qquad (3.205)$$

We assume that the functions $(x, y, v, w, z) \to L(x, y, v, w, z)$ and $(x, y, v, w) \to l(x, y, v, w)$ are of class C^1, and the trajectories $y : [a, b] \to \mathbb{R}$ are absolute continuous functions, $y \in AC([a, b]; \mathbb{R})$, such that ${}_a^C D_x^\alpha y(x)$ and ${}_a I_x^\beta y(x)$ exist and are continuous on $[a, b]$. We denote such class of functions by $\mathcal{F}([a, b]; \mathbb{R})$. Also, to simplify, by $[\cdot]$ and $\{\cdot\}$ we denote the operators

$$[y](x) = (x, y(x), {}_a^C D_x^\alpha y(x), {}_a I_x^\beta y(x), z(x))$$

and

$$\{y\}(x) = (x, y(x), {}_a^C D_x^\alpha y(x), {}_a I_x^\beta y(x)).$$

Theorem 3.46. *Let $y \in \mathcal{F}([a,b];\mathbb{R})$ be a minimizer of J as in (3.204), subject to the boundary conditions (3.205). Then, for all $x \in [a,b]$, y is a solution of the fractional Euler–Lagrange equation*

$$\frac{\partial L}{\partial y}[y](x) + {}_xD_b^\alpha\left(\frac{\partial L}{\partial v}[y](x)\right) + {}_xI_b^\beta\left(\frac{\partial L}{\partial w}[y](x)\right)$$

$$+ \int_x^b \frac{\partial L}{\partial z}[y](t)dt \cdot \frac{\partial l}{\partial y}\{y\}(x) + {}_xD_b^\alpha\left(\int_x^b \frac{\partial L}{\partial z}[y](t)dt \cdot \frac{\partial l}{\partial v}\{y\}(x)\right)$$

$$+ {}_xI_b^\beta\left(\int_x^b \frac{\partial L}{\partial z}[y](t)dt \cdot \frac{\partial l}{\partial w}\{y\}(x)\right) = 0. \quad (3.206)$$

Proof. Let $h \in \mathcal{F}([a,b];\mathbb{R})$ be such that $h(a) = 0 = h(b)$, and ϵ be a real number with $|\epsilon| \ll 1$. If we define j as $j(\epsilon) = J(y + \epsilon h)$, then $j'(0) = 0$. Differentiating j at $\epsilon = 0$, we find

$$\int_a^b \left[\frac{\partial L}{\partial y}[y](x)h(x) + \frac{\partial L}{\partial v}[y](x){}_a^C D_x^\alpha h(x) + \frac{\partial L}{\partial w}[y](x){}_aI_x^\beta h(x) + \frac{\partial L}{\partial z}[y](x)\right.$$

$$\left.\times \int_a^x \left(\frac{\partial l}{\partial y}\{y\}(t)h(t) + \frac{\partial l}{\partial v}\{y\}(t){}_a^C D_t^\alpha h(t) + \frac{\partial l}{\partial w}\{y\}(t){}_aI_t^\beta h(t)\right)dt\right]dx = 0.$$

The necessary optimality condition (3.206) follows from the next relations and the fundamental lemma of the calculus of variations:

$$\int_a^b \frac{\partial L}{\partial v}[y](x){}_a^C D_x^\alpha h(x)dx$$

$$= \int_a^b {}_xD_b^\alpha\left(\frac{\partial L}{\partial v}[y](x)\right)h(x)dx + \left[{}_xI_b^{1-\alpha}\left(\frac{\partial L}{\partial v}[y](x)\right)h(x)\right]_a^b,$$

$$\int_a^b \frac{\partial L}{\partial w}[y](x){}_aI_x^\beta h(x)dx = \int_a^b {}_xI_b^\beta\left(\frac{\partial L}{\partial w}[y](x)\right)h(x)dx,$$

$$\int_a^b \frac{\partial L}{\partial z}[y](x) \left(\int_a^x \frac{\partial l}{\partial y}\{y\}(t)h(t)dt \right) dx$$

$$= \int_a^b \left(-\frac{d}{dx} \int_x^b \frac{\partial L}{\partial z}[y](t)dt \right) \left(\int_a^x \frac{\partial l}{\partial y}\{y\}(t)h(t)dt \right) dx$$

$$= \left[-\left(\int_x^b \frac{\partial L}{\partial z}[y](t)dt \right) \left(\int_a^x \frac{\partial l}{\partial y}\{y\}(t)h(t)dt \right) \right]_a^b$$

$$+ \int_a^b \left(\int_x^b \frac{\partial L}{\partial z}[y](t)dt \right) \frac{\partial l}{\partial y}\{y\}(x)h(x)\, dx$$

$$= \int_a^b \left(\int_x^b \frac{\partial L}{\partial z}[y](t)dt \right) \frac{\partial l}{\partial y}\{y\}(x)h(x)\, dx,$$

$$\int_a^b \frac{\partial L}{\partial z}[y](x) \left(\int_a^x \frac{\partial l}{\partial v}\{y\}(t)\,_a^C D_t^\alpha h(t)dt \right) dx$$

$$= \int_a^b \left(-\frac{d}{dx} \int_x^b \frac{\partial L}{\partial z}[y](t)dt \right) \left(\int_a^x \frac{\partial l}{\partial v}\{y\}(t)\,_a^C D_t^\alpha h(t)dt \right) dx$$

$$= \left[-\left(\int_x^b \frac{\partial L}{\partial z}[y](t)dt \right) \left(\int_a^x \frac{\partial l}{\partial v}\{y\}(t)\,_a^C D_t^\alpha h(t)dt \right) \right]_a^b$$

$$+ \int_a^b \left(\int_x^b \frac{\partial L}{\partial z}[y](t)dt \right) \frac{\partial l}{\partial v}\{y\}(x)\,_a^C D_x^\alpha h(x)\, dx$$

$$= \int_a^b {}_x D_b^\alpha \left(\int_x^b \frac{\partial L}{\partial z}[y](t)dt \frac{\partial l}{\partial v}\{y\}(x) \right) h(x)\, dx$$

$$+ \left[{}_x I_b^{1-\alpha} \left(\int_x^b \frac{\partial L}{\partial z}[y](t)dt \frac{\partial l}{\partial v}\{y\}(x) \right) h(x) \right]_a^b,$$

and

$$\int_a^b \frac{\partial L}{\partial z}[y](x) \left(\int_a^x \frac{\partial l}{\partial w}\{y\}(t)\,_a I_t^\beta h(t)dt \right) dx$$

$$= \int_a^b {}_x I_b^\beta \left(\int_x^b \frac{\partial L}{\partial z}[y](t)dt \frac{\partial l}{\partial w}\{y\}(x) \right) h(x)\, dx. \qquad \square$$

The fractional Euler–Lagrange equation (3.206) involves not only fractional integrals and fractional derivatives, but also indefinite integrals. Theorem 3.46 gives a necessary condition to determine the possible choices for extremizers.

Definition 3.31. *Solutions to the fractional Euler–Lagrange equation (3.206) are called extremals for J defined by (3.204).*

Example 3.19. Consider the functional

$$J(y) = \int_0^1 \left[\left({}_0^C D_x^\alpha y(x) - \Gamma(\alpha + 2)x \right)^2 + z(x) \right] dx, \qquad (3.207)$$

where $\alpha \in (0, 1]$ and

$$z(x) = \int_0^x \left(y(t) - t^{\alpha+1} \right)^2 dt,$$

defined on the set

$$\{ y \in \mathcal{F}([0, 1]; \mathbb{R}) : y(0) = 0 \text{ and } y(1) = 1 \}.$$

Let

$$y_\alpha(x) = x^{\alpha+1}, \quad x \in [0, 1]. \qquad (3.208)$$

Then,

$${}_0^C D_x^\alpha y_\alpha(x) = \Gamma(\alpha + 2)x.$$

Since $J(y) \geq 0$ for all admissible functions y, and $J(y_\alpha) = 0$, we have that y_α is a minimizer of J. The Euler–Lagrange equation applied to (3.207) gives

$${}_x D_1^\alpha ({}_0^C D_x^\alpha y(x) - \Gamma(\alpha + 2)x) + \int_x^1 1 dt \, (y(x) - x^{\alpha+1}) = 0. \qquad (3.209)$$

Obviously, y_α is a solution of the fractional differential equation (3.209).

The extremizer (3.208) of Example 3.19 is smooth on the closed interval $[0, 1]$. This is not always the case. As the next example shows, minimizers of (3.204)–(3.205) are not necessarily C^1 functions.

Example 3.20. Consider the following fractional variational problem: to minimize the functional

$$J(y) = \int_0^1 \left[\left({}_0^C D_x^\alpha y(x) - 1 \right)^2 + z(x) \right] dx \qquad (3.210)$$

on

$$\left\{ y \in \mathcal{F}([0, 1]; \mathbb{R}) : y(0) = 0 \quad \text{and} \quad y(1) = \frac{1}{\Gamma(\alpha + 1)} \right\},$$

where z is given by

$$z(x) = \int_0^x \left(y(t) - \frac{t^\alpha}{\Gamma(\alpha + 1)} \right)^2 dt.$$

Since $_0^C D_x^\alpha x^\alpha = \Gamma(\alpha + 1)$, we deduce easily that function

$$\overline{y}(x) = \frac{x^\alpha}{\Gamma(\alpha + 1)} \qquad (3.211)$$

is the global minimizer to the problem. Indeed, $J(y) \geq 0$ for all y, and $J(\overline{y}) = 0$. Let us see that $\overline{y}$ is an extremal for J. The fractional Euler–Lagrange equation (3.206) becomes

$$2 \, _xD_1^\alpha (_0^C D_x^\alpha y(x) - 1) + \int_x^1 1 \, dt \cdot 2 \left(y(x) - \frac{x^\alpha}{\Gamma(\alpha + 1)} \right) = 0. \qquad (3.212)$$

Obviously, $\overline{y}$ is a solution of equation (3.212).

Remark 3.37. The minimizer (3.211) of Example 3.20 is not differentiable at 0, as $0 < \alpha < 1$. However, $\overline{y}(0) = 0$ and $_0^C D_x^\alpha \overline{y}(x) = {}_0D_x^\alpha \overline{y}(x) = \Gamma(\alpha + 1)$ for any $x \in [0, 1]$.

Corollary 3.9. *If y is a minimizer of*

$$J(y) = \int_a^b L(x, y(x), {}_a^C D_x^\alpha y(x)) dx, \qquad (3.213)$$

subject to the boundary conditions (3.205), then y is a solution of the fractional Euler–Lagrange equation

$$\frac{\partial L}{\partial y}[y](x) + {}_xD_b^\alpha \left(\frac{\partial L}{\partial v}[y](x) \right) = 0.$$

Proof.   Follows from Theorem 3.46 with an L that does not depend on $_aI_x^\beta y$ and z.   □

We now derive the Euler–Lagrange equations for functionals containing several dependent variables, i.e., for functionals of type

$$J(y_1, \ldots, y_n) = \int_a^b \Big(x, y_1(x), \ldots, y_n(x), {}_a^C D_x^\alpha y_1(x),$$

$$\ldots, {}_a^C D_x^\alpha y_n(x), {}_aI_x^\beta y_1(x), \ldots, {}_aI_x^\beta y_n(x), z(x) \Big) dx, \qquad (3.214)$$

where $n \in \mathbb{N}$ and z is defined by

$$z(x) = \int_a^x l \Big(t, y_1(t), \ldots, y_n(t), {}_a^C D_t^\alpha y_1(t),$$

$$\ldots, {}_a^C D_t^\alpha y_n(t), {}_aI_t^\beta y_1(t), \ldots, {}_aI_t^\beta y_n(t) \Big) dt,$$

subject to the boundary conditions

$$y_k(a) = y_{a,k} \quad \text{and} \quad y_k(b) = y_{b,k}, \quad k \in \{1, \ldots, n\}. \qquad (3.215)$$

To simplify, we consider y as the vector $y = (y_1, \ldots, y_n)$. Consider a family of variations $y + \epsilon h$, where $|\epsilon| \ll 1$ and $h = (h_1, \ldots, h_n)$. The boundary conditions (3.215) imply that $h_k(a) = 0 = h_k(b)$, for $k \in \{1, \ldots, n\}$. The following theorem can be easily proved.

Theorem 3.47. *Let y be a minimizer of J as in* (3.214), *subject to the boundary conditions* (3.215). *Then, for all $k \in \{1, \ldots, n\}$ and for all $x \in [a, b]$, y is a solution of the fractional Euler–Lagrange equation*

$$\frac{\partial L}{\partial y_k}[y](x) + {_xD_b^\alpha}\left(\frac{\partial L}{\partial v_k}[y](x)\right) + {_xI_b^\beta}\left(\frac{\partial L}{\partial w_k}[y](x)\right)$$

$$+ \int_x^b \frac{\partial L}{\partial z}[y](t)dt \cdot \frac{\partial l}{\partial y_k}\{y\}(x) + {_xD_b^\alpha}\left(\int_x^b \frac{\partial L}{\partial z}[y](t)dt \cdot \frac{\partial l}{\partial v_k}\{y\}(x)\right)$$

$$+ {_xI_b^\beta}\left(\int_x^b \frac{\partial L}{\partial z}[y](t)dt \cdot \frac{\partial l}{\partial w_k}\{y\}(x)\right) = 0.$$

Natural boundary conditions

In this section we consider a more general question. Not only is the unknown function y a variable in the problem, but also the terminal time T is an unknown. For $T \in [a, b]$, consider the functional

$$J(y, T) = \int_a^T L[y](x)dx, \tag{3.216}$$

where

$$[y](x) = (x, y(x), {_a^C D_x^\alpha} y(x), {_a I_x^\beta} y(x), z(x)).$$

The problem consists in finding a pair $(y, T) \in \mathcal{F}([a, b]; \mathbb{R}) \times [a, b]$ for which the functional J attains a minimum value. First we give a remark that will be used later in the proof of Theorem 3.48.

Remark 3.38. If ϕ is a continuous function, then (cf. p. 46 of Miller and Ross, 1993)

$$\lim_{x \to T} {_xI_T^{1-\alpha}}\phi(x) = 0$$

for any $\alpha \in (0, 1]$.

Theorem 3.48. *Let (y, T) be a minimizer of J as in (3.216). Then, for all $x \in [a, T]$, (y, T) is a solution of the fractional Euler–Lagrange equation*

$$\frac{\partial L}{\partial y}[y](x) + {}_xD_T^\alpha \left(\frac{\partial L}{\partial v}[y](x) \right) + {}_xI_T^\beta \left(\frac{\partial L}{\partial w}[y](x) \right) + \int_x^T \frac{\partial L}{\partial z}[y](t)dt \cdot \frac{\partial l}{\partial y}\{y\}(x)$$

$$+ {}_xD_T^\alpha \left(\int_x^T \frac{\partial L}{\partial z}[y](t)dt \cdot \frac{\partial l}{\partial v}\{y\}(x) \right) + {}_xI_T^\beta \left(\int_x^T \frac{\partial L}{\partial z}[y](t)dt \cdot \frac{\partial l}{\partial w}\{y\}(x) \right) = 0$$

and satisfies the transversality conditions

$$\left[{}_xI_T^{1-\alpha} \left(\frac{\partial L}{\partial v}[y](x) + \int_x^T \frac{\partial L}{\partial z}[y](t)\, dt \cdot \frac{\partial l}{\partial v}\{y\}(x) \right) \right]_{x=a} = 0$$

and

$$L[y](T) = 0.$$

Proof. Let $h \in \mathcal{F}([a, b]; \mathbb{R})$ be a variation, and let ΔT be a real number. Define the function

$$j(\epsilon) = J(y + \epsilon h, T + \epsilon \Delta T)$$

with $|\epsilon| \ll 1$. Differentiating j at $\epsilon = 0$, and using the same procedure as in Theorem 3.46, we deduce that

$$\Delta T \cdot L[y](T) + \int_a^T \left[\frac{\partial L}{\partial y}[y](x) + {}_xD_T^\alpha \left(\frac{\partial L}{\partial v}[y](x) \right) + {}_xI_T^\beta \left(\frac{\partial L}{\partial w}[y](x) \right) \right.$$

$$+ \int_x^T \frac{\partial L}{\partial z}[y](t)dt \cdot \frac{\partial l}{\partial y}\{y\}(x) + {}_xD_T^\alpha \left(\int_x^T \frac{\partial L}{\partial z}[y](t)dt \cdot \frac{\partial l}{\partial v}\{y\}(x) \right)$$

$$\left. + {}_xI_T^\beta \left(\int_x^T \frac{\partial L}{\partial z}[y](t)dt \cdot \frac{\partial l}{\partial w}\{y\}(x) \right) \right] h(x)dx$$

$$+ \left[{}_xI_T^{1-\alpha} \left(\frac{\partial L}{\partial v}[y](x) \right) h(x) \right]_a^T$$

$$+ \left[{}_xI_T^{1-\alpha} \left(\int_x^T \frac{\partial L}{\partial z}[y](t)dt \cdot \frac{\partial l}{\partial v}\{y\}(x) \right) h(x) \right]_a^T = 0.$$

The theorem follows from the arbitrariness of h and ΔT. □

Remark 3.39. If T is fixed, say $T = b$, then $\Delta T = 0$ and the transversality conditions reduce to

$$\left[{}_xI_b^{1-\alpha} \left(\frac{\partial L}{\partial v}[y](x) + \int_x^b \frac{\partial L}{\partial z}[y](t)\, dt \cdot \frac{\partial l}{\partial v}\{y\}(x) \right) \right]_a = 0. \qquad (3.217)$$

Example 3.21. Consider the problem of minimizing the functional J as in (3.210), but without given boundary conditions. Besides equation (3.212), extremals must also satisfy

$$\left[_x I_1^{1-\alpha} \left(_0^C D_x^\alpha y(x) - 1 \right) \right]_0 = 0. \tag{3.218}$$

Again, $\bar{y}$ given by (3.211) is a solution of (3.212) and (3.218).

As a particular case, the following result is deduced.

Corollary 3.10. *If y is a minimizer of J as in (3.213), then y is a solution of*

$$\frac{\partial L}{\partial y}[y](x) + {_x D_b^\alpha} \left(\frac{\partial L}{\partial v}[y](x) \right) = 0$$

and satisfies the transversality condition

$$\left[_x I_b^{1-\alpha} \left(\frac{\partial L}{\partial v}[y](x) \right) \right]_a = 0.$$

Proof. The Lagrangian L in (3.213) does not depend on $_a I_x^\beta y$ and z, and the result follows from Theorem 3.48. □

Remark 3.40. Observe that the condition

$$\left[_x I_b^{1-\alpha} \left(\frac{\partial L}{\partial v}[y](x) \right) \right]_b = 0$$

is implicitly satisfied in Corollary 3.10 (cf. Remark 3.38).

Fractional isoperimetric problems

Here, we state the isoperimetric problem in the following way. Determine the minimizers of a given functional

$$J(y) = \int_a^b L(x, y(x), {_a^C D_x^\alpha y(x)}, {_a I_x^\beta y(x)}, z(x)) dx \tag{3.219}$$

subject to the boundary conditions

$$y(a) = y_a \quad \text{and} \quad y(b) = y_b \tag{3.220}$$

and the fractional integral constraint

$$I(y) = \int_a^b G(x, y(x), {_a^C D_x^\alpha y(x)}, {_a I_x^\beta y(x)}, z(x)) dx = \gamma, \quad \gamma \in \mathbb{R}, \tag{3.221}$$

where z is defined by

$$z(x) = \int_a^x l(t, y(t), {_a^C D_t^\alpha y(t)}, {_a I_t^\beta y(t)}) dt.$$

As usual, we assume that all the functions $(x, y, v, w, z) \to L(x, y, v, w, z)$, $(x, y, v, w) \to l(x, y, v, w)$, and $(x, y, v, w, z) \to G(x, y, v, w, z)$ are C^1.

Theorem 3.49. *Let y be a minimizer of J as in* (3.219), *under the boundary conditions* (3.220) *and isoperimetric constraint* (3.221). *Suppose that y is not an extremal for I in* (3.221). *Then there exists a constant λ such that y is a solution of the fractional Euler–Lagrange equation*

$$\frac{\partial F}{\partial y}[y](x) + {}_x D_b^\alpha \left(\frac{\partial F}{\partial v}[y](x) \right) + {}_x I_b^\beta \left(\frac{\partial F}{\partial w}[y](x) \right) + \int_x^b \frac{\partial F}{\partial z}[y](t) dt \cdot \frac{\partial l}{\partial y}\{y\}(x)$$

$$+ {}_x D_b^\alpha \left(\int_x^b \frac{\partial F}{\partial z}[y](t) dt \cdot \frac{\partial l}{\partial v}\{y\}(x) \right) + {}_x I_b^\beta \left(\int_x^b \frac{\partial F}{\partial z}[y](t) dt \cdot \frac{\partial l}{\partial w}\{y\}(x) \right) = 0,$$

where $F = L - \lambda G$, for all $x \in [a, b]$.

Proof. Let $\epsilon_1, \epsilon_2 \in \mathbb{R}$ be two real numbers such that $|\epsilon_1| \ll 1$ and $|\epsilon_2| \ll 1$, with ϵ_1 free and ϵ_2 to be determined later, and let h_1 and h_2 be two functions satisfying

$$h_1(a) = h_1(b) = h_2(a) = h_2(b) = 0.$$

Define functions j and i by

$$j(\epsilon_1, \epsilon_2) = J(y + \epsilon_1 h_1 + \epsilon_2 h_2)$$

and

$$i(\epsilon_1, \epsilon_2) = I(y + \epsilon_1 h_1 + \epsilon_2 h_2) - \gamma.$$

Analogous calculations as in the proof of Theorem 3.46 give

$$\left. \frac{\partial i}{\partial \epsilon_2} \right|_{(0,0)} = \int_a^b \left[\frac{\partial G}{\partial y}[y](x) + {}_x D_b^\alpha \left(\frac{\partial G}{\partial v}[y](x) \right) + {}_x I_b^\alpha \left(\frac{\partial G}{\partial w}[y](x) \right) \right.$$

$$+ \int_x^b \frac{\partial G}{\partial z}[y](t) dt \cdot \frac{\partial l}{\partial y}\{y\}(x)$$

$$+ {}_x D_b^\alpha \left(\int_x^b \frac{\partial G}{\partial z}[y](t) dt \cdot \frac{\partial l}{\partial v}\{y\}(x) \right)$$

$$\left. + {}_x I_b^\beta \left(\int_x^b \frac{\partial G}{\partial z}[y](t) dt \cdot \frac{\partial l}{\partial w}\{y\}(x) \right) \right] h_2(x) \, dx.$$

By hypothesis, y is not an extremal for I and therefore a function h_2 must exist for which

$$\left. \frac{\partial i}{\partial \epsilon_2} \right|_{(0,0)} \neq 0.$$

Since $i(0,0) = 0$, by the implicit function theorem there exists a function $\epsilon_2(\cdot)$, defined in some neighborhood of zero, such that

$$i(\epsilon_1, \epsilon_2(\epsilon_1)) = 0. \tag{3.222}$$

On the other hand, j subject to the constraint (3.222) attains a minimum value at $(0,0)$. Because $\nabla i(0,0) \neq (0,0)$, by the Lagrange multiplier rule (see, e.g., p. 77 of van Brunt, 2004), there exists a constant λ such that

$$\nabla(j(0,0) - \lambda i(0,0)) = (0,0).$$

So

$$\left.\frac{\partial j}{\partial \epsilon_1}\right|_{(0,0)} - \lambda \left.\frac{\partial i}{\partial \epsilon_1}\right|_{(0,0)} = 0.$$

Differentiating j and i at zero, and doing the same calculations as before, we get the desired result. $\qquad\square$

Using the abnormal Lagrange multiplier rule (cf. p. 82 of van Brunt, 2004), the previous result can be generalized to include the case when the minimizer is an extremal of I.

Theorem 3.50. *Let y be a minimizer of J as in* (3.219)*, subject to the constraints* (3.220) *and* (3.221)*. Then there exist two constants λ_0 and λ, not both zero, such that y is a solution of the fractional Euler–Lagrange equation*

$$\frac{\partial K}{\partial y}[y](x) + {}_xD_b^\alpha\left(\frac{\partial K}{\partial v}[y](x)\right) + {}_xI_b^\beta\left(\frac{\partial K}{\partial w}[y](x)\right)$$

$$+ \int_x^b \frac{\partial K}{\partial z}[y](t)dt \cdot \frac{\partial l}{\partial y}\{y\}(x) + {}_xD_b^\alpha\left(\int_x^b \frac{\partial K}{\partial z}[y](t)dt \cdot \frac{\partial l}{\partial v}\{y\}(x)\right)$$

$$+ {}_xI_b^\beta\left(\int_x^b \frac{\partial K}{\partial z}[y](t)dt \cdot \frac{\partial l}{\partial w}\{y\}(x)\right) = 0$$

for all $x \in [a,b]$, where $K = \lambda_0 L - \lambda G$.

Corollary 3.11. *Let y be a minimizer of*

$$J(y) = \int_a^b L(x, y(x), {}_a^C D_x^\alpha y(x))dx$$

subject to the boundary conditions

$$y(a) = y_a \quad \text{and} \quad y(b) = y_b$$

and the isoperimetric constraint

$$I(y) = \int_a^b G(x, y(x), {}_a^C D_x^\alpha y(x)) dx = \gamma, \quad \gamma \in \mathbb{R}.$$

Then, there exist two constants λ_0 and λ, not both zero, such that y is a solution of the fractional Euler–Lagrange equation

$$\frac{\partial K}{\partial y}(x, y(x), {}_a^C D_x^\alpha y(x)) + {}_x D_b^\alpha \left(\frac{\partial K}{\partial v}(x, y(x), {}_a^C D_x^\alpha y(x)) \right) = 0$$

for all $x \in [a, b]$, where $K = \lambda_0 L - \lambda G$. Moreover, if y is not an extremal for I, then we may take $\lambda_0 = 1$.

Holonomic constraints

In this section we consider the following problem. Minimize the functional

$$J(y_1, y_2) = \int_a^b L(x, y_1(x), y_2(x), {}_a^C D_x^\alpha y_1(x), {}_a^C D_x^\alpha y_2(x),$$

$$\qquad\qquad {}_a I_x^\beta y_1(x), {}_a I_x^\beta y_2(x), z(x)) dx, \quad (3.223)$$

where z is defined by

$$z(x) = \int_a^x l(t, y_1(t), y_2(t), {}_a^C D_t^\alpha y_1(t), {}_a^C D_t^\alpha y_2(t), {}_a I_t^\beta y_1(t), {}_a I_t^\beta y_2(t)) dt,$$

when restricted to the boundary conditions

$$(y_1(a), y_2(a)) = (y_1^a, y_2^a) \text{ and } (y_1(b), y_2(b)) = (y_1^b, y_2^b), \quad y_1^a, y_2^a, y_1^b, y_2^b \in \mathbb{R},$$
$$(3.224)$$

and the holonomic constraint

$$g(x, y_1(x), y_2(x)) = 0. \quad (3.225)$$

As usual, here

$$(x, y_1, y_2, v_1, v_2, w_1, w_2, z) \to L(x, y_1, y_2, v_1, v_2, w_1, w_2, z),$$

$$(x, y_1, y_2, v_1, v_2, w_1, w_2) \to l(x, y_1, y_2, v_1, v_2, w_1, w_2)$$

and

$$(x, y_1, y_2) \to g(x, y_1, y_2)$$

are all smooth. In what follows we make use of the operator $[\cdot, \cdot]$ given by

$$[y_1, y_2](x)$$
$$= (x, y_1(x), y_2(x), {}_a^C D_x^\alpha y_1(x), {}_a^C D_x^\alpha y_2(x), {}_a I_x^\beta y_1(x), {}_a I_x^\beta y_2(x), z(x)),$$

we denote $(x, y_1(x), y_2(x))$ by $(x, \mathbf{y}(x))$, and the Euler–Lagrange equation obtained in (3.206) with respect to y_i by (ELE_i), $i = 1, 2$.

Remark 3.41. For simplicity, we are considering functionals depending only on two functions y_1 and y_2. Theorem 3.51 is, however, easily generalized for n variables $y_1, \ldots, y_n$.

Theorem 3.51. *Let the pair (y_1, y_2) be a minimizer of J as in* (3.223), *subject to the constraints* (3.224)–(3.225). *If $\frac{\partial g}{\partial y_2} \neq 0$, then there exists a continuous function $\lambda : [a, b] \to \mathbb{R}$ such that (y_1, y_2) is a solution of the fractional Euler–Lagrange equation*

$$
\frac{\partial F}{\partial y_i}[y_1, y_2](x) + {}_xD_b^\alpha\left(\frac{\partial F}{\partial v_i}[y_1, y_2](x)\right) + {}_xI_b^\beta\left(\frac{\partial F}{\partial w_i}[y_1, y_2](x)\right)
$$

$$
+ \int_x^b \frac{\partial F}{\partial z}[y_1, y_2](t)dt \cdot \frac{\partial l}{\partial y_i}\{y_1, y_2\}(x)
$$

$$
+ {}_xD_b^\alpha\left(\int_x^b \frac{\partial F}{\partial z}[y_1, y_2](t)dt \cdot \frac{\partial l}{\partial v_i}\{y_1, y_2\}(x)\right) \tag{3.226}
$$

$$
+ {}_xI_b^\beta\left(\int_x^b \frac{\partial F}{\partial z}[y_1, y_2](t)dt \cdot \frac{\partial l}{\partial w_i}\{y_1, y_2\}(x)\right) = 0
$$

for all $x \in [a, b]$ and $i = 1, 2$, where

$$
F[y_1, y_2](x) = L[y_1, y_2](x) - \lambda(x)g(x, \mathbf{y}(x)).
$$

Proof. Consider a variation of the optimal solution of type

$$
(\overline{y}_1, \overline{y}_2) = (y_1 + \epsilon h_1, y_2 + \epsilon h_2),
$$

where h_1, h_2 are two functions defined on $[a, b]$ satisfying

$$
h_1(a) = h_1(b) = h_2(a) = h_2(b) = 0,
$$

and ϵ is a sufficiently small real parameter. Since $\frac{\partial g}{\partial y_2}(x, \overline{y}_1(x), \overline{y}_2(x)) \neq 0$ for all $x \in [a, b]$, we can solve equation $g(x, \overline{y}_1(x), \overline{y}_2(x)) = 0$ with respect to h_2, $h_2 = h_2(\epsilon, h_1)$. Differentiating $J(\overline{y}_1, \overline{y}_2)$ at $\epsilon = 0$, and proceeding similarly as in the proof of Theorem 3.46, we deduce that

$$
\int_a^b (ELE_1)h_1(x) + (ELE_2)h_2(x)\, dx = 0. \tag{3.227}
$$

Besides, since $g(x, \overline{y}_1(x), \overline{y}_2(x)) = 0$, differentiating at $\epsilon = 0$ we get

$$
h_2(x) = -\frac{\frac{\partial g}{\partial y_1}(x, \mathbf{y}(x))}{\frac{\partial g}{\partial y_2}(x, \mathbf{y}(x))}h_1(x). \tag{3.228}
$$

Define the function λ on $[a, b]$ as

$$
\lambda(x) = \frac{(ELE_2)}{\frac{\partial g}{\partial y_2}(x, \mathbf{y}(x))}. \tag{3.229}
$$

Combining (3.228) and (3.229), equation (3.227) can be written as

$$
\int_a^b \left[(ELE_1) - \lambda(x)\frac{\partial g}{\partial y_1}(x, \mathbf{y}(x))\right] h_1(x)\, dx = 0.
$$

By the arbitrariness of h_1, if follows that

$$(ELE_1) - \lambda(x)\frac{\partial g}{\partial y_1}(x, \mathbf{y}(x)) = 0.$$

Define F as

$$F[y_1, y_2](x) = L[y_1, y_2](x) - \lambda(x)g(x, \mathbf{y}(x)).$$

Then, equations (3.226) follow.   □

Higher-order Caputo derivatives

In this section we consider fractional variational problems in the presence of higher-order Caputo derivatives.

Let $n \in \mathbb{N}$, $\beta > 0$, and $\alpha_k \in \mathbb{R}$ be such that $\alpha_k \in (k-1, k)$ for $k \in \{1, \ldots, n\}$. Admissible functions y belong to $AC^n([a, b]; \mathbb{R})$ and are such that $_a^C D_x^{\alpha_k} y$, $k = 1, \ldots, n$, and $_a I_x^\beta y$ exist and are continuous on $[a, b]$. We denote such a class of functions by $\mathcal{F}^n([a, b]; \mathbb{R})$. For $\alpha = (\alpha_1, \ldots, \alpha_n)$, define the vector

$$_a^C D_x^\alpha y(x) = (_a^C D_x^{\alpha_1} y(x), \ldots, _a^C D_x^{\alpha_n} y(x)). \tag{3.230}$$

The optimization problem is the following: to minimize or maximize the functional

$$J(y) = \int_a^b L(x, y(x), {}_a^C D_x^\alpha y(x), {}_a I_x^\beta y(x), z(x))dx, \tag{3.231}$$

$y \in \mathcal{F}^n([a, b]; \mathbb{R})$, subject to the boundary conditions

$$y^{(k)}(a) = y_{a,k} \quad \text{and} \quad y^{(k)}(b) = y_{b,k}, \quad k \in \{0, \ldots, n-1\}, \tag{3.232}$$

where $z : [a, b] \to \mathbb{R}$ is defined by

$$z(x) = \int_a^x l(t, y(t), {}_a^C D_t^\alpha y(t), {}_a I_t^\beta y(t))dt.$$

Theorem 3.52. *If $y \in \mathcal{F}^n([a, b]; \mathbb{R})$ is a minimizer of J as in (3.231), subject to the boundary conditions (3.232), then y is a solution of the fractional Euler–Lagrange equation*

$$\frac{\partial L}{\partial y}[y](x) + \sum_{k=1}^n {}_x D_b^{\alpha_k}\left(\frac{\partial L}{\partial v_k}[y](x)\right) + {}_x I_b^\beta\left(\frac{\partial L}{\partial w}[y](x)\right)$$

$$+ \int_x^b \frac{\partial L}{\partial z}[y](t)dt \cdot \frac{\partial l}{\partial y}\{y\}(x) + \sum_{k=1}^n {}_x D_b^{\alpha_k}\left(\int_x^b \frac{\partial L}{\partial z}[y](t)dt \cdot \frac{\partial l}{\partial v_k}\{y\}(x)\right)$$

$$+ {}_x I_b^\beta\left(\int_x^b \frac{\partial L}{\partial z}[y](t)dt \cdot \frac{\partial l}{\partial w}\{y\}(x)\right) = 0$$

for all $x \in [a, b]$, where $[y](x) = (x, y(x), {}_a^C D_x^\alpha y(x), {}_a I_x^\beta y(x), z(x))$ with $_a^C D_x^\alpha y(x)$ as in (3.230).

Proof. Let $h \in \mathcal{F}^n([a,b]; \mathbb{R})$ be such that $h^{(k)}(a) = h^{(k)}(b) = 0$, for $k \in \{0, \ldots, n-1\}$. Define the new function j as $j(\epsilon) = J(y + \epsilon h)$. Then

$$\int_a^b \left[\frac{\partial L}{\partial y}[y](x)h(x) + \sum_{k=1}^n \frac{\partial L}{\partial v_k}[y](x) {}_a^C D_x^{\alpha_k} h(x) + \frac{\partial L}{\partial w}[y](x) {}_a I_x^\beta h(x) \right.$$

$$+ \frac{\partial L}{\partial z}[y](x) \int_a^x \left(\frac{\partial l}{\partial y}\{y\}(t)h(t) + \sum_{k=1}^n \frac{\partial l}{\partial v_k}\{y\}(t) {}_a^C D_t^{\alpha_k} h(t) \right.$$

$$\left. \left. + \frac{\partial l}{\partial w}\{y\}(t) {}_a I_t^\beta h(t) \right) dt \right] dx = 0. \quad (3.233)$$

Integrating by parts, we obtain

$$\int_a^b \frac{\partial L}{\partial v_k}[y](x) {}_a^C D_x^{\alpha_k} h(x) dx$$

$$= \int_a^b {}_x D_b^{\alpha_k} \left(\frac{\partial L}{\partial v_k}[y](x) \right) h(x) dx$$

$$+ \sum_{m=0}^{k-1} \left[{}_x D_b^{\alpha_k + m - k} \left(\frac{\partial L}{\partial v_k}[y](x) \right) h^{(k-1-m)}(x) \right]_a^b$$

$$= \int_a^b {}_x D_b^{\alpha_k} \left(\frac{\partial L}{\partial v_k}[y](x) \right) h(x) dx$$

for all $k \in \{1, \ldots, n\}$. Moreover, one has

$$\int_a^b \frac{\partial L}{\partial w}[y](x) {}_a I_x^\beta h(x) dx = \int_a^b {}_x I_b^\beta \left(\frac{\partial L}{\partial w}[y](x) \right) h(x) dx,$$

$$\int_a^b \frac{\partial L}{\partial z}[y](x) \int_a^x \frac{\partial l}{\partial y}\{y\}(t)h(t) dt \, dx$$

$$= \int_a^b \left(\int_x^b \frac{\partial L}{\partial z}[y](t) dt \right) \frac{\partial l}{\partial y}\{y\}(x)h(x) \, dx,$$

$$\int_a^b \frac{\partial L}{\partial z}[y](x) \left(\int_a^x \frac{\partial l}{\partial v_k}\{y\}(t) \,{}^C_a D_t^{\alpha_k} h(t)dt \right) dx$$

$$= \int_a^b \left(\int_x^b \frac{\partial L}{\partial z}[y](t)dt \right) \frac{\partial l}{\partial v_k}\{y\}(x) \,{}^C_a D_x^{\alpha_k} h(x) \, dx$$

$$= \int_a^b {}_x D_b^{\alpha_k} \left(\int_x^b \frac{\partial L}{\partial z}[y](t)dt \frac{\partial l}{\partial v_k}\{y\}(x) \right) h(x) \, dx$$

$$+ \sum_{m=0}^{k-1} \left[{}_x D_b^{\alpha_k+m-k} \left(\int_x^b \frac{\partial L}{\partial z}[y](t)dt \frac{\partial l}{\partial v_k}\{y\}(x) \right) h^{(k-1-m)}(x) \right]_a^b$$

$$= \int_a^b {}_x D_b^{\alpha_k} \left(\int_x^b \frac{\partial L}{\partial z}[y](t)dt \frac{\partial l}{\partial v_k}\{y\}(x) \right) h(x) \, dx,$$

and

$$\int_a^b \frac{\partial L}{\partial z}[y](x) \left(\int_a^x \frac{\partial l}{\partial w}\{y\}(t) \,{}_a I_t^{\beta} h(t)dt \right) dx$$

$$= \int_a^b {}_x I_b^{\beta} \left(\int_x^b \frac{\partial L}{\partial z}[y](t)dt \frac{\partial l}{\partial w}\{y\}(x) \right) h(x) \, dx.$$

Replacing these last relations into equation (3.233), and applying the fundamental lemma of the calculus of variations, we obtain the intended necessary optimality condition. □

We now consider the higher-order problem without the presence of boundary conditions (3.232).

Theorem 3.53. *If* $y \in \mathcal{F}^n([a,b];\mathbb{R})$ *is a minimizer of J as in* (3.231), *then y is a solution of the fractional Euler–Lagrange equation*

$$\frac{\partial L}{\partial y}[y](x) + \sum_{k=1}^{n} {}_x D_b^{\alpha_k} \left(\frac{\partial L}{\partial v_k}[y](x) \right) + {}_x I_b^{\beta} \left(\frac{\partial L}{\partial w}[y](x) \right)$$

$$+ \int_x^b \frac{\partial L}{\partial z}[y](t)dt \cdot \frac{\partial l}{\partial y}\{y\}(x) + \sum_{k=1}^{n} {}_x D_b^{\alpha_k} \left(\int_x^b \frac{\partial L}{\partial z}[y](t)dt \cdot \frac{\partial l}{\partial v_k}\{y\}(x) \right)$$

$$+ {}_x I_b^{\beta} \left(\int_x^b \frac{\partial L}{\partial z}[y](t)dt \cdot \frac{\partial l}{\partial w}\{y\}(x) \right) = 0$$

for all $x \in [a,b]$, *and satisfies the natural boundary conditions*

$$\sum_{m=k}^{n} \left[{}_x D_b^{\alpha_m-k} \left(\frac{\partial L}{\partial v_k}[y](x) + \int_x^b \frac{\partial L}{\partial z}[y](t)dt \frac{\partial l}{\partial v_k}\{y\}(x) \right) \right]_a^b = 0, \quad (3.234)$$

for all $k \in \{1, \ldots, n\}$.

Proof. The proof follows the same pattern as the proof of Theorem 3.52. Since admissible functions y are not required to satisfy given boundary conditions, the variation functions h may also take any value at the boundaries, and thus the condition

$$h^{(k)}(a) = h^{(k)}(b) = 0, \quad \text{for } k \in \{0, \ldots, n-1\}, \tag{3.235}$$

is no longer imposed *a priori*. If we consider the first variation of J for variations h satisfying condition (3.235), we obtain the Euler–Lagrange equation. Replacing it in the expression of the first variation, we conclude that

$$\sum_{k=1}^{n} \sum_{m=0}^{k-1} \left[{}_x D_b^{\alpha_k + m - k} \left(\frac{\partial L}{\partial v_k}[y](x) + \int_x^b \frac{\partial L}{\partial z}[y](t)dt \frac{\partial l}{\partial v_k}\{y\}(x) \right) \right.$$
$$\left. \times h^{(k-1-m)}(x) \right]_a^b = 0.$$

To obtain the transversality condition with respect to k, for $k \in \{1, \ldots, n\}$, we consider variations satisfying the condition

$$h^{(k-1)}(a) \neq 0 \neq h^{(k-1)}(b) \quad \text{and } h^{(j-1)}(a) = 0 = h^{(j-1)}(b),$$

for all $j \in \{0, \ldots, n\} \setminus \{k\}$. □

Remark 3.42. Some of the terms that appear in the natural boundary conditions (3.234) are equal to zero (cf. Remarks 3.38 and 3.40).

Fractional Lagrange problems

We now prove a necessary optimality condition for a fractional Lagrange problem, when the Lagrangian depends again on an indefinite integral. Consider the cost functional defined by

$$J(y, u) = \int_a^b L\left(x, y(x), u(x), {}_aI_x^\beta y(x), z(x)\right) dx, \tag{3.236}$$

to be minimized or maximized subject to the fractional dynamical system

$${}_a^C D_x^\alpha y(x) = f(x, y(x), u(x), {}_aI_x^\beta y(x), z(x)) \tag{3.237}$$

and the boundary conditions

$$y(a) = y_a \quad \text{and} \quad y(b) = y_b, \tag{3.238}$$

where

$$z(x) = \int_a^x l\left(t, y(t), {}_a^C D_t^\alpha y(t), {}_aI_t^\beta y(t)\right) dt.$$

We assume the functions $(x, y, v, w, z) \to f(x, y, v, w, z)$, $(x, y, v, w, z) \to L(x, y, v, w, z)$, and $(x, y, v, w) \to l(x, y, v, w)$, to be of class C^1 with respect to all their arguments.

Remark 3.43. If $f(x, y(x), u(x), {}_aI_x^\beta y(x), z(x)) = u(x)$, the Lagrange problem (3.236)–(3.238) reduces to the fractional variational problem (3.204)–(3.205) studied in Section 3.11.3.

An optimal solution is a pair of functions (y, u) that minimizes J as in (3.236), subject to the fractional dynamic equation (3.237) and the boundary conditions (3.238).

Theorem 3.54. *If (y, u) is an optimal solution to the fractional Lagrange problem (3.236)–(3.238), then there exists a function p for which the triplet (y, u, p) satisfies the Hamiltonian system*

$$
\begin{cases}
{}_a^C D_x^\alpha y(x) = \dfrac{\partial H}{\partial p}\lceil y, u, p\rceil(x), \\[2mm]
{}_xD_b^\alpha p(x) = \dfrac{\partial H}{\partial y}\lceil y, u, p\rceil(x) + {}_xI_b^\beta \left(\dfrac{\partial H}{\partial w}\lceil y, u, p\rceil(x)\right) \\[2mm]
\qquad + \displaystyle\int_x^b \dfrac{\partial H}{\partial z}\lceil y, u, p\rceil(t)dt \cdot \dfrac{\partial l}{\partial y}\{y\}(x) \\[2mm]
\qquad + {}_xD_b^\alpha \left(\displaystyle\int_x^b \dfrac{\partial H}{\partial z}\lceil y, u, p\rceil(t)dt \cdot \dfrac{\partial l}{\partial v}\{y\}(x)\right) \\[2mm]
\qquad + {}_xI_b^\beta \left(\displaystyle\int_x^b \dfrac{\partial H}{\partial z}\lceil y, u, p\rceil(t)dt \cdot \dfrac{\partial l}{\partial w}\{y\}(x)\right)
\end{cases}
$$

and the stationary condition

$$
\frac{\partial H}{\partial u}\lceil y, u, p\rceil(x) = 0,
$$

where the Hamiltonian H is defined by

$$
H\lceil y, u, p\rceil(x) = L(x, y(x), u(x), {}_aI_x^\beta y(x), z(x))
$$
$$
+ p(x)f(x, y(x), u(x), {}_aI_x^\beta y(x), z(x))
$$

and

$$
\lceil y, u, p\rceil(x) = (x, y(x), u(x), {}_aI_x^\beta y(x), z(x), p(x)),
$$
$$
\{y\}(x) = (x, y(x), {}_a^C D_x^\alpha y(x), {}_aI_x^\beta y(x)).
$$

Proof.   The result follows applying Theorem 3.47 to

$$
J^*(y, u, p) = \int_a^b H\lceil y, u, p\rceil(x) - p(x){}_a^C D_x^\alpha y(x)dx
$$

with respect to y, u and p. $\square$

In the particular case when L does not depend on $_aI_x^\beta y$ and z, we obtain the following corollary.

Corollary 3.12. *Let $(y(x), u(x))$ be a solution of*

$$J(y, u) = \int_a^b L(x, y(x), u(x))dx \longrightarrow \min$$

subject to the fractional control system $_a^C D_x^\alpha y(x) = f(x, y(x), u(x))$ and the boundary conditions $y(a) = y_a$ and $y(b) = y_b$. Define the Hamiltonian by $H(x, y, u, p) = L(x, y, u) + pf(x, y, u)$. Then there exists a function p for which the triplet (y, u, p) fulfill the Hamiltonian system

$$\begin{cases} _a^C D_x^\alpha y(x) = \frac{\partial H}{\partial p}(x, y(x), u(x), p(x)), \\ _x D_b^\alpha p(x) = \frac{\partial H}{\partial y}(x, y(x), u(x), p(x)), \end{cases}$$

and the stationary condition $\frac{\partial H}{\partial u}(x, y(x), u(x), p(x)) = 0$.

Sufficient conditions of optimality

Now we present a sufficient optimality condition using the notion of convexity (concavity) given in Definition 1.6.

Theorem 3.55. *Consider the functional J as in (3.204), and let $y \in \mathcal{F}([a, b]; \mathbb{R})$ be a solution of the fractional Euler–Lagrange equation (3.206) satisfying the boundary conditions (3.205). Assume that L is convex in (y, v, w, z). If one of the two following conditions is satisfied,*

(i) l is convex in (y, v, w) and $\frac{\partial L}{\partial z}[y](x) \geq 0$ for all $x \in [a, b]$;

(ii) l is concave in (y, v, w) and $\frac{\partial L}{\partial z}[y](x) \leq 0$ for all $x \in [a, b]$;

then y is a (global) minimizer of problem (3.204)–(3.205).

Proof.   Consider h of class $\mathcal{F}([a,b];\mathbb{R})$ such that $h(a) = h(b) = 0$. Then,

$J(y + h) - J(y)$

$$= \int_a^b L\left(x, y(x) + h(x), {}_a^C D_x^\alpha y(x) + {}_a^C D_x^\alpha h(x), {}_a I_x^\beta y(x) + {}_a I_x^\beta h(x), \right.$$

$$\int_a^x l(t, y(t) + h(t), {}_a^C D_t^\alpha y(t) + {}_a^C D_t^\alpha h(t), {}_a I_t^\beta y(t) + {}_a I_t^\beta h(t))dt \Bigg) dx$$

$$- \int_a^b L\left(x, y(x), {}_a^C D_x^\alpha y(x), {}_a I_x^\beta y(x), \int_a^x l(t, y(t), {}_a^C D_t^\alpha y(t), {}_a I_t^\beta y(t))dt \right) dx$$

$$\geq \int_a^b \left[\frac{\partial L}{\partial y}[y](x)h(x) + \frac{\partial L}{\partial v}[y](x){}_a^C D_x^\alpha h(x) + \frac{\partial L}{\partial w}[y](x){}_a I_x^\beta h(x) + \frac{\partial L}{\partial z}[y](x) \right.$$

$$\times \int_a^x \left(\frac{\partial l}{\partial y}\{y\}(t)h(t) + \frac{\partial l}{\partial v}\{y\}(t){}_a^C D_t^\alpha h(t) + \frac{\partial l}{\partial w}\{y\}(t){}_a I_t^\beta h(t) \right) dt \Bigg] dx$$

$$= \int_a^b \left[\frac{\partial L}{\partial y}[y](x) + {}_x D_b^\alpha \left(\frac{\partial L}{\partial v}[y](x) \right) + {}_x I_b^\beta \left(\frac{\partial L}{\partial w}[y](x) \right) \right.$$

$$+ \int_x^b \frac{\partial L}{\partial z}[y](t)dt \cdot \frac{\partial l}{\partial y}\{y\}(x) + {}_x D_b^\alpha \left(\int_x^b \frac{\partial L}{\partial z}[y](t)dt \cdot \frac{\partial l}{\partial v}\{y\}(x) \right)$$

$$+ {}_x I_b^\beta \left(\int_x^b \frac{\partial L}{\partial z}[y](t)dt \cdot \frac{\partial l}{\partial w}\{y\}(x) \right) \Bigg] h(x)dx = 0.$$

$\square$

One can easily include the case when the boundary conditions (3.205) are not given.

Theorem 3.56. *Consider functional J as in (3.204) and let $y \in \mathcal{F}([a,b];\mathbb{R})$ be a solution of the fractional Euler–Lagrange equation (3.206) and the fractional natural boundary condition (3.217). Assume that L is convex in (y, v, w, z). If one of the two following conditions is satisfied,*

(i) l is convex in (y, v, w) and $\frac{\partial L}{\partial z}[y](x) \geq 0$ for all $x \in [a,b]$;

(ii) l is concave in (y, v, w) and $\frac{\partial L}{\partial z}[y](x) \leq 0$ for all $x \in [a,b]$;

then y is a (global) minimizer of (3.204).

Numerical simulations

Solving a variational problem usually means solving Euler–Lagrange differential equations subject to some boundary conditions. It turns out that most fractional Euler–Lagrange equations cannot be solved analytically.

Therefore, in practical terms, numerical methods need to be developed and used in order to solve the fractional variational problems. A numerical scheme to solve fractional Lagrange problems has been presented in Agrawal, 2004. The method is based on approximating the problem to a set of algebraic equations using some basis functions. A more general approach can be found in Tricaud and Chen, 2010, that uses the Oustaloup recursive approximation of the fractional derivative, and reduces the problem to an integer order (classical) optimal control problem. A similar approach is presented in Jelicic and Petrovacki, 2009, using an expansion formula for the left Riemann–Liouville fractional derivative developed in Atanackovic and Stankovic, 2008. Here we use a modified approximation (see Remark 3.44) based on the same expansion, to reduce a given fractional problem to a classical one. The expansion formula is given in the following lemma.

Lemma 3.5 (cf. Eq. (12) of Atanackovic and Stankovic, 2008).
Suppose that $f \in AC^2[0,b]$, $f'' \in L_1[0,b]$ and $0 < \alpha \le 1$. Then the left Riemann–Liouville fractional derivative can be expanded as

$$_0D_x^\alpha f(x) = A(\alpha)x^{-\alpha}f(x) + B(\alpha)x^{1-\alpha}f'(x) - \sum_{k=2}^\infty C(k,\alpha)x^{1-k-\alpha}v_k(x),$$

where

$$v'_k(x) = (1-k)x^{k-2}f(x), \qquad v_k(0) = 0, \qquad k = 2,3,\dots,$$

$$A(\alpha) = \frac{1}{\Gamma(1-\alpha)} - \frac{1}{\Gamma(2-\alpha)\Gamma(\alpha-1)}\sum_{k=2}^\infty \frac{\Gamma(k-1+\alpha)}{(k-1)!},$$

$$B(\alpha) = \frac{1}{\Gamma(2-\alpha)}\left[1 + \sum_{k=1}^\infty \frac{\Gamma(k-1+\alpha)}{\Gamma(\alpha-1)k!}\right],$$

$$C(k,\alpha) = \frac{1}{\Gamma(2-\alpha)\Gamma(\alpha-1)}\frac{\Gamma(k-1+\alpha)}{(k-1)!}.$$

In practice, we only keep a finite number of terms in the series. We use the approximation

$$_0D_x^\alpha f(x) \simeq A(\alpha,N)x^{-\alpha}f(x) + B(\alpha,N)x^{1-\alpha}f'(x) - \sum_{k=2}^N C(k,\alpha)x^{1-k-\alpha}v_k(x)$$

$$(3.239)$$

for some fixed number N, where

$$A(\alpha, N) = \frac{1}{\Gamma(1-\alpha)} - \frac{1}{\Gamma(2-\alpha)\Gamma(\alpha-1)} \sum_{k=2}^{N} \frac{\Gamma(k-1+\alpha)}{(k-1)!},$$

$$B(\alpha, N) = \frac{1}{\Gamma(2-\alpha)} \left[1 + \sum_{k=1}^{N} \frac{\Gamma(k-1+\alpha)}{\Gamma(\alpha-1)k!} \right].$$

Remark 3.44. In Atanackovic and Stankovic, 2008, the authors use the fact that $1 + \sum_{k=1}^{\infty} \frac{\Gamma(k-1+\alpha)}{\Gamma(\alpha-1)k!} = 0$, and apply in their method the approximation

$$_0D_x^\alpha f(x) \simeq A(\alpha, N)x^{-\alpha}f(x) - \sum_{k=2}^{N} C(k,\alpha)x^{1-k-\alpha}v_k(x).$$

Regarding the value of $B(\alpha, N)$ for some values of N (see Table 3.1), we take the first derivative in the expansion into account and keep the approximation in the form of equation (3.239).

N	4	7	15	30	70	120	170
$B(0.3, N)$	0.1357	0.0928	0.0549	0.0339	0.0188	0.0129	0.0101
$B(0.5, N)$	0.3085	0.2364	0.1630	0.1157	0.0760	0.0581	0.0488
$B(0.7, N)$	0.5519	0.4717	0.3783	0.3083	0.2396	0.2040	0.1838
$B(0.9, N)$	0.8470	0.8046	0.7481	0.6990	0.6428	0.6092	0.5884

Table 3.1 $B(\alpha, N)$ for $\alpha \in \{0.3, 0.5, 0.7, 0.9\}$ and different values of N.

We illustrate with Examples 3.19 and 3.20 how the approximation (3.239) provides an accurate and efficient numerical method to solve fractional variational problems.

Example 3.22. We obtain an approximated solution to the problem considered in Example 3.19. Since $y(0) = 0$, the Caputo derivative coincides with the Riemann–Liouville derivative and we can approximate the fractional problem using (3.239). We reformulate the problem using the Hamiltonian formalism by letting $_0^C D_x^\alpha y(x) = u(x)$. Then,

$$A(\alpha, N)x^{-\alpha}y(x) + B(\alpha, N)x^{1-\alpha}y'(x) - \sum_{k=2}^{N} C(k,\alpha)x^{1-k-\alpha}v_k(x) = u(x).$$

$$(3.240)$$

We also include the variable $z(x)$ with

$$z'(x) = \left(y(x) - x^{\alpha+1}\right)^2.$$

In summary, one has the following Lagrange problem:

$$\tilde{J}(y) = \int_0^1 [(u(x) - \Gamma(\alpha+2)x)^2 + z(x)]dx \longrightarrow \min$$

$$\begin{cases} y'(x) = -AB^{-1}x^{-1}y(x) + \sum_{k=2}^N B^{-1}C_k x^{-k}v_k(x) + B^{-1}x^{\alpha-1}u(x) \\ v'_k(x) = (1-k)x^{k-2}y(x), \qquad k = 1, 2, \ldots \\ z'(x) = \left(y(x) - x^{\alpha+1}\right)^2 \end{cases}$$

(3.241)

subject to the boundary conditions $y(0) = 0$, $v_k(0) = 0$, $k = 1, 2, \ldots$, and $z(0) = 0$. Setting $N = 2$, the Hamiltonian is given by

$$\begin{aligned} H = &-[(u(x) - \Gamma(\alpha+2)x)^2 + z(x)] \\ &+ p_1(x)\left(-AB^{-1}x^{-1}y(x) + B^{-1}C_2 x^{-2}v_2(x) + B^{-1}x^{\alpha-1}u(x)\right) \\ &- p_2(x)y(x) + p_3(x)\left(y(x) - x^{\alpha+1}\right)^2. \end{aligned}$$

Using the classical necessary optimality condition for problem (3.241), we end up with the following two point boundary value problem:

$$\begin{cases} y'(x) = -AB^{-1}x^{-1}y(x) + B^{-1}C_2 x^{-2}v_2(x) + \frac{1}{2}B^{-2}x^{2\alpha-2}p_1(x) \\ \qquad + \Gamma(\alpha+2)B^{-1}x^{\alpha} \\ v'_2(x) = -y(x) \\ z'(x) = (y(x) - x^{\alpha+1})^2 \\ p'_1(x) = AB^{-1}x^{-1}p_1(x) + p_2(x) - 2p_3(x)(y(x) - x^{\alpha+1}) \\ p'_2(x) = -B^{-1}C_2 x^{-2}p_1(x) \\ p'_3(x) = 1 \end{cases}$$

(3.242)

subject to the boundary conditions

$$\begin{cases} y(0) = 0 \\ v_2(0) = 0 \\ z(0) = 0 \end{cases} \qquad \begin{cases} y(1) = 1 \\ p_2(1) = 0 \\ p_3(1) = 0. \end{cases} \qquad (3.243)$$

We solved system (3.242) subject to (3.243) using the MATLAB built-in function bvp4c. The resulting graph of $y(x)$ for $\alpha = \frac{1}{2}$, together with the corresponding value of J, is given in Fig. 3.3.

Our numerical method works well, even in the case where the minimizer is not a Lipschitz function.

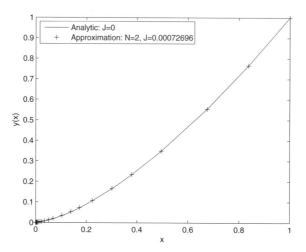

Fig. 3.3   Analytic versus numerical solution to problem of Example 3.19.

Example 3.23. An approximated solution to the problem considered in Example 3.20 can be obtained following exactly the same steps as in Example 3.22. Recall that the minimizer (3.211) to that problem is not a Lipschitz function. As before, one has $y(0) = 0$ and the Caputo derivative coincides with the Riemann–Liouville derivative. We approximate the fractional problem using (3.239). Let $^C_0D^\alpha_x y(x) = u(x)$. Then (3.240) holds. In this case the variable $z(x)$ satisfies

$$z'(x) = \left(y(x) - \frac{x^\alpha}{\Gamma(\alpha + 1)} \right)^2$$

and we approximate the fractional variational problem with the following classical one:

$$\tilde{J}(y) = \int_0^1 [(u(x) - 1)^2 + z(x)]dx \longrightarrow \min$$

$$\begin{cases} y'(x) = -AB^{-1}x^{-1}y(x) + \sum_{k=2}^N B^{-1}C_k x^{-k}v_k(x) + B^{-1}x^{\alpha-1}u(x) \\ v'_k(x) = (1-k)x^{k-2}y(x), \qquad k = 1,2,\dots \\ z'(x) = \left(y(x) - \frac{x^\alpha}{\Gamma(\alpha+1)} \right)^2 \end{cases}$$

subject to the boundary conditions

$$y(0) = 0, \quad z(0) = 0, \quad v_k(0) = 0, \quad k = 1,2,\dots$$

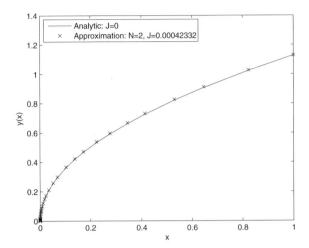

Fig. 3.4   Analytic versus numerical solution to problem of Example 3.20.

Setting $N = 2$, the Hamiltonian is given by

$$H = p_1(x)\left(-AB^{-1}x^{-1}y(x) + B^{-1}C_2x^{-2}v_2(x) + B^{-1}x^{\alpha-1}u(x)\right)$$
$$- p_2(x)y(x) + p_3(x)\left(y(x) - \frac{x^\alpha}{\Gamma(\alpha+1)}\right)^2 - \left[(u(x)-1)^2 + z(x)\right].$$

The classical theory (Pontryagin *et al.*, 1962) enables us to solve the system

$$\begin{cases} y'(x) = B^{-1}C_2x^{-2}v_2(x) + \frac{1}{2}B^{-2}x^{2\alpha-2}p_1(x) + B^{-1}x^{\alpha-1} - AB^{-1}x^{-1}y(x) \\ v_2'(x) = -y(x) \\ z'(x) = \left(y(x) - \frac{x^\alpha}{\Gamma(\alpha+1)}\right)^2 \\ p_1'(x) = AB^{-1}x^{-1}p_1(x) + p_2(x) - 2p_3(x)\left(y(x) - \frac{x^\alpha}{\Gamma(\alpha+1)}\right) \\ p_2'(x) = -B^{-1}C_2x^{-2}p_1(x) \\ p_3'(x) = 1 \end{cases}$$

$$(3.244)$$

subject to boundary conditions

$$\begin{cases} y(0) = 0 \\ v_2(0) = 0 \\ z(0) = 0 \end{cases} \qquad \begin{cases} y(1) = \dfrac{1}{\Gamma(\alpha+1)} \\ p_2(1) = 0 \\ p_3(1) = 0. \end{cases} \qquad (3.245)$$

As in Example 3.22, we solved (3.244)–(3.245) using the MATLAB built-in function **bvp4c**. The resulting graph of $y(x)$ for $\alpha = \frac{1}{2}$, together with the corresponding value of J, is given in Fig. 3.4 in contrast with the exact minimizer (3.211).

Chapter 4

Other Approaches to the Fractional Calculus of Variations

To better model non-conservative dynamical systems, a novel approach entitled Fractional Action-Like Variational Approach (FALVA) has been introduced by El-Nabulsi. This approach is based on the concept of fractional integration. Section 4.1 is devoted to the study of such problems. In Section 4.2 we extend FALVA formalism and previous results by considering Cresson's Riemann–Liouville fractional derivatives of order (α, β). Finally, in Section 4.3 we consider Jumarie's approach, based on a simple modification of the Riemann–Liouville derivative. The results of this chapter appeared in El-Nabulsi and Torres, 2007, 2008; Frederico and Torres, 2006b, 2007b, 2008c; Almeida and Torres, 2011b; Malinowska, 2012a; Malinowska, Sidi Ammi and Torres, 2010; Almeida, Malinowska and Torres, 2010; Odzijewicz and Torres, 2011.

4.1 Fractional Action-Like Variational Approach (FALVA)

This section is devoted to the study of fractional action-like variational problems with intrinsic and observer times. In Subsection 4.1.1 we extend Noether's symmetry theorem to non-conservative FALVA problems. We prove higher-order Euler–Lagrange (Subsection 4.1.2), DuBois–Reymond stationary conditions (Subsection 4.1.3), and Noether's Theorem (Subsection 4.1.5) for higher-order FALVA problems. More general fractional action-like optimal control problems are considered in Subsection 4.1.4.

4.1.1   Fractional Action-Like Noether's Theorem

We consider the fundamental problem of the calculus of variations with Riemann–Liouville fractional integral:

$$I[q(\cdot)] = \frac{1}{\Gamma(\alpha)} \int_a^b L\left(\theta, q(\theta), \dot{q}(\theta)\right)(t-\theta)^{\alpha-1} d\theta \longrightarrow \min, \qquad (4.1)$$

under given boundary conditions $q(a) = q_a$ and $q(b) = q_b$, where $\dot{q} = \frac{dq}{d\theta}$, $0 < \alpha \le 1$, θ is the intrinsic time, t is the observer time, $t \ne \theta$, and the Lagrangian $L : [a, b] \times \mathbb{R}^n \times \mathbb{R}^n \to \mathbb{R}$ is a C^2 function with respect to its arguments. We will denote by $\partial_i L$ the partial derivative of L with respect to the ith argument, $i = 1, 2, 3$. Admissible functions $q(\cdot)$ are assumed to be C^2.

Theorem 4.1. *If q is a minimizer to problem* (4.1), *then q satisfies the following Euler–Lagrange equation:*

$$\partial_2 L\left(\theta, q(\theta), \dot{q}(\theta)\right) - \frac{d}{d\theta} \partial_3 L\left(\theta, q(\theta), \dot{q}(\theta)\right) = \frac{1-\alpha}{t-\theta} \partial_3 L\left(\theta, q(\theta), \dot{q}(\theta)\right). \quad (4.2)$$

We now introduce the following definition of variational quasi-invariance up to a gauge term (cf. Torres, 2004b).

Definition 4.1 (Quasi-invariance of (4.1) up to a gauge term Λ). *Functional* (4.1) *is said to be quasi-invariant under the infinitesimal ε-parameter transformations*

$$\begin{cases} \bar{\theta} = \theta + \varepsilon\tau(\theta, q) + o(\varepsilon) \\ \bar{q}(\bar{\theta}) = q(\theta) + \varepsilon\xi(\theta, q) + o(\varepsilon) \end{cases} \qquad (4.3)$$

up to the gauge term Λ if, and only if,

$$L\left(\bar{\theta}, \bar{q}(\bar{\theta}), \bar{q}'(\bar{\theta})\right)(t-\bar{\theta})^{\alpha-1} \frac{d\bar{\theta}}{d\theta}$$

$$= L\left(\theta, q(\theta), \dot{q}(\theta)\right)(t-\theta)^{\alpha-1} + \varepsilon(t-\theta)^{\alpha-1} \frac{d\Lambda}{d\theta}\left(\theta, q(\theta), \dot{q}(\theta)\right) + o(\varepsilon). \quad (4.4)$$

Lemma 4.1 (Necessary condition for quasi-invariance). *If functional* (4.1) *is quasi-invariant up to Λ under the infinitesimal transformations* (4.3), *then*

$$\partial_1 L\left(\theta, q, \dot{q}\right)\tau + \partial_2 L\left(\theta, q, \dot{q}\right) \cdot \xi + \partial_3 L\left(\theta, q, \dot{q}\right) \cdot \left(\dot{\xi} - \dot{q}\dot{\tau}\right)$$

$$+ L\left(\theta, q, \dot{q}\right)\left(\dot{\tau} + \frac{1-\alpha}{t-\theta}\tau\right) = \dot{\Lambda}\left(\theta, q, \dot{q}\right). \quad (4.5)$$

Proof. Condition (4.4) is equivalent to

$$\left[L\left(\theta + \varepsilon\tau + o(\varepsilon), q + \varepsilon\xi + o(\varepsilon), \frac{\dot{q} + \varepsilon\dot{\xi} + o(\varepsilon)}{1 + \varepsilon\dot{\tau} + o(\varepsilon)} \right) \right]$$

$$(t - \theta - \varepsilon\tau - o(\varepsilon))^{\alpha-1} (1 + \varepsilon\dot{\tau} + o(\varepsilon))$$

$$= L\left(\theta, q, \dot{q} \right)(t - \theta)^{\alpha-1} + \varepsilon(t-\theta)^{\alpha-1}\frac{d}{d\theta}\Lambda\left(\theta, q, \dot{q} \right) + o(\varepsilon). \quad (4.6)$$

Equation (4.5) is obtained differentiating both sides of condition (4.6) with respect to ε and then putting $\varepsilon = 0$. □

Definition 4.2 (Constant of motion). *A quantity* $C\left(\theta, q(\theta), \dot{q}(\theta) \right)$, $\theta \in [a, b]$, *is said to be a constant of motion if, and only if,* $\frac{d}{d\theta}C\left(\theta, q(\theta), \dot{q}(\theta) \right) = 0$ *for all the solutions* q *of the Euler–Lagrange equation (4.2).*

Theorem 4.2 (Noether's theorem). *If the fractional integral (4.1) is quasi-invariant up to* Λ, *in the sense of Definition 4.1, and functions* $\tau(\theta, q)$ *and* $\xi(\theta, q)$ *satisfy the condition*

$$L\left(\theta, q, \dot{q} \right)\tau = -\partial_3 L\left(\theta, q, \dot{q} \right) \cdot (\xi - \dot{q}\tau), \quad (4.7)$$

then

$$\partial_3 L\left(\theta, q, \dot{q} \right) \cdot \xi(\theta, q) + [L(\theta, q, \dot{q}) - \partial_3 L\left(\theta, q, \dot{q} \right) \cdot \dot{q}]\tau(\theta, q) - \Lambda\left(\theta, q, \dot{q} \right) \quad (4.8)$$

is a constant of motion.

Remark 4.1. Under our hypothesis (4.7) the necessary and sufficient condition of quasi-invariance (4.5) is reduced to

$$\partial_1 L\left(\theta, q, \dot{q} \right)\tau + \partial_2 L\left(\theta, q, \dot{q} \right) \cdot \xi + \partial_3 L\left(\theta, q, \dot{q} \right) \cdot \left(\dot{\xi} - \dot{q}\dot{\tau} \right)$$

$$+ L\left(\theta, q, \dot{q} \right)\dot{\tau} - \frac{1-\alpha}{t-\theta}\partial_3 L\left(\theta, q, \dot{q} \right) \cdot (\xi - \dot{q}\tau) = \dot{\Lambda}\left(\theta, q, \dot{q} \right). \quad (4.9)$$

Conditions (4.7) and (4.9) correspond to the generalized equations of Noether–Bessel–Hagen of a non-conservative mechanical system.

Proof. We can write (4.9) in the form

$$\left[\partial_1 L\left(\theta, q, \dot{q} \right) + \frac{1-\alpha}{t-\theta}\partial_3 L\left(\theta, q, \dot{q} \right) \cdot \dot{q} \right]\tau + [L\left(\theta, q, \dot{q} \right) - \partial_3 L\left(\theta, q, \dot{q} \right) \cdot \dot{q}]\dot{\tau}$$

$$+ \left[\partial_2 L\left(\theta, q, \dot{q} \right) - \frac{1-\alpha}{t-\theta}\partial_3 L\left(\theta, q, \dot{q} \right) \right] \cdot \xi + \partial_3 L\left(\theta, q, \dot{q} \right) \cdot \dot{\xi} - \dot{\Lambda} = 0. \quad (4.10)$$

Using the Euler–Lagrange equation (4.2), equality (4.10) is equivalent to

$$\frac{d}{d\theta} \left[L\left(\theta, q, \dot{q}\right) - \partial_3 L\left(\theta, q, \dot{q}\right) \cdot \dot{q} \right] \tau + \left[L\left(\theta, q, \dot{q}\right) - \partial_3 L\left(\theta, q, \dot{q}\right) \cdot \dot{q} \right] \dot{\tau}$$

$$+ \frac{d}{d\theta} \left[\partial_3 L\left(\theta, q, \dot{q}\right) \right] \cdot \xi + \partial_3 L\left(\theta, q, \dot{q}\right) \cdot \dot{\xi} - \dot{\Lambda} = 0$$

and the intended conclusion follows:

$$\frac{d}{d\theta} \left[\partial_3 L\left(\theta, q, \dot{q}\right) \cdot \xi + \left(L(\theta, q, \dot{q}) - \partial_3 L\left(\theta, q, \dot{q}\right) \cdot \dot{q} \right) \tau - \Lambda\left(\theta, q, \dot{q}\right) \right] = 0 . \quad \square$$

Example 4.1. Conservation of momentum, when L is not a function of q, or conservation of energy, when L has no explicit dependence on time θ, are no more true for a fractional order of integration α, $\alpha \neq 1$. Indeed, as we shall see, these facts follow from Theorem 4.2. Moreover, our Noether's theorem gives new explicit formulas for the fractional constants of motion. For the particular case $\alpha = 1$ we recover the classical constants of motion of momentum and energy.

Let us first consider an arbitrary fractional action-like problem (4.1) with an autonomous L: $L\left(\theta, q, \dot{q}\right) = L\left(q, \dot{q}\right)$. In this case $\partial_1 L = 0$, and it is a simple exercise to check that (4.9) is satisfied with $\tau = 1$, $\xi = 0$ and Λ given by

$$\dot{\Lambda} = \frac{1 - \alpha}{t - \theta} \frac{\partial L}{\partial \dot{q}} \cdot \dot{q} .$$

It follows from the Noether theorem given by Theorem 4.2 that

$$L\left(q, \dot{q}\right) - \frac{\partial L}{\partial \dot{q}}\left(q, \dot{q}\right) \cdot \dot{q} - (1 - \alpha) \int \frac{1}{t - \theta} \frac{\partial L}{\partial \dot{q}}\left(q, \dot{q}\right) \cdot \dot{q} \, d\theta \equiv \text{constant} . \quad (4.11)$$

In the classical framework $\alpha = 1$ and we then get from our expression (4.11) the well-known constant of motion:

$$L\left(q, \dot{q}\right) - \frac{\partial L}{\partial \dot{q}}\left(q, \dot{q}\right) \cdot \dot{q} \equiv \text{constant},$$

which corresponds in mechanics to conservation of energy.

When L is not a function of q one has $\frac{\partial L}{\partial q} = 0$ and (4.9) holds true with $\tau = 0$, $\xi = 1$ and Λ given by

$$\dot{\Lambda} = -\frac{1 - \alpha}{t - \theta} \frac{\partial L}{\partial \dot{q}}\left(\theta, \dot{q}\right) .$$

The constant of motion (4.8) takes the form

$$\frac{\partial L}{\partial \dot{q}}\left(\theta, \dot{q}\right) + (1 - \alpha) \int \frac{1}{t - \theta} \frac{\partial L}{\partial \dot{q}}\left(\theta, \dot{q}\right) d\theta . \quad (4.12)$$

For $\alpha = 1$ (4.12) implies conservation of momentum: $\frac{\partial L}{\partial \dot{q}} = \text{const}$.

4.1.2 Higher-Order Euler–Lagrange Equations

We prove Euler–Lagrange equations to higher-order problems of the calculus of variations with fractional integrals of Riemann–Liouville, i.e., to FALVA problems with higher-order derivatives.

Problem 4.1. *The higher-order FALVA problem consists to find stationary values of an integral functional*

$$I^m[q(\cdot)] = \frac{1}{\Gamma(\alpha)} \int_a^t L\Big(\theta, q(\theta), \dot{q}(\theta), \ldots, q^{(m)}(\theta)\Big)(t-\theta)^{\alpha-1}d\theta, \quad (4.13)$$

$m \geq 1$, *subject to initial conditions*

$$q^{(i)}(a) = q_a^i, \quad i = 0, \ldots, m-1, \quad (4.14)$$

where $q^{(0)}(\theta) = q(\theta)$, $q^{(i)}(\theta)$ *is the ith derivative,* $i \geq 1$; $0 < \alpha \leq 1$; θ *is the intrinsic time; t the observer's time,* $t \neq \theta$; *and the Lagrangian* $L : [a,b] \times \mathbb{R}^{n \times (m+1)} \to \mathbb{R}$ *is a function of class* C^{2m} *with respect to all the arguments.*

Remark 4.2. In the particular case when $m = 1$, functional (4.13) reduces to (4.1).

To establish the Euler–Lagrange stationary condition for Problem 4.1, we follow the standard steps used to derive the necessary conditions in the calculus of variations.

Let us suppose $q(\cdot)$ a solution to Problem 4.1. The variation $\delta I^m[q(\cdot)]$ of the integral functional (4.13) is given by

$$\frac{1}{\Gamma(\alpha)} \int_a^t \left(\sum_{i=0}^m \partial_{i+2} L \cdot \delta q^{(i)} \right)(t-\theta)^{\alpha-1} d\theta, \quad (4.15)$$

where $\delta q^{(i)} \in C^{2m}([a,b]; \mathbb{R}^n)$ represents the variation of $q^{(i)}$, $i = 1, \ldots, m$, and satisfy

$$\delta q^{(i)}(a) = 0. \quad (4.16)$$

Having in account conditions (4.16), repeated integration by parts of each integral containing $\delta q^{(i)}$ in (4.15) leads to

$$m = 1: \quad \delta I[q(\cdot)] = \frac{1}{\Gamma(\alpha)} \int_a^t \left[\left(\partial_2 L - \frac{d}{d\theta} \partial_3 L \right) \right.$$
$$\left. - \frac{1-\alpha}{t-\theta} \partial_3 L \right](t-\theta)^{\alpha-1} \cdot \delta q \, d\theta; \quad (4.17)$$

$$m = 2: \quad \delta I^2[q(\cdot)] = \frac{1}{\Gamma(\alpha)} \int_a^t \left[\left(\partial_2 L - \frac{d}{d\theta} \partial_3 L \right. \right.$$

$$\left. + \frac{d^2}{d\theta^2} \partial_4 L \right) \left(\frac{1-\alpha}{t-\theta} \left(\partial_3 L - 2 \frac{d}{d\theta} \partial_4 L \right) \right.$$

$$\left. \left. - \frac{(1-\alpha)(2-\alpha)}{(t-\theta)^2} \partial_4 L \right) \right] (t-\theta)^{\alpha-1} \cdot \delta q \, d\theta \, ; \quad (4.18)$$

and, in general,

$$\delta I^m[q(\cdot)] = \frac{1}{\Gamma(\alpha)} \int_a^t \left[\left(\partial_2 L + \sum_{i=1}^m (-1)^i \frac{d^i}{d\theta^i} \partial_{i+2} L \right) \right.$$

$$- \frac{1-\alpha}{t-\theta} \sum_{i=1}^m i (-1)^{i-1} \frac{d^{i-1}}{d\theta^{i-1}} \partial_{i+2} L$$

$$\left. - \sum_{k=2}^m \sum_{i=2}^k (-1)^{i-1} \frac{\Gamma(i-\alpha+1)}{(t-\theta)^i \Gamma(1-\alpha)} \binom{k}{k-i} \frac{d^{k-i}}{d\theta^{k-i}} \partial_{k+2} L \right]$$

$$\cdot (t-\theta)^{\alpha-1} \cdot \delta q \, d\theta \, .$$

The integral functional $I^m[\cdot]$ has, by hypothesis, a stationary value for $q(\cdot)$, so that $\delta I^m[q(\cdot)] = 0$. The fundamental lemma of the calculus of variations asserts that all the coefficients of δq must vanish.

Theorem 4.3 (Higher-order Euler–Lagrange equations).
If $q(\cdot)$ gives a stationary value to functional (4.13), then $q(\cdot)$ satisfy the higher-order Euler–Lagrange equations

$$\sum_{i=0}^m (-1)^i \frac{d^i}{d\theta^i} \partial_{i+2} L \left(\theta, q(\theta), \dot{q}(\theta), \dots, q^{(m)}(\theta) \right)$$

$$= F \left(\theta, q(\theta), \dot{q}(\theta), \dots, q^{(2m-1)}(\theta) \right) , \quad (4.19)$$

where $m \geq 1$ and

$$F \left(\theta, q(\theta), \dot{q}(\theta), \dots, q^{(2m-1)}(\theta) \right)$$

$$= \frac{1-\alpha}{t-\theta} \sum_{i=1}^m i (-1)^{i-1} \frac{d^{i-1}}{d\theta^{i-1}} \partial_{i+2} L$$

$$+ \sum_{k=2}^m \sum_{i=2}^k (-1)^{i-1} \frac{\Gamma(i-\alpha+1)}{(t-\theta)^i \Gamma(1-\alpha)} \binom{k}{k-i} \frac{d^{k-i}}{d\theta^{k-i}} \partial_{k+2} L \quad (4.20)$$

with the partial derivatives of the Lagrangian L evaluated at $\left(\theta, q(\theta), \dot{q}(\theta), \ldots, q^{(m)}(\theta)\right)$.

Remark 4.3. Function F in (4.19) may be viewed as an external non-conservative friction force acting on the system. If $\alpha = 1$, then $F = 0$ and equation (4.19) is nothing more than the standard Euler–Lagrange equation for the classical problem of the calculus of variations with higher-order derivatives:

$$\sum_{i=0}^{m}(-1)^i \frac{d^i}{d\theta^i}\partial_{i+2}L\left(\theta, q(\theta), \dot{q}(\theta), \ldots, q^{(m)}(\theta)\right) = 0\,.$$

Remark 4.4. If $m = 1$, the Euler–Lagrange equations (4.19) coincide with the Euler–Lagrange equations (4.2).

Remark 4.5. For $m = 2$, the Euler–Lagrange equations (4.19) reduce to

$$\left(\partial_2 L\left(\theta, q, \dot{q}, \ddot{q}\right) - \frac{d}{d\theta}\partial_3 L\left(\theta, q, \dot{q}, \ddot{q}\right)\right.$$

$$\left. + \frac{d^2}{d\theta^2}\partial_4 L\left(\theta, q, \dot{q}, \ddot{q}\right)\right) = F\left(\theta, q, \dot{q}, \ddot{q}, \dddot{q}\right) \quad (4.21)$$

where

$$F\left(\theta, q, \dot{q}, \ddot{q}, \dddot{q}\right) = \frac{1-\alpha}{t-\theta}\left(\partial_3 L\left(\theta, q, \dot{q}, \ddot{q}\right)\right.$$

$$\left. - 2\frac{d}{d\theta}\partial_4 L\left(\theta, q, \dot{q}, \ddot{q}\right)\right)$$

$$- \frac{\Gamma(3-\alpha)}{(t-\theta)^2\Gamma(1-\alpha)}\binom{2}{0}\partial_4 L\left(\theta, q, \dot{q}, \ddot{q}\right)$$

$$= \frac{1-\alpha}{t-\theta}\left(\partial_3 L\left(\theta, q, \dot{q}, \ddot{q}\right) - 2\frac{d}{d\theta}\partial_4 L\left(\theta, q, \dot{q}, \ddot{q}\right)\right)$$

$$- \frac{(1-\alpha)(2-\alpha)}{(t-\theta)^2}\partial_4 L\left(\theta, q, \dot{q}, \ddot{q}\right)\,. \quad (4.22)$$

Proof. Theorem 4.3 is proved by induction. For $m = 1$ and $m = 2$, the Euler–Lagrange equations (4.2) and (4.21)–(4.22) are obtained applying the fundamental lemma of the calculus of variations respectively to (4.17) and (4.18). From the induction hypothesis,

$$\sum_{i=0}^{j}(-1)^i \frac{d^i}{d\theta^i}\partial_{i+2}L\left(\theta, q(\theta), \dot{q}(\theta), \ldots, q^{(m)}(\theta)\right)$$

$$= F\left(\theta, q(\theta), \dot{q}(\theta), \ldots, q^{(2j-1)}(\theta)\right)\,,\ m = j > 2. \quad (4.23)$$

We need to prove that equations (4.19)–(4.20) hold for $m = j + 1$. For simplicity, let us focus our attention on the variation of $q^{(j+1)}(\theta)$. From hypotheses (variation of $q^{(j+1)}(\theta)$ up to order $m = j$), and having in mind that $C_i^j + C_{i+1}^j = C_{i+1}^{j+1}$ and $m\Gamma(m) = \Gamma(m+1)$, we obtain equations (4.19) for $m = j + 1$ using integration by parts followed by the application of the fundamental lemma of the calculus of variations. $\qquad\square$

It is convenient to introduce the following quantity:

$$\psi^j = \sum_{i=0}^{m-j} (-1)^i \frac{d^i}{d\theta^i} \partial_{i+j+2} L\left(\theta, q(\theta), \dot{q}(\theta), \dots, q^{(m)}(\theta)\right), \qquad (4.24)$$

$j = 1, \dots, m$. This notation is useful for our purposes because of the following property:

$$\frac{d}{d\theta} \psi^j = \partial_{j+1} L\left(\theta, q(\theta), \dot{q}(\theta), \dots, q^{(m)}(\theta)\right) - \psi^{j-1}, \qquad (4.25)$$

$j = 1, \dots, m$.

Remark 4.6. Equation (4.19) can be written in the following form:

$$\partial_2 L\left(\theta, q(\theta), \dot{q}(\theta), \dots, q^{(m)}(\theta)\right) - \frac{d}{d\theta} \psi^1$$
$$= F\left(\theta, q(\theta), \dot{q}(\theta), \dots, q^{(2m-1)}(\theta)\right). \qquad (4.26)$$

4.1.3 *Higher-Order DuBois–Reymond Condition*

We now prove a DuBois–Reymond condition for FALVA problems.

Theorem 4.4 (Higher-order DuBois–Reymond condition). *A necessary condition for $q(\cdot)$ to be a solution to Problem 4.1 is given by the following higher-order DuBois–Reymond condition:*

$$\frac{d}{d\theta} \left\{ L\left(\theta, q(\theta), \dot{q}(\theta), \dots, q^{(m)}(\theta)\right) - \sum_{j=1}^{m} \psi^j \cdot q^{(j)}(\theta) \right\}$$
$$= \partial_1 L\left(\theta, q(\theta), \dot{q}(\theta), \dots, q^{(m)}(\theta)\right)$$
$$+ F\left(\theta, q(\theta), \dot{q}(\theta), \dots, q^{(2m-1)}(\theta)\right) \cdot \dot{q}(\theta), \qquad (4.27)$$

where F and ψ^j are defined by (4.20) and (4.24), respectively.

Remark 4.7. If $\alpha = 1$, then $F = 0$ and condition (4.27) is reduced to the classical higher-order DuBois–Reymond condition:

$$\partial_1 L\left(\theta, q(\theta), \dot{q}(\theta), \ldots, q^{(m)}(\theta)\right)$$
$$= \frac{d}{d\theta}\left\{ L\left(\theta, q(\theta), \dot{q}(\theta), \ldots, q^{(m)}(\theta)\right) - \sum_{j=1}^{m} \psi^j \cdot q^{(j)}(\theta) \right\}.$$

Proof. The total derivative of

$$L\left(\theta, q(\theta), \dot{q}(\theta), \ldots, q^{(m)}(\theta)\right) - \sum_{j=1}^{m} \psi^j \cdot q^{(j)}(\theta)$$

with respect to θ is:

$$\frac{d}{d\theta}\left\{ L\left(\theta, q(\theta), \dot{q}(\theta), \ldots, q^{(m)}(\theta)\right) - \sum_{j=1}^{m} \psi^j \cdot q^{(j)}(\theta) \right\}$$
$$= \frac{\partial L}{\partial \theta}\left(\theta, q(\theta), \dot{q}(\theta), \ldots, q^{(m)}(\theta)\right)$$
$$+ \sum_{j=0}^{m} \frac{\partial L}{\partial q^{(j)}}\left(\theta, q(\theta), \dot{q}(\theta), \ldots, q^{(m)}(\theta)\right) \cdot q^{(j+1)}(\theta) \qquad (4.28)$$
$$- \sum_{j=1}^{m} \left(\dot{\psi}^j \cdot q^{(j)}(\theta) + \psi^j \cdot q^{(j+1)}(\theta)\right).$$

From (4.25) it follows that (4.28) is equivalent to

$$\frac{d}{d\theta}\left\{ L\left(\theta, q(\theta), \dot{q}(\theta), \ldots, q^{(m)}(\theta)\right) - \sum_{j=1}^{m} \psi^j \cdot q^{(j)}(\theta) \right\}$$
$$= \frac{\partial L}{\partial \theta}\left(\theta, q(\theta), \dot{q}(\theta), \ldots, q^{(m)}(\theta)\right)$$
$$+ \sum_{j=0}^{m} \frac{\partial L}{\partial q^{(j)}}\left(\theta, q(\theta), \dot{q}(\theta), \ldots, q^{(m)}(\theta)\right) \cdot q^{(j+1)}(\theta) \qquad (4.29)$$
$$- \sum_{j=1}^{m} \Big[\big(\frac{\partial L}{\partial q^{(j-1)}}\left(\theta, q(\theta), \dot{q}(\theta), \ldots, q^{(m)}(\theta)\right)$$
$$- \psi^{j-1} \big) \cdot q^{(j)}(\theta) + \psi^j \cdot q^{(j+1)}(\theta) \Big].$$

We now simplify the last term on the right-hand side of (4.29):

$$\sum_{j=1}^{m} \left[\left(\frac{\partial L}{\partial q^{(j-1)}} - \psi^{j-1} \right) \cdot q^{(j)}(\theta) + \psi^j \cdot q^{(j+1)}(\theta) \right]$$

$$= \sum_{j=0}^{m-1} \left[\frac{\partial L}{\partial q^{(j)}} \cdot q^{(j+1)}(\theta) - \psi^j \cdot q^{(j+1)}(\theta) + \psi^{j+1} \cdot q^{(j+2)}(\theta) \right] \quad (4.30)$$

$$= \sum_{j=0}^{m-1} \left[\frac{\partial L}{\partial q^{(j)}} \cdot q^{(j+1)}(\theta) \right] - \psi^0 \cdot \dot{q}(\theta) + \psi^m \cdot q^{(m+1)}(\theta),$$

where the partial derivatives of the Lagrangian L are evaluated at $\left(\theta, q(\theta), \dot{q}(\theta), \ldots, q^{(m)}(\theta) \right)$. Substituting (4.30) into (4.29), and using the higher-order Euler–Lagrange equations (4.19), we obtain the intended result, that is,

$$\frac{d}{d\theta} \left\{ L\left(\theta, q(\theta), \dot{q}(\theta), \ldots, q^{(m)}(\theta) \right) - \sum_{j=1}^{m} \psi^j \cdot q^{(j)}(\theta) \right\}$$

$$= \frac{\partial L}{\partial \theta} \left(\theta, q(\theta), \dot{q}(\theta), \ldots, q^{(m)}(\theta) \right)$$

$$+ \frac{\partial L}{\partial q^{(m)}} \left(\theta, q(\theta), \dot{q}(\theta), \ldots, q^{(m)}(\theta) \right) \cdot q^{(m+1)}(\theta) + \psi^0 \cdot \dot{q}(\theta)$$

$$- \psi^m \cdot q^{(m+1)}(\theta)$$

$$= \partial_1 L\left(\theta, q(\theta), \dot{q}(\theta), \ldots, q^{(m)}(\theta) \right)$$

$$+ F\left(\theta, q(\theta), \dot{q}(\theta), \ldots, q^{(2m-1)}(\theta) \right) \cdot \dot{q}(\theta),$$

since, by definition,

$$\psi^m = \frac{\partial L}{\partial q^{(m)}} \left(\theta, q(\theta), \dot{q}(\theta), \ldots, q^{(m)}(\theta) \right)$$

and

$$\psi^0 = \sum_{i=0}^{m} (-1)^i \frac{d^i}{d\theta^i} \partial_{i+2} L\left(\theta, q(\theta), \dot{q}(\theta), \ldots, q^{(m)}(\theta) \right). \qquad \square$$

Corollary 4.1 (DuBois–Reymond condition). *If $q(\cdot)$ is a solution to Problem 4.1 with $m = 1$, then the following (first-order) DuBois–Reymond condition holds:*

$$\frac{d}{d\theta} \left\{ L\left(\theta, q(\theta), \dot{q}(\theta) \right) - \partial_3 L\left(\theta, q(\theta), \dot{q}(\theta) \right) \cdot \dot{q}(\theta) \right\}$$

$$= \partial_1 L\left(\theta, q(\theta), \dot{q}(\theta) \right) + \frac{1-\alpha}{t-\theta} \partial_3 L\left(\theta, q(\theta), \dot{q}(\theta) \right) \cdot \dot{q}(\theta). \quad (4.31)$$

Proof. For $m = 1$, condition (4.27) is reduced to

$$\frac{d}{d\theta} \left\{ L\left(\theta, q(\theta), \dot{q}(\theta)\right) - \psi^1 \cdot \dot{q}(\theta) \right\}$$
$$= \partial_1 L\left(\theta, q(\theta), \dot{q}(\theta)\right) + F\left(\theta, q(\theta), \dot{q}(\theta)\right) \cdot \dot{q}(\theta) . \quad (4.32)$$

Having in mind (4.20) and (4.24), we obtain that

$$\psi^1 = \partial_3 L\left(\theta, q(\theta), \dot{q}(\theta)\right) , \quad (4.33)$$

$$F\left(\theta, q(\theta), \dot{q}(\theta)\right) = \frac{1-\alpha}{t-\theta} \partial_3 L\left(\theta, q(\theta), \dot{q}(\theta)\right) . \quad (4.34)$$

One finds the intended equality (4.31) by substituting the quantities (4.33) and (4.34) into (4.32). $\qquad \square$

4.1.4 *Stationary Conditions for Optimal Control*

Here we obtain stationary conditions for two-time FALVA problems of optimal control. We begin by defining the problem.

Problem 4.2. *The two-time optimal control FALVA problem consists in finding the stationary values of the integral functional*

$$I[q(\cdot), u(\cdot)] = \frac{1}{\Gamma(\alpha)} \int_a^t L\left(\theta, q(\theta), u(\theta)\right)(t-\theta)^{\alpha-1} d\theta , \quad (4.35)$$

when subject to the control system

$$\dot{q}(\theta) = \varphi\left(\theta, q(\theta), u(\theta)\right) \quad (4.36)$$

and the initial condition $q(a) = q_a$. The Lagrangian $L : [a,b] \times \mathbb{R}^n \times \mathbb{R}^r \to \mathbb{R}$ and the velocity vector $\varphi : [a,b] \times \mathbb{R}^n \times \mathbb{R}^r \to \mathbb{R}^n$ are assumed to be C^1 functions with respect to all their arguments. In accordance with the calculus of variations, we suppose that the control functions $u(\cdot)$ take values on an open set of $\mathbb{R}^r$.

Remark 4.8. The problem defined by equation (4.1) is a particular case of Problem 4.2 where $\varphi(\theta, q, u) = u$. FALVA problems of the calculus of variations with higher-order derivatives are also easily written in the optimal control form (4.35)–(4.36). For example, the integral functional of the second-order FALVA problem of the calculus of variations,

$$I^2[q(\cdot)] = \frac{1}{\Gamma(\alpha)} \int_a^t L\left(\theta, q(\theta), \dot{q}(\theta), \ddot{q}(\theta)\right)(t-\theta)^{\alpha-1} d\theta ,$$

is equivalent to

$$\frac{1}{\Gamma(\alpha)} \int_a^t L\left(\theta, q^0(\theta), q^1(\theta), u(\theta)\right) (t-\theta)^{\alpha-1} d\theta,$$

$$\begin{cases} \dot{q}^0(\theta) = q^1(\theta), \\ \dot{q}^1(\theta) = u(\theta). \end{cases}$$

We now adopt the Hamiltonian formalism. We reduce (4.35)–(4.36) to the form (4.1) by considering the augmented functional:

$$J[q(\cdot), u(\cdot), p(\cdot)]$$

$$= \frac{1}{\Gamma(\alpha)} \int_a^t \left[\mathcal{H}\left(\theta, q(\theta), u(\theta), p(\theta)\right) - p(\theta) \cdot \dot{q}(\theta)\right] d\theta, \quad (4.37)$$

where the Hamiltonian $\mathcal{H}$ is defined by

$$\mathcal{H}\left(\theta, q, u, p\right) = L\left(\theta, q, u\right)(t-\theta)^{\alpha-1} + p \cdot \varphi\left(\theta, q, u\right). \quad (4.38)$$

Definition 4.3 (Process). *A pair $(q(\cdot), u(\cdot))$ that satisfies the control system $\dot{q}(\theta) = \varphi\left(\theta, q(\theta), u(\theta)\right)$ and the initial condition $q(a) = q_a$ of Problem 4.2 is said to be a process.*

The next theorem gives the weak Pontryagin maximum principle for Problem 4.2.

Theorem 4.5. *If $(q(\cdot), u(\cdot))$ is a stationary process for Problem 4.2, then there exists a vectorial function $p(\cdot) \in C^1([a, b]; \mathbb{R}^n)$ such that for all θ the tuple $(q(\cdot), u(\cdot), p(\cdot))$ satisfy the following conditions:*

(i) the Hamiltonian system

$$\begin{cases} \dot{q}(\theta) = \partial_4 \mathcal{H}(\theta, q(\theta), u(\theta), p(\theta)), \\ \dot{p}(\theta) = -\partial_2 \mathcal{H}(\theta, q(\theta), u(\theta), p(\theta)); \end{cases} \quad (4.39)$$

(ii) the stationary condition

$$\partial_3 \mathcal{H}(\theta, q(\theta), u(\theta), p(\theta)) = 0; \quad (4.40)$$

where $\mathcal{H}$ is given by (4.38).

Proof. We begin by remarking that the first equation in the Hamiltonian system, $\dot{q} = \partial_4 \mathcal{H}$, is nothing more than the control system (4.36). We write the augmented functional (4.37) in the following form:

$$\frac{1}{\Gamma(\alpha)} \int_a^t \left[\frac{\mathcal{H} - p(\theta) \cdot \dot{q}(\theta)}{(t-\theta)^{\alpha-1}}\right] (t-\theta)^{\alpha-1} d\theta, \quad (4.41)$$

where $\mathcal{H}$ is evaluated at $(\theta, q(\theta), u(\theta), p(\theta))$. Intended conditions are obtained by applying the stationary condition (4.2) to (4.41):

$$\begin{cases} \frac{d}{d\theta} \frac{\partial}{\partial \dot{q}} \left[\frac{\mathcal{H} - p \cdot \dot{q}}{(t-\theta)^{\alpha-1}} \right] = \frac{\partial}{\partial q} \left[\frac{\mathcal{H} - p \cdot \dot{q}}{(t-\theta)^{\alpha-1}} \right] - \frac{1-\alpha}{t-\theta} \frac{\partial}{\partial \dot{q}} \left[\frac{\mathcal{H} - p \cdot \dot{q}}{(t-\theta)^{\alpha-1}} \right] \\ \frac{d}{d\theta} \frac{\partial}{\partial \dot{u}} \left[\frac{\mathcal{H} - p \cdot \dot{q}}{(t-\theta)^{\alpha-1}} \right] = \frac{\partial}{\partial u} \left[\frac{\mathcal{H} - p \cdot \dot{q}}{(t-\theta)^{\alpha-1}} \right] - \frac{1-\alpha}{t-\theta} \frac{\partial}{\partial \dot{u}} \left[\frac{\mathcal{H} - p \cdot \dot{q}}{(t-\theta)^{\alpha-1}} \right] \end{cases}$$

$$\Leftrightarrow \begin{cases} -\dot{p} = \partial_2 \mathcal{H} \\ 0 = \partial_3 \mathcal{H} \end{cases}$$

$\square$

Remark 4.9. For FALVA problems of the calculus of variations, Theorem 4.5 takes the form of Theorem 4.3.

Definition 4.4 (Pontryagin FALVA extremal). *We call any tuple $(q(\cdot), u(\cdot), p(\cdot))$ satisfying Theorem 4.5, a Pontryagin FALVA extremal.*

The next theorem generalizes the DuBois–Reymond condition (4.27) to Problem 4.2.

Theorem 4.6. *The following property holds along the Pontryagin FALVA extremals:*

$$\frac{d\mathcal{H}}{d\theta}(\theta, q(\theta), u(\theta), p(\theta)) = \partial_1 \mathcal{H}(\theta, q(\theta), u(\theta), p(\theta)). \qquad (4.42)$$

Proof. Condition (4.42) is a simple consequence of Theorem 4.5. $\square$

Remark 4.10. In the classical framework, i.e., for $\alpha = 1$, the Hamiltonian $\mathcal{H}$ does not depend explicitly on θ when the Lagrangian L and the velocity vector φ are autonomous. In that case, it follows from (4.42) that the Hamiltonian $\mathcal{H}$ (interpreted as energy in mechanics) is conserved. In the FALVA setting, i.e., for $\alpha \neq 1$, this is no longer true: condition (4.42) holds but we have no conservation of energy since, by definition (cf. (4.38)), the Hamiltonian $\mathcal{H}$ is never autonomous ($\mathcal{H}$ always depend explicitly on θ for $\alpha \neq 1$, thus $\partial_1 \mathcal{H} \neq 0$).

4.1.5 Higher-Order Noether's Theorem

In order to generalize the Noether's theorem to Problem 4.1 (see Theorem 4.8 below) we use the DuBois–Reymond necessary stationary condition (4.27) and the following invariance definition.

Definition 4.5 (Invariance of (4.13)). *The functional* (4.13) *is said to be invariant under the infinitesimal transformations*

$$\begin{cases} \bar{\theta} = \theta + \varepsilon\tau(\theta, q) + o(\varepsilon) \\ \bar{q}(\bar{\theta}) = q(\theta) + \varepsilon\xi(\theta, q) + o(\varepsilon) \end{cases} \tag{4.43}$$

if

$$L\left(\bar{\theta}, \bar{q}(\bar{\theta}), \bar{q}'(\bar{\theta}), \dots, \bar{q}'^{(m)}(\bar{\theta})\right)(t - \bar{\theta})^{\alpha-1}\frac{d\bar{\theta}}{d\theta}$$

$$= L\left(\theta, q(\theta), \dot{q}(\theta), \dots, q^{(m)}(\theta)\right)(t - \theta)^{\alpha-1}$$

$$+ \varepsilon(t - \theta)^{\alpha-1}\frac{d\Lambda}{d\theta}\left(\theta, q(\theta), \dot{q}(\theta), \dots, q^{(2m-1)}(\theta)\right) + o(\varepsilon). \tag{4.44}$$

Remark 4.11. Expressions $\bar{q}'^{(i)}$ in equation (4.44), $i = 1, \dots, m$, are interpreted as

$$\bar{q}' = \frac{d\bar{q}}{d\bar{\theta}} = \frac{\frac{d\bar{q}}{d\theta}}{\frac{d\bar{\theta}}{d\theta}}, \quad \bar{q}'^{(i)} = \frac{d^i\bar{q}}{d\bar{\theta}^i} = \frac{\frac{d}{d\theta}\left(\frac{d^{i-1}}{d\bar{\theta}^{i-1}}\bar{q}\right)}{\frac{d\bar{\theta}}{d\theta}} \quad (i = 2, \dots, m). \tag{4.45}$$

The next theorem gives a necessary and sufficient condition for invariance of (4.13). Theorem 4.7 is useful to check invariance and also to compute the infinitesimal generators τ and ξ.

Theorem 4.7. *The integral functional* (4.13) *is invariant in the sense of Definition 4.5 if, and only if,*

$$\partial_1 L\left(\theta, q(\theta), \dot{q}(\theta), \dots, q^{(m)}(\theta)\right)\tau + \sum_{i=0}^{m}\partial_{i+2}L\left(\theta, q(\theta), \dot{q}(\theta), \dots, q^{(m)}(\theta)\right)\cdot\rho^i$$

$$+ L\left(\theta, q(\theta), \dot{q}(\theta), \dots, q^{(m)}(\theta)\right)\left(\dot{\tau} + \frac{1-\alpha}{t-\theta}\tau\right)$$

$$= \dot{\Lambda}\left(\theta, q(\theta), \dot{q}(\theta), \dots, q^{(2m-1)}(\theta)\right), \tag{4.46}$$

where

$$\begin{cases} \rho^0 = \xi, \\ \rho^i = \frac{d}{d\theta}\left(\rho^{i-1}\right) - q^{(i)}(\theta)\dot{\tau}, \quad i = 1, \dots, m. \end{cases} \tag{4.47}$$

Proof. Differentiating equation (4.44) with respect to ε, then setting $\varepsilon = 0$, we obtain:

$$\partial_1 L\tau + \sum_{i=0}^{m}\partial_{i+2}L\cdot\frac{\partial}{\partial\varepsilon}\left(\frac{d^i\bar{q}}{d\bar{\theta}^i}\right)\Bigg|_{\varepsilon=0} + L\left(\dot{\tau} + \frac{1-\alpha}{t-\theta}\tau\right) = \dot{\Lambda}.$$

The intended conclusion follows from (4.45):

$$\frac{\partial}{\partial \varepsilon} \left(\frac{d\bar{q}}{d\bar{\theta}} \right) \Bigg|_{\varepsilon=0} = \dot{\xi} - \dot{q}\dot{\tau} \, ,$$

$$\frac{\partial}{\partial \varepsilon} \left(\frac{d^i \bar{q}}{d\bar{\theta}^i} \right) \Bigg|_{\varepsilon=0} = \frac{d}{d\theta} \left[\frac{\partial}{\partial \varepsilon} \left(\frac{d^{i-1}\bar{q}}{d\bar{\theta}^{i-1}} \right) \Bigg|_{\varepsilon=0} \right] - q^{(i)}\dot{\tau} \, , \quad i = 2, \ldots, m \, .$$

$\square$

Remark 4.12. If $\alpha = 1$, condition (4.46) gives the higher-order necessary and sufficient condition of invariance:

$$\partial_1 L \left(\theta, q(\theta), \dot{q}(\theta), \ldots, q^{(m)}(\theta) \right) \tau + \sum_{i=0}^{m} \partial_{i+2} L \left(\theta, q(\theta), \dot{q}(\theta), \ldots, q^{(m)}(\theta) \right) \cdot \rho^i$$

$$+ L \left(\theta, q(\theta), \dot{q}(\theta), \ldots, q^{(m)}(\theta) \right) \dot{\tau} = \dot{\Lambda} \left(\theta, q(\theta), \dot{q}(\theta), \ldots, q^{(2m-1)}(\theta) \right) \, .$$

Definition 4.6 (Higher-order conservation law). *A quantity* $C(\theta, q(\theta), \dot{q}(\theta), \ldots, q^{(2m-1)}(\theta))$ *is said to be a conservation law if*

$$\frac{d}{d\theta} C \left(\theta, q(\theta), \dot{q}(\theta), \ldots, q^{(2m-1)}(\theta) \right) = 0$$

along all the solutions $q(\cdot)$ *of the higher-order Euler–Lagrange equation* (4.19).

Theorem 4.8 (Higher-order Noether's theorem). *If the integral functional* (4.13) *is invariant, in the sense of Definition 4.5, and* $\tau(\theta, q)$ *and* $\xi(\theta, q)$ *satisfy the condition*

$$G \left(\theta, q(\theta), \dot{q}(\theta), \ldots, q^{(2m-1)}(\theta) \right) \cdot \Omega = -L \left(\theta, q(\theta), \dot{q}(\theta), \ldots, q^{(m)}(\theta) \right) \tau \, ,$$
(4.48)

where

$$G \left(\theta, q(\theta), \dot{q}(\theta), \ldots, q^{(2m-1)}(\theta) \right)$$

$$= \sum_{i=1}^{m} (-1)^{i-1} i \frac{d^{i-1}}{d\theta^{i-1}} \partial_{i+2} L \left(\theta, q(\theta), \dot{q}(\theta), \ldots, q^{(m)}(\theta) \right)$$

$$+ \sum_{k=2}^{m} \sum_{i=2}^{k} (-1)^i \frac{\Gamma(i - \alpha + 1)}{\Gamma(2 - \alpha)(t - \theta)^{i-1}}$$
(4.49)

$$\times \binom{k}{k-i} \frac{d^{k-i}}{d\theta^{k-i}} \partial_{k+2} L \left(\theta, q(\theta), \dot{q}(\theta), \ldots, q^{(m)}(\theta) \right)$$

and $\Omega = \xi - \dot{q}\tau$, then

$$C\left(\theta, q(\theta), \dot{q}(\theta), \ldots, q^{(2m-1)}(\theta)\right)$$

$$= \sum_{j=1}^{m} \psi^j \cdot \rho^{j-1} + \left(L\left(\theta, q(\theta), \dot{q}(\theta), \ldots, q^{(m)}(\theta)\right) - \sum_{j=1}^{m} \psi^j \cdot q^{(j)}(\theta)\right)\tau$$

$$- \Lambda\left(\theta, q(\theta), \dot{q}(\theta), \ldots, q^{(2m-1)}(\theta)\right) \quad (4.50)$$

is a conservation law.

Proof.   We begin by writing the Noether's conservation law (4.50) in the form

$$C = \psi^1 \cdot \rho^0 + \sum_{j=2}^{m} \psi^j \cdot \rho^{j-1} + \left(L - \sum_{j=1}^{m} \psi^j \cdot q^{(j)}(\theta)\right)\tau - \Lambda. \quad (4.51)$$

Differentiation of equation (4.51) with respect to θ gives

$$\dot{\Lambda} = \rho^0 \cdot \frac{d}{d\theta}\psi^1 + \psi^1 \cdot \frac{d}{d\theta}\rho^0 + \sum_{j=2}^{m}\left(\rho^{j-1} \cdot \frac{d}{d\theta}\psi^j + \psi^j \cdot \frac{d}{d\theta}\left(\rho^{j-1}\right)\right)$$

$$+ \tau\frac{d}{d\theta}\left(L - \sum_{j=1}^{m} \psi^j \cdot q^{(j)}(\theta)\right) + \left(L - \sum_{j=1}^{m} \psi^j \cdot q^{(j)}(\theta)\right)\frac{d}{d\theta}\tau. \quad (4.52)$$

Using the Euler–Lagrange equation (4.19), the DuBois–Reymond condition (4.27), and relations (4.25) and (4.47) in (4.52), we obtain:

$$\dot{\Lambda} = (\partial_2 L - F) \cdot \xi + \psi^1 \cdot (\rho^1 + \dot{q}\dot{\tau})$$

$$+ \sum_{j=2}^{m}\left[\left(\partial_{j+1}L - \psi^{j-1}\right) \cdot \rho^{j-1} + \psi^j \cdot \left(\rho^j + q^{(j)}(\theta)\dot{\tau}\right)\right]$$

$$+ (\partial_1 L + F \cdot \dot{q})\tau + \left(L - \sum_{j=1}^{m} \psi^j \cdot q^{(j)}(\theta)\right)\dot{\tau}$$

$$= \partial_1 L\tau + L\dot{\tau} + \partial_2 L \cdot \xi + \psi^1 \cdot (\rho^1 + \dot{q}\dot{\tau}) - \psi^1 \cdot \rho^1$$

$$- \psi^1 \cdot \dot{q}\dot{\tau} + \psi^m \cdot \rho^m + \sum_{j=2}^{m} \partial_{j+1}L \cdot \rho^{j-1}. \quad (4.53)$$

Simplification of (4.53) leads us to the necessary and sufficient condition of invariance (4.54).   □

Remark 4.13. Under hypothesis (4.48), the necessary and sufficient condition of invariance (4.46) takes the following form:

$$\partial_1 L\left(\theta, q(\theta), \dot{q}(\theta), \ldots, q^{(m)}(\theta)\right) \tau + \sum_{i=0}^{m} \partial_{i+2} L\left(\theta, q(\theta), \dot{q}(\theta), \ldots, q^{(m)}(\theta)\right) \cdot \rho^i$$

$$+ L\left(\theta, q(\theta), \dot{q}(\theta), \ldots, q^{(m)}(\theta)\right) \dot{\tau} - F\left(\theta, q(\theta), \dot{q}(\theta), \ldots, q^{(2m-1)}(\theta)\right) \cdot \Omega$$

$$= \dot{\Lambda}\left(\theta, q(\theta), \dot{q}(\theta), \ldots, q^{(2m-1)}(\theta)\right). \quad (4.54)$$

In the particular case $m = 1$ we obtain from Theorem 4.8 the following result:

Corollary 4.2. *If the integral functional* (4.1) *is invariant under the infinitesimal transformations* (4.43), *and* $\tau(\theta, q)$ *and* $\xi(\theta, q)$ *satisfy the condition*

$$\partial_3 L(\theta, q, \dot{q}) \cdot \Omega = -L(\theta, q, \dot{q}) \tau, \quad (4.55)$$

then

$$C(\theta, q, \dot{q}) = \partial_3 L(\theta, q, \dot{q}) \cdot \xi(\theta, q) + (L(\theta, q, \dot{q}) - \partial_3 L(\theta, q, \dot{q}) \cdot \dot{q}) \tau(\theta, q)$$

$$- \Lambda(\theta, q, \dot{q}) \quad (4.56)$$

is a conservation law (i.e., (4.56) *is constant along all the solutions* $q(\cdot)$ *of the Euler–Lagrange equation* (4.2)*).*

Proof. For $m = 1$ we obtain from (4.49) and (4.50) that

$$G(\theta, q, \dot{q}) = \partial_3 L(\theta, q, \dot{q}) \quad (4.57)$$

and

$$C(\theta, q, \dot{q}) = \psi^1 \cdot \rho^0 + \left(L(\theta, q, \dot{q}) - \psi^1 \cdot \dot{q}(\theta)\right) \tau - \Lambda(\theta, q, \dot{q}). \quad (4.58)$$

Having in mind the equations (4.2) and (4.47), we conclude that

$$\begin{cases} \psi^1 = \partial_3 L(\theta, q, \dot{q}), \\ \rho^0 = \xi. \end{cases} \quad (4.59)$$

We obtain the intended result substituting (4.59) into (4.58). $\quad \square$

In order to illustrate our result, we consider an example for which the Lagrangian L do not depend explicitly on the intrinsic time θ.

Example 4.2. Let us consider the following second-order ($m = 2$) FALVA problem: to find a stationary function $q(\cdot)$ for the integral functional

$$I^2[q(\cdot)] = \frac{1}{2} \int_0^t \left(aq^2 + b\dot{q}^2 + \ddot{q}^2\right)(t - \theta)^{\alpha-1} d\theta, \quad (4.60)$$

where a and b are arbitrary constants. In this case the Euler–Lagrange equation (4.19) reads

$$-aq + \frac{b(1-\alpha)}{t-\theta}\dot{q} + \left(b - \frac{(1-\alpha)(2-\alpha)}{(t-\theta)^2}\right)\ddot{q} - \left(1 + \frac{2(1-\alpha)}{(t-\theta)}\right)\dddot{q} = 0.$$
(4.61)

Since the Lagrangian L do not depend explicitly on the independent variable θ, the necessary and sufficient invariance condition (4.54) is satisfied with

$$\tau = 1,$$
(4.62)

$$\xi = 0,$$
(4.63)

$$\dot{\Lambda} = F\dot{q} \Rightarrow \Lambda = \int F\dot{q}\,d\theta,$$
(4.64)

where

$$F = \frac{1-\alpha}{t-\theta}(b\dot{q} - 2\dddot{q}) - \frac{(1-\alpha)(2-\alpha)}{(t-\theta)^2}\ddot{q}.$$
(4.65)

The conservation law (4.50) with $m = 2$ takes the following form:

$$C(\theta, q, \dot{q}, \ddot{q}, \dddot{q}) = L(\theta, q, \dot{q}, \ddot{q})\tau + \left(\partial_3 L(\theta, q, \dot{q}, \ddot{q}) - \frac{d}{d\theta}\partial_4 L(\theta, q, \dot{q}, \ddot{q})\right) \cdot \Omega$$

$$+ \partial_4 L(\theta, q, \dot{q}, \ddot{q}) \cdot \dot{\Omega} - \Lambda(\theta, q, \dot{q}, \ddot{q}, \dddot{q}). \quad (4.66)$$

Substituting the quantities $L = \frac{1}{2}\left(aq^2 + b\dot{q}^2 + \ddot{q}^2\right)$, (4.62), (4.63) and (4.64) into (4.66), we conclude that

$$\frac{1}{2}\left(aq^2 - b\dot{q}^2 + 3\ddot{q}^2\right) - \dot{q}\dddot{q} - \int F\dot{q}\,d\theta$$
(4.67)

is constant along any solution q of (4.61).

If $\alpha = 1$, then one can see from (4.65) that $F = 0$, and (4.67) gives the classical result:

$$\frac{1}{2}\left(aq^2 - b\dot{q}^2 + 3\ddot{q}^2\right) - \dot{q}\dddot{q}$$

is a conservation law.

4.2 Cresson's Approach

In this section we study fractional action-like integrals of the calculus of variations which depend on the Riemann–Liouville derivatives of order (α, β), $\alpha > 0$, $\beta > 0$, introduced by J. Cresson. Subsection 4.2.2 presents Euler–Lagrange type equations. In Subsection 4.2.3 the concept of the fractional constant of motion is introduced and illustrated. The last two subsections are devoted to multidimensional fractional action-like problems of the calculus of variations with fractional derivatives of Cresson.

4.2.1 The Fractional Derivative of Cresson

In this section we consider the fractional derivative operator of order (α, β) (Cresson, 2007).

Definition 4.7. *Given a, $b \in \mathbb{R}$, $a < b$, and $\gamma \in \mathbb{C}$, the fractional derivative operator of order (α, β), α, $\beta > 0$, is defined by*

$$D_\gamma^{\alpha,\beta} = \frac{1}{2} \left[{}_aD_t^\alpha - {}_tD_b^\beta \right] + \frac{i\gamma}{2} \left[{}_aD_t^\alpha + {}_tD_b^\beta \right]$$

where $i = \sqrt{-1}$.

Remark 4.14. The operator $D_\gamma^{\alpha,\beta}$ extends the classical Riemann–Liouville fractional derivatives: for $\gamma = -i$ we have $D_\gamma^{\alpha,\beta} = {}_aD_t^\alpha$; for $\gamma = i$ we obtain $D_\gamma^{\alpha,\beta} = -{}_tD_b^\beta$.

The following lemma is useful: we make use of (4.68) to prove the fractional Euler–Lagrange equations associated with our problem (cf. proof of Theorem 4.9).

Lemma 4.2 (Cresson, 2007). *If f, $g \in C^1$ with $f(a) = f(b) = 0$ or $g(a) = g(b) = 0$, then*

$$\int_a^b D_\gamma^{\alpha,\beta} f(t) g(t)\, dt = - \int_a^b f(t) D_{-\gamma}^{\beta,\alpha} g(t)\, dt \,. \tag{4.68}$$

4.2.2 Fractional Euler–Lagrange Equations

We now have the conditions to formulate the (α, β) fractional action-like variational problem.

Definition 4.8. *Consider a smooth manifold M and let L be a smooth Lagrangian function $L : \mathbb{C}^d \times \mathbb{R}^d \times \mathbb{R} \to \mathbb{R}$, $d \geq 1$. For any piecewise smooth path $q : [a, b] \to M$ satisfying fixed boundary conditions $q(a) = q_a$ and $q(b) = q_b$ we define the following fractional action integral:*

$$S_{\gamma,(a,b)}^{\alpha,\beta}[q] = \frac{1}{\Gamma(\alpha)} \int_a^b L\left(D_\gamma^{\alpha,\beta} q(\tau), q(\tau), \tau\right)(t - \tau)^{\alpha-1}\, d\tau \,. \tag{4.69}$$

The (α, β) fractional action-like variational problem consists in finding an admissible $q(\cdot)$ which minimizes (4.69).

Remark 4.15. The fractional action integral (4.69) is a generalization of the FALVA action integral.

Theorem 4.9 gives a necessary optimality condition for $q(\cdot)$ to be a solution of the (α, β) fractional action-like variational problem.

Theorem 4.9 ((α, β) fractional Euler–Lagrange equations). *If q : $[a, b] \to M$ is a minimizer of the (α, β) fractional action-like variational problem (cf. Definition 4.8), then*

$$\frac{\partial L}{\partial q}\left(D_\gamma^{\alpha,\beta}q\left(\tau\right), q\left(\tau\right), \tau\right) - D_{-\gamma;\tau}^{\beta,\alpha}\frac{\partial L}{\partial \dot{q}}\left(D_\gamma^{\alpha,\beta}q\left(\tau\right), q\left(\tau\right), \tau\right)$$

$$= \frac{1-\alpha}{t-\tau}\frac{\partial L}{\partial \dot{q}}\left(D_\gamma^{\alpha,\beta}q\left(\tau\right), q\left(\tau\right), \tau\right) \quad (4.70)$$

where $D_{-\gamma;\tau}^{\beta,\alpha}$ represents the fractional derivative with respect to time τ.

Proof. We perform a small perturbation of the generalized coordinates as $q \to q + \varepsilon h$, $\varepsilon \ll 1$. As a result, $D_\gamma^{\alpha,\beta}\left(q + \varepsilon h\right) = D_\gamma^{\alpha,\beta}q + \varepsilon D_\gamma^{\alpha,\beta}h$ and

$$S_{\gamma,(a,b)}^{\alpha,\beta}\left[q + \varepsilon h\right] = \frac{1}{\Gamma\left(\alpha\right)}\int_a^b L\left(D_\gamma^{\alpha,\beta}q + \varepsilon D_\gamma^{\alpha,\beta}h, q + \varepsilon h, \tau\right)\left(t - \tau\right)^{\alpha-1} d\tau$$

which, using a Taylor expansion of $L\left(D_\gamma^{\alpha,\beta}q + \varepsilon D_\gamma^{\alpha,\beta}h, q + \varepsilon h, \tau\right)$ in ε around zero, and integrating by parts, imply that

$$S_{\gamma,(a,b)}^{\alpha,\beta}\left[q + \varepsilon h\right] = S_{\gamma,(a,b)}^{\alpha,\beta}\left[q\right]$$

$$- \frac{\varepsilon}{\Gamma\left(\alpha\right)}\int_a^b \left[-\frac{\partial L}{\partial q}\left(D_\gamma^{\alpha,\beta}q\left(\tau\right), q\left(\tau\right), \tau\right)\left(t - \tau\right)^{\alpha-1} h\left(\tau\right)\right.$$

$$+ \left(t - \tau\right)^{\alpha-1}\frac{\partial L}{\partial \dot{q}}\left(D_\gamma^{\alpha,\beta}q\left(\tau\right) D_\gamma^{\alpha,\beta}h\left(\tau\right), q\left(\tau\right), \tau\right) D_{\gamma;\tau}^{\alpha,\beta}h\left(\tau\right)$$

$$+ \left.\left(t - \tau\right)^{\alpha-1}\frac{\partial L}{\partial \dot{q}}\left(D_\gamma^{\alpha,\beta}q\left(\tau\right) D_\gamma^{\alpha,\beta}h\left(\tau\right), q\left(\tau\right), \tau\right) D_\gamma^{\alpha,\beta}h\left(\tau\right)\right] d\tau + O\left(\varepsilon\right).$$

Using Lemma 4.2 and the least action principle, we arrive at (4.70). □

Remark 4.16. Theorem 4.9 is an extension of the Euler–Lagrange equations derived in FALVA: one just needs to choose $\beta = 1$ in (4.70) to obtain the FALVA Euler–Lagrange equations.

Definition 4.9. *A path $q : [a, b] \to M$ satisfying equation (4.70) is said to be a fractional extremal associated to the Lagrangian L.*

Definition 4.10. *The right-hand side of* (4.70),

$$F_{\gamma,\tau}^{\alpha,\beta} = \frac{\alpha-1}{T} \frac{\partial L}{\partial \dot{q}} \left(D_\gamma^{\alpha,\beta} q\left(\tau\right), q\left(\tau\right), \tau\right), \tag{4.71}$$

$T = \tau - t$, *defines the fractional decaying friction force.*

Remark 4.17. The fractional decaying friction force satisfies the asymptotic property $\lim\limits_{T\to\infty} F_{\gamma,\tau}^{\alpha,\beta} = 0$.

4.2.3 The Fractional Hamiltonian Formalism

We now consider a more general class of fractional optimal control problems:

$$S_{\gamma,(a,b)}^{\alpha,\beta}[q,u] = \frac{1}{\Gamma(\alpha)} \int_a^b L\left(u\left(\tau\right), q\left(\tau\right), \tau\right) \left(t-\tau\right)^{\alpha-1} d\tau \longrightarrow \min,$$
$$D_\gamma^{\alpha,\beta} q\left(\tau\right) = \varphi\left(u\left(\tau\right), q\left(\tau\right), \tau\right), \tag{4.72}$$

$a, b \in \mathbb{R}$, $a < b$. In the particular case where $\varphi\left(u, q, \tau\right) = u$ (4.72) reduces to (4.69). To obtain a necessary optimality condition to problem (4.72) we introduce the augmented action integral:

$$S_{\gamma,(a,b)}^{\alpha,\beta}\left[q,u,p^{\alpha,\beta}\right] = \frac{1}{\Gamma(\alpha)} \int_a^b \left[\mathcal{H}^{\alpha,\beta}\left(u\left(\tau\right), q\left(\tau\right), p^{\alpha,\beta}\left(\tau\right), \tau\right) \right.$$
$$\left. - p^{\alpha,\beta}\left(\tau\right) D_\gamma^{\alpha,\beta} q\left(\tau\right) \right] d\tau \quad (4.73)$$

where $p^{\alpha,\beta}$ is the fractional Lagrange multiplier and the fractional Hamiltonian $\mathcal{H}^{\alpha,\beta}$ is defined by

$$\mathcal{H}^{\alpha,\beta}\left(u, q, p^{\alpha,\beta}, \tau\right) = L\left(u, q, \tau\right)\left(t-\tau\right)^{\alpha-1} + p^{\alpha,\beta} \varphi\left(u, q, \tau\right).$$

Theorem 4.10. *If* (q, u) *is a minimizer to problem* (4.72)*, then there exists a co-vector function* $p^{\alpha,\beta}$ *such that the following conditions hold:*

(i) the fractional Hamiltonian system

$$\begin{cases} D_\gamma^{\alpha,\beta} q\left(\tau\right) = \frac{\partial \mathcal{H}^{\alpha,\beta}}{\partial p^{\alpha,\beta}}\left(u\left(\tau\right), q\left(\tau\right), p^{\alpha,\beta}\left(\tau\right), \tau\right), \\ D_{-\gamma;\tau}^{\beta,\alpha} p^{\alpha,\beta}\left(\tau\right) = -\frac{\partial \mathcal{H}^{\alpha,\beta}}{\partial q}\left(u\left(\tau\right), q\left(\tau\right), p^{\alpha,\beta}\left(\tau\right), \tau\right); \end{cases} \tag{4.74}$$

(ii) the fractional stationary condition

$$\frac{\partial \mathcal{H}^{\alpha,\beta}}{\partial u}\left(u\left(\tau\right), q\left(\tau\right), p^{\alpha,\beta}\left(\tau\right), \tau\right) = 0. \tag{4.75}$$

Proof. Theorem 4.10 is proved applying the (α, β) fractional Euler–Lagrange equations to the augmented action integral (4.73), i.e., applying (4.70) to

$$S_{\gamma,(a,b)}^{\alpha,\beta}\left[q, u, p^{\alpha,\beta}\right] = \frac{1}{\Gamma(\alpha)} \int_a^b \left[\frac{\mathcal{H}^{\alpha,\beta} - p^{\alpha,\beta}(\tau) D_\gamma^{\alpha,\beta} q(\tau)}{(t - \tau)^{\alpha-1}} \right] (t - \tau)^{\alpha-1} d\tau,$$

where $\mathcal{H}^{\alpha,\beta} = \mathcal{H}^{\alpha,\beta}\left(u(\tau), q(\tau), p^{\alpha,\beta}(\tau), \tau\right)$. The Euler–Lagrange equation with respect to q gives the second equation of the fractional Hamiltonian system (4.74); the Euler–Lagrange equation with respect to u gives the fractional stationary condition (4.75); and, finally, the Euler–Lagrange equation with respect to $p^{\alpha,\beta}$ gives the first equation of (4.74). $\square$

Remark 4.18. For the fractional problem of the calculus of variations (4.69) one has

$$\mathcal{H}^{\alpha,\beta}\left(u, q, p^{\alpha,\beta}, \tau\right) = L(u, q, \tau)(t - \tau)^{\alpha-1} + p^{\alpha,\beta} u.$$

The fractional stationary condition (4.75) reduces to

$$p^{\alpha,\beta}(\tau) = -\frac{\partial L}{\partial u}(u(\tau), q(\tau), \tau)(t - \tau)^{\alpha-1} ; \qquad (4.76)$$

the first equation in (4.74) to

$$u(\tau) = D_\gamma^{\alpha,\beta} q(\tau) ; \qquad (4.77)$$

while the second equation in (4.74) takes the form

$$D_{-\gamma;\tau}^{\beta,\alpha} p^{\alpha,\beta}(\tau) = -\frac{\partial L}{\partial q}(u(\tau), q(\tau), \tau)(t - \tau)^{\alpha-1} . \qquad (4.78)$$

Using (4.76) and (4.77) in (4.78) we arrive at:

$$D_{-\gamma;\tau}^{\beta,\alpha} \left[\frac{\partial L}{\partial u}\left(D_\gamma^{\alpha,\beta} q(\tau), q(\tau), \tau\right)(t - \tau)^{\alpha-1} \right]$$

$$= \frac{\partial L}{\partial q}\left(D_\gamma^{\alpha,\beta} q(\tau), q(\tau), \tau\right)(t - \tau)^{\alpha-1} . \qquad (4.79)$$

Simple calculations show that (4.79) is equivalent to (4.70), that is, Theorem 4.10 is a generalization of Theorem 4.9 to the fractional optimal control problem (4.72).

Remark 4.19. Let us define the Poisson bracket of two dynamical quantities f and g, with respect to coordinates q and fractional momentum $p^{\alpha,\beta}$, by

$$\{f, g\} = \frac{\partial f}{\partial p^{\alpha,\beta}} \cdot \frac{\partial g}{\partial q} - \frac{\partial f}{\partial q} \cdot \frac{\partial g}{\partial p^{\alpha,\beta}} .$$

The fractional Hamiltonian system (4.74) can be written in the following form:

$$\begin{cases} D^{\alpha,\beta}_{\gamma} q\left(\tau\right) = \left\{ \mathcal{H}^{\alpha,\beta}, q \right\}, \\ D^{\beta,\alpha}_{-\gamma;\tau} p^{\alpha,\beta}\left(\tau\right) = \left\{ \mathcal{H}^{\alpha,\beta}, p^{\alpha,\beta} \right\}. \end{cases}$$

In classical mechanics, constants of motion are derived from the first integrals of the Euler–Lagrange equations. In the fractional case, as we saw, it is necessary to change the definition of constant of motion in a proper way. Here, in order to account the presence of the fractional decaying friction force $F^{\alpha,\beta}_{\gamma,\tau}$ (4.71), we propose the following definition of the fractional constant of motion.

Definition 4.11. *We say that a function C of τ is a fractional constant of motion if, and only if, $D^{\beta,\alpha}_{-\gamma;\tau} C = 0$.*

Corollary 4.3. *If L and φ do not depend on q, then it follows from the fractional Hamiltonian system (cf. Theorem 4.10) that $D^{\beta,\alpha}_{-\gamma;\tau} p^{\alpha,\beta}\left(\tau\right) = 0$, i.e., $p^{\alpha,\beta}$ is a fractional constant of motion.*

Definition 4.12 (Fractional momentum of order (α, β)).
Associated with an (α, β) fractional action-like variational problem (4.69) we define the fractional momentum by (4.76)–(4.77):

$$p^{\alpha,\beta}\left(\tau\right) = -\frac{\partial L}{\partial \dot{q}}\left(D^{\alpha,\beta}_{\gamma} q\left(\tau\right), q\left(\tau\right), \tau \right)\left(t - \tau\right)^{\alpha-1}.$$

Corollary 4.4. *For the fractional problem of the calculus of variations (4.69) with $\frac{\partial L}{\partial q} = 0$, the fractional momentum of order (α, β) is a fractional constant of motion.*

4.2.4 Double-Weighted Fractional Variational Principles

We denote by M a smooth n-dimensional manifold; the admissible paths are smooth functions $q : \Omega \subset \mathbb{R}^2 \to M$ satisfying fixed Dirichlet boundary conditions on $\partial\Omega$; the Lagrangian function $(q_x, q_y, q, x, y) \to L(q_x, q_y, q, x, y)$ is supposed to be sufficiently smooth with respect to all its arguments; α and β are two real parameters taking values on the interval $(0, 1)$.

Definition 4.13. *The double-weighted fractional action integral is defined by*

$$\frac{1}{\Gamma\left(\alpha\right)\Gamma\left(\beta\right)} \iint_{\Omega(\xi,\lambda)} L(D^{\alpha,\delta}_{\gamma;x} q, D^{\beta,\chi}_{\gamma;y} q, q, x, y)\left(\xi - x\right)^{\alpha-1}\left(\lambda - y\right)^{\beta-1} dxdy,$$

$$(4.80)$$

where ξ and λ are the observer times, $(\xi, \lambda) \in \Omega$; x and y are the intrinsic times, $(x, y) \in \Omega(\xi, \lambda) \subseteq \Omega$, $x \neq \xi$ and $y \neq \lambda$; $q = q(x, y)$; $D_{\gamma;x}^{\delta,\alpha}$ and $D_{\gamma;y}^{\chi,\beta}$ are the fractional derivative operators (see Definition 4.7), respectively of order (δ, α) with respect to x and of order (χ, β) with respect to y. We denote (4.80) by $S_{\gamma}^{\alpha,\beta,\delta,\chi}[q](\xi, \lambda)$.

The primary objective is to find functions $q = q(x, y)$ that make the fractional action $S_{\gamma}^{\alpha,\beta,\delta,\chi}[q](\xi, \lambda)$ stationary for every $(\xi, \lambda) \in \Omega$.

Theorem 4.11. *Given a smooth Lagrangian*

$$(q_x, q_y, q, x, y) \to L(q_x, q_y, q, x, y),$$

if $q = q(x, y)$ makes the fractional action $S_{\gamma}^{\alpha,\beta,\delta,\chi}[q](\xi, \lambda)$ defined by (4.80) stationary for every $(\xi, \lambda) \in \Omega$, then the following double-weighted Euler–Lagrange equation holds for every $(x, y) \in \Omega(\xi, \lambda)$:

$$D_{-\gamma;x}^{\delta,\alpha}\left(\frac{\partial L(D_{\gamma;x}^{\alpha,\delta}q, D_{\gamma;y}^{\beta,\chi}q, q, x, y)}{\partial q_x}\right) + D_{-\gamma;y}^{\chi,\beta}\left(\frac{\partial L(D_{\gamma;x}^{\alpha,\delta}q, D_{\gamma;y}^{\beta,\chi}q, q, x, y)}{\partial q_y}\right)$$

$$+ \frac{1-\alpha}{\xi-x}\left(\frac{\partial L(D_{\gamma;x}^{\alpha,\delta}q, D_{\gamma;y}^{\beta,\chi}q, q, x, y)}{\partial q_x}\right) + \frac{1-\beta}{\lambda-y}\left(\frac{\partial L(D_{\gamma;x}^{\alpha,\delta}q, D_{\gamma;y}^{\beta,\chi}q, q, x, y)}{\partial q_y}\right)$$

$$- \frac{\partial L(D_{\gamma;x}^{\alpha,\delta}q, D_{\gamma;y}^{\beta,\chi}q, q, x, y)}{\partial q} = 0. \quad (4.81)$$

Proof. Let q be a stationary solution, $h \ll 1$ a small real parameter, and $w(x, y)$ an admissible variation, i.e., an arbitrary smooth function satisfying $w(x, y) = 0$ for all $(x, y) \in \partial\Omega$ so that $q + hw$ satisfies the given Dirichlet boundary conditions for all h. The fractional action $S_{\gamma}^{\alpha,\beta,\delta,\chi}[q + hw]$ can be written as

$$\frac{1}{\Gamma(\alpha)\Gamma(\beta)}\iint_{\Omega(\xi,\lambda)} L(D_{\gamma;x}^{\alpha,\delta}q + hD_{\gamma;x}^{\alpha,\delta}w, D_{\gamma;y}^{\beta,\chi}q + hD_{\gamma;y}^{\beta,\chi}w, q + hw, x, y)$$

$$\times (\xi-x)^{\alpha-1}(\lambda-y)^{\beta-1}\,dxdy,$$

and the stationary condition $\frac{d}{dh}S_{\gamma}^{\alpha,\beta,\delta,\chi}[q + hw]\big|_{h=0} = 0$ gives

$$\frac{1}{\Gamma(\alpha)\Gamma(\beta)}\iint_{\Omega(\xi,\lambda)}\left(w\frac{\partial L}{\partial q} + D_{\gamma;x}^{\alpha,\delta}w\frac{\partial L}{\partial q_x} + D_{\gamma;y}^{\beta,\chi}w\frac{\partial L}{\partial q_y}\right)$$

$$\times (\xi-x)^{\alpha-1}(\lambda-y)^{\beta-1}\,dxdy = 0, \quad (4.82)$$

where the partial derivatives of function $(q_x, q_y, q, x, y) \to L(q_x, q_y, q, x, y)$ are evaluated at $(D^{\alpha,\delta}_{\gamma;x} q(x,y), D^{\beta,\chi}_{\gamma;y} q(x,y), q(x,y), x, y)$. Using integration by parts and Green's theorem, we know that

$$\iint_{\Omega(\xi,\lambda)} \left(\frac{\partial P}{\partial \xi} G_1 + \frac{\partial P}{\partial \lambda} G_2 \right) d\bar{\xi} d\bar{\lambda}$$

$$= \oint_{\partial \Omega} P \left(-G_2 d\bar{\xi} + G_1 d\bar{\lambda} \right) - \iint_{\Omega(\xi,\lambda)} \left(P \left(\frac{\partial G_1}{\partial \xi} + \frac{\partial G_2}{\partial \lambda} \right) \right) d\bar{\xi} d\bar{\lambda}$$

for any smooth functions G_1 and G_2, where

$$\Gamma(1+\alpha)\bar{\xi} = \xi^{\alpha} - (\xi - x)^{\alpha},$$
$$\Gamma(1+\alpha)\bar{\lambda} = \lambda^{\beta} - (\lambda - y)^{\beta}.$$

We conclude from (4.82) that

$$0 = -\frac{1}{\Gamma(\alpha)\Gamma(\beta)} \iint_{\Omega(\xi,\lambda)} w \left[(\xi - x)^{\alpha-1} (\lambda - y)^{\beta-1} \right.$$

$$\times \left(D^{\alpha,\delta}_{\gamma;x} \left(\frac{\partial L}{\partial q_x} \right) + D^{\beta,\chi}_{\gamma;y} \left(\frac{\partial L}{\partial q_y} \right) \right)$$

$$+ (1-\alpha) \left(\frac{\partial L}{\partial q_x} \right) (\xi - x)^{\alpha-2} (\lambda - y)^{\beta-1}$$

$$+ (1-\beta) \left(\frac{\partial L}{\partial q_y} \right) (\xi - x)^{\alpha-1} (\lambda - y)^{\beta-2} - \frac{\partial L}{\partial q} (\xi - x)^{\alpha-1} (\lambda - y)^{\beta-1} \right] dx dy$$

and then, because of the arbitrariness of $w(x,y)$ inside $\Omega(\xi,\lambda)$, it follows (4.81), which is the 2D-FELE. $\square$

We expect that fractional variational problems involving multiple integrals may have important consequences in mechanical problems involving dissipative systems with infinitely many degrees of freedom.

4.2.5 *N-Weighted Fractional Variational Principles*

All the arguments of Section 4.2.4 can be repeated, *mutatis mutandis*, to the N-dimensional situation when the admissible paths are smooth functions $q : \Omega \subset \mathbb{R}^N \to M$ satisfying given Dirichlet boundary conditions on $\partial \Omega$.

Definition 4.14. *Consider a smooth manifold* M *and let*

$$(q_{x_1}, \ldots, q_{x_N}, q, x_1, \ldots, x_N) \to L(q_{x_1}, \ldots, q_{x_N}, q, x_1, \ldots, x_N)$$

be a sufficiently smooth Lagrangian function. The N*-weighted fractional functional is defined by*

$$S_\gamma^{\alpha,\delta}[q](\xi) = \frac{1}{\displaystyle\prod_{i=1}^N \Gamma(\alpha_i)} \int \cdots \int_{\Omega(\xi)} L\left(\nabla_\gamma^{\alpha,\delta} q(x), q(x), x\right) \prod_{i=1}^N (\xi_i - x_i)^{\alpha_i - 1} dx,$$

where $x = (x_1, \ldots, x_N)$ *is the intrinsic time vector,* $\xi = (\xi_1, \ldots, \xi_N) \in \Omega$ *the observer time vector,* $x \in \Omega(\xi) \subseteq \Omega$ *with* $x_i \neq \xi_i$ *(i = 1, \ldots, N), dx =* $dx_1 \cdots dx_N$, $\alpha = (\alpha_1, \ldots, \alpha_N)$, $\delta = (\delta_1, \ldots, \delta_N)$, $0 < \alpha_i < 1$ *(i = 1, \ldots N),* *and* $\nabla_\gamma^{\alpha,\delta} = (D_{\gamma;x_1}^{\alpha_1,\delta_1}, \ldots, D_{\gamma;x_N}^{\alpha_N,\delta_N})$.

Theorem 4.12. *The* N*-dimensional Euler–Lagrange equation associated with the fractional functional* $S_\gamma^{\alpha,\delta}[q](\xi)$, $\xi \in \Omega$, *is given by*

$$\sum_{i=1}^N \left[D_{-\gamma;x_i}^{\delta_i,\alpha_i} \left(\frac{\partial L}{\partial q_{x_i}} \right) + \frac{1 - \alpha_i}{\xi_i - x_i} \left(\frac{\partial L}{\partial q_{x_i}} \right) \right] - \frac{\partial L}{\partial q} = 0,$$

where all partial derivatives of L *are evaluated at* $\left(\nabla_\gamma^{\alpha,\delta} q(x), q(x), x \right)$, $x \in \Omega(\xi)$.

4.3   Jumarie's Approach

We now prove necessary optimality conditions, in the class of continuous functions, for variational problems defined with Jumarie's modified Riemann–Liouville derivative. The fractional basic problem of the calculus of variations with free boundary conditions is considered, as well as problems with isoperimetric and holonomic constraints. In Subsection 4.3.1 we state the assumptions, notations, and the results of the literature needed in the sequel. Subsection 4.3.2 reviews Jumarie's fractional Euler–Lagrange equations (Jumarie, 2009b). In Subsection 4.3.3 we consider the case when no boundary conditions are imposed on the problem, and we prove associated transversality (natural boundary) conditions; optimization with constraints (integral or not) are studied in Subsections 4.3.4 and 4.3.5. Necessary and sufficient optimality conditions for fractional problems of the calculus of variations with a Lagrangian depending on the free end-points are proved in Subsection 4.3.6. In Subsection 4.3.7 we consider fractional problems of the calculus of variations which are given by a composition of functionals (4.106). We obtain Euler–Lagrange equations and natural boundary conditions for the general problem (Theorem 4.23), which are then applied to the product (Corollary 4.6) and the quotient (Corollary 4.7). Finally we introduce a fractional theory of the calculus of variations for multiple integrals. Main results provide fractional versions of the theorems of Green and Gauss; fractional Euler–Lagrange equations; fractional natural boundary conditions; and necessary optimality conditions for fractional isoperimetric problems of calculus of variations with double integrals (Theorem 4.29), as well as a sufficient condition under appropriate convexity assumptions (Theorem 4.31) and fractional natural boundary conditions (Theorem 4.30). As an application, we discuss the fractional equation of motion of a vibrating string in Subsection 4.3.10.

4.3.1 *Jumarie's Riemann–Liouville Derivative*

Throughout $f : [0, 1] \to \mathbb{R}$ is a continuous function and α a real number on the interval $(0, 1)$. Jumarie's modified Riemann–Liouville fractional derivative is defined by

$$f^{(\alpha)}(x) = \frac{1}{\Gamma(1 - \alpha)} \frac{d}{dx} \int_0^x (x - t)^{-\alpha} (f(t) - f(0)) \, dt.$$

Definition 4.15. *Let $f : [a, b] \to \mathbb{R}$ be a continuous function. The Jumarie fractional derivative of f is defined by*

$$f^{(\alpha)}(t) := \frac{1}{\Gamma(-\alpha)} \int_0^t (t - \tau)^{-\alpha - 1} (f(\tau) - f(a)) \, d\tau, \quad \alpha < 0,$$

where $\Gamma(z) = \int_0^\infty t^{z-1} e^{-t} \, dt$. For positive α, one will set

$$f^{(\alpha)}(t) = (f^{(\alpha-1)}(t))' = \frac{1}{\Gamma(1 - \alpha)} \frac{d}{dt} \int_0^t (t - \tau)^{-\alpha} (f(\tau) - f(a)) \, d\tau,$$

for $0 < \alpha < 1$, and $f^{(\alpha)}(t) := (f^{(\alpha-n)}(t))^{(n)}$, $n \leq \alpha < n + 1$, $n \geq 1$.

If $f(0) = 0$, then $f^{(\alpha)}$ is equal to the Riemann–Liouville fractional derivative of f of order α. We remark that the fractional derivative of a constant is zero, as desired. Moreover, $f(0) = 0$ is no longer a necessary condition for the fractional derivative of f to be continuous on $[0, 1]$.

The Jumarie fractional derivative satisfies the following properties:

(a) The αth derivative of a constant is zero.

(b) If $0 < \alpha \leq 1$, then the Laplace transform of $f^{(\alpha)}$ is

$$\mathcal{L}\{f^{(\alpha)}(t)\} = s^{\alpha}\mathcal{L}\{f(t)\} - s^{\alpha-1}f(0).$$

(c) $(g(t)f(t))^{(\alpha)} = g^{(\alpha)}(t)f(t) + g(t)f^{(\alpha)}(t), \quad 0 < \alpha < 1.$

Example 4.3. Let $f(t) = t^{\gamma}$. Then $f^{(\alpha)}(x) = \Gamma(\gamma+1)\Gamma^{-1}(\gamma+1-\alpha)t^{\gamma-\alpha}$, where $0 < \alpha < 1$ and $\gamma > 0$.

Example 4.4. The solution of the fractional differential equation

$$x^{(\alpha)}(t) = c, \quad x(0) = x_0, \quad c = constant,$$

is

$$x(t) = \frac{c}{\alpha!}t^{\alpha} + x_0,$$

with the notation $\alpha! := \Gamma(1 + \alpha)$.

The $(dt)^{\alpha}$ integral of f is given by

$$\int_0^x f(t)(dt)^{\alpha} = \alpha \int_0^x (x - t)^{\alpha-1}f(t)dt.$$

A motivation of this definition is given in Jumarie, 2005. The integral with respect to $(dt)^{\alpha}$ is defined as the solution of the fractional differential equation

$$dy = f(x)(dx)^{\alpha}, \quad x \geq 0, \quad y(0) = y_0, \quad 0 < \alpha \leq 1 \qquad (4.83)$$

which is provided by the following result:

Lemma 4.3. *Let $f(t)$ denote a continuous function. The solution of the equation* (4.83) *is defined by the equality*

$$\int_0^t f(\tau)(d\tau)^{\alpha} = \alpha \int_0^t (t - \tau)^{\alpha-1}f(\tau)d\tau, \quad 0 < \alpha \leq 1.$$

Example 4.5. Let $f(t) = 1$. Then $\int_0^t (d\tau)^{\alpha} = t^{\alpha}$, $0 < \alpha \leq 1$.

Example 4.6. The solution of the fractional differential equation

$$x^{(\alpha)}(t) = f(t), \quad x(0) = x_0$$

is

$$x(t) = x_0 + \Gamma^{-1}(\alpha) \int_0^t (t - \tau)^{\alpha-1}f(\tau)d\tau.$$

One can easily generalize the previous definitions and results for functions with a domain $[a, b]$:

$$f^{(\alpha)}(t) = \frac{1}{\Gamma(1-\alpha)} \frac{d}{dt} \int_a^t (t-\tau)^{-\alpha} (f(\tau) - f(a)) \, d\tau$$

and

$$\int_a^t f(\tau)(d\tau)^\alpha = \alpha \int_a^t (t-\tau)^{\alpha-1} f(\tau) d\tau.$$

Remark 4.20. This type of fractional derivative and integral has found applications in some physical phenomena. The definition of the fractional derivative via difference reads

$$f^{(\alpha)}(x) = \lim_{h \downarrow 0} \frac{\triangle^\alpha f(x)}{h^\alpha}, \quad 0 < \alpha < 1,$$

and obviously this contributes some questions on the sign of h, as it is emphasized by the fractional Rolle's formula $f(x+h) \cong f(x) + h^\alpha f^{(\alpha)}(x)$. In a first approach, in a realm of physics, when h denotes time, then this feature could show the irreversibility of time. The fractional derivative is quite suitable to describe dynamics evolving in space which exhibit coarse-grained phenomenon. When the point in this space is not infinitely thin but rather a thickness, then it would be better to replace dx by $(dx)^\alpha$, $0 < \alpha < 1$, where α characterizes the grade of the phenomenon. The fractal feature of the space is transported on time, and so both space and time are fractal. Thus, the increment of time of the dynamics of the system is not dx but $(dx)^\alpha$. For more on the subject see, e.g., Almeida, Malinowska and Torres, 2010; Filatova, Grzywaczewski and Osmolovskii, 2010; Jumarie, 2009b, 2010a,b.

Our results make use of the formula of integration by parts for the $(dx)^\alpha$ integral. This formula follows from the fractional Leibniz rule and the fractional Barrow's formula.

Theorem 4.13 (Fractional Leibniz rule (Jumarie, 2008a)).
If f and g are two continuous functions on $[0, 1]$, then

$$(f(x)g(x))^{(\alpha)} = (f(x))^{(\alpha)} g(x) + f(x)(g(x))^{(\alpha)}. \tag{4.84}$$

Kolwankar obtained the same formula (4.84) by using an approach on Cantor space (Kolwankar, 2004).

Theorem 4.14 (Fractional Barrow's formula (Jumarie, 2009a)).
For a continuous function f, we have

$$\int_0^x f^{(\alpha)}(t)(dt)^{(\alpha)} = \alpha!(f(x) - f(0)),$$

where $\alpha! = \Gamma(1 + \alpha)$.

From Theorems 4.13 and 4.14 we deduce the following formula of integration by parts:

$$\int_0^1 u^{(\alpha)}(x)v(x)\,(dx)^\alpha = \int_0^1 (u(x)v(x))^{(\alpha)}\,(dx)^\alpha - \int_0^1 u(x)v^{(\alpha)}(x)\,(dx)^\alpha$$
$$= \alpha![u(x)v(x)]_0^1 - \int_0^1 u(x)v^{(\alpha)}(x)\,(dx)^\alpha.$$

More generally, one has

$$\int_a^b u^{(\alpha)}(t)v(t)\,(dt)^\alpha = \alpha![u(t)v(t)]_a^b - \int_a^b u(t)v^{(\alpha)}(t)\,(dt)^\alpha. \tag{4.85}$$

It has been proved that the fractional Taylor series holds for nondifferentiable functions. See, for instance, Jumarie, 2008b. Another approach is to check that this formula holds for the Mittag–Leffler function, and then to consider functions which can be approximated by the former. The first term of this series is the Rolle's fractional formula which has been obtained by Kolwankar and Jumarie and provides the equality $d^\alpha x(t) = \alpha! dx(t)$.

It is a simple exercise to verify that the fundamental lemma of the calculus of variations is valid for the $(dx)^\alpha$ integral.

Lemma 4.4. *Let g be a continuous function and assume that*

$$\int_a^b g(x)h(x)\,(dx)^\alpha = 0$$

for every continuous function h satisfying

$$h(a) = h(b) = 0.$$

Then $g \equiv 0$.

4.3.2   The Euler–Lagrange Equations

Let $0 < \alpha < 1$. Consider functionals

$$\mathcal{J}(y) = \int_0^1 L\left(x, y(x), y^{(\alpha)}(x)\right)(dx)^\alpha \tag{4.86}$$

defined on the set of continuous curves $y : [0, 1] \to \mathbb{R}$, where $L(\cdot, \cdot, \cdot)$ has continuous partial derivatives with respect to the second and third variable. Jumarie has addressed (in Jumarie, 2009b) the basic problem of calculus of variations: to minimize (or maximize) $\mathcal{J}$, when restricted to the class of continuous curves satisfying prescribed boundary conditions $y(0) = y_0$ and $y(1) = y_1$. Let us denote this problem by (P). A necessary condition for problem (P) is given by the next result.

Theorem 4.15. *If y is a solution to the basic fractional problem of the calculus of variations (P), then*

$$\frac{\partial L}{\partial y} - \frac{d^\alpha}{dx^\alpha} \frac{\partial L}{\partial y^{(\alpha)}} = 0 \tag{4.87}$$

is satisfied along y, for all $x \in [0, 1]$.

Definition 4.16. *A curve that satisfies equation (4.87) for all $x \in [0, 1]$ is said to be an extremal for $\mathcal{J}$.*

Example 4.7. Consider the following problem:

$$\mathcal{J}(y) = \int_0^1 \left[\frac{x^\alpha}{\Gamma(\alpha + 1)} (y^{(\alpha)})^2 - 2x^\alpha y^{(\alpha)} \right]^2 (dx)^\alpha \longrightarrow \text{extremize}$$

subject to the boundary conditions

$$y(0) = 1 \text{ and } y(1) = 2.$$

The Euler–Lagrange equation associated to this problem is

$$-\frac{d^\alpha}{dx^\alpha} \left(2 \left[\frac{x^\alpha}{\Gamma(\alpha + 1)} (y^{(\alpha)})^2 - 2x^\alpha y^{(\alpha)} \right] \cdot \left[\frac{2x^\alpha}{\Gamma(\alpha + 1)} y^{(\alpha)} - 2x^\alpha \right] \right) = 0. \tag{4.88}$$

Let $y = x^\alpha + 1$. Since $y^{(\alpha)} = \Gamma(\alpha + 1)$, it follows that y is a solution of (4.88). We remark that the extremal curve is not differentiable in $[0, 1]$.

Remark 4.21. When $\alpha = 1$, we obtain the variational functional

$$\mathcal{J}(y) = \int_0^1 \left[x(y')^2 - 2xy' \right]^2 dx.$$

It is easy to verify that $y = x + 1$ satisfies the (standard) Euler–Lagrange equation.

We now present the Euler–Lagrange equation for functionals containing several dependent variables.

Theorem 4.16.   *Consider a functional $\mathcal{J}$, defined on the set of curves satisfying the boundary conditions $\boldsymbol{y}(0) = \boldsymbol{y}_0$ and $\boldsymbol{y}(1) = \boldsymbol{y}_1$, of the form*

$$\mathcal{J}(\boldsymbol{y}) = \int_0^1 L\left(x, \boldsymbol{y}(x), \boldsymbol{y}^{(\alpha)}(x)\right) (dx)^\alpha,$$

where $\boldsymbol{y} = (y_1, \dots, y_n)$, $\boldsymbol{y}^{(\alpha)} = (y_1^{(\alpha)}, \dots, y_n^{(\alpha)})$, and y_k, $k = 1, \dots, n$, are continuous real valued functions defined on $[0,1]$. Let $\boldsymbol{y}$ be an extremizer of $\mathcal{J}$. Then,

$$\frac{\partial L}{\partial y_k}\left(x, \boldsymbol{y}(x), \boldsymbol{y}^{(\alpha)}(x)\right) - \frac{d^\alpha}{dx^\alpha}\frac{\partial L}{\partial y_k^{(\alpha)}}\left(x, \boldsymbol{y}(x), \boldsymbol{y}^{(\alpha)}(x)\right) = 0,$$

$k = 1, \dots, n$, for all $x \in [0,1]$.

4.3.3   Natural Boundary Conditions

The problem is stated as follows. Given a functional

$$\mathcal{J}(y) = \int_0^1 L\left(x, y(x), y^{(\alpha)}(x)\right) (dx)^\alpha,$$

where the Lagrangian $L(\cdot, \cdot, \cdot)$ has continuous partial derivatives with respect to the second and third variables, determine continuous curves $y : [0,1] \to \mathbb{R}$ such that $\mathcal{J}$ has an extremum at y. Note that no boundary conditions are now imposed.

Theorem 4.17.   *Let y be an extremizer for $\mathcal{J}$. Then y satisfies the Euler–Lagrange equation*

$$\frac{\partial L}{\partial y} - \frac{d^\alpha}{dx^\alpha}\frac{\partial L}{\partial y^{(\alpha)}} = 0 \tag{4.89}$$

for all $x \in [0,1]$ and the natural boundary conditions

$$\left.\frac{\partial L}{\partial y^{(\alpha)}}\right|_{x=0} = 0 \quad and \quad \left.\frac{\partial L}{\partial y^{(\alpha)}}\right|_{x=1} = 0. \tag{4.90}$$

Proof.   Let h be any continuous curve and let $j(\epsilon) = \mathcal{J}(y + \epsilon h)$. It follows that

$$0 = \int_0^1 \left(\frac{\partial L}{\partial y} \cdot h(x) + \frac{\partial L}{\partial y^{(\alpha)}} \cdot h^{(\alpha)}(x)\right) (dx)^\alpha$$

$$= \int_0^1 \left(\frac{\partial L}{\partial y} - \frac{d^\alpha}{dx^\alpha}\frac{\partial L}{\partial y^{(\alpha)}}\right) \cdot h(x) \, (dx)^\alpha + \alpha! \left[\frac{\partial L}{\partial y^{(\alpha)}} h(x)\right]_{x=0}^{x=1}.$$

If we choose curves such that $h(0) = h(1) = 0$, we deduce by Lemma 4.4 the Euler–Lagrange equation (4.89). Then condition

$$\left[\frac{\partial L}{\partial y^{(\alpha)}} h(x) \right]_{x=0}^{x=1} = 0$$

must be verified. Picking curves such that $h(0) = 0$ and $h(1) \neq 0$, and others such that $h(1) = 0$ and $h(0) \neq 0$, we deduce the natural boundary conditions (4.90). □

If one of the end-points is specified, say $y(0) = y_0$, then the necessary conditions become

$$\frac{\partial L}{\partial y} - \frac{d^\alpha}{dx^\alpha} \frac{\partial L}{\partial y^{(\alpha)}} = 0$$

and

$$\left. \frac{\partial L}{\partial y^{(\alpha)}} \right|_{x=1} = 0.$$

Example 4.8. Let $\mathcal{J}$ be given by the expression

$$\mathcal{J}(y) = \int_0^1 \sqrt{1 + y^{(\alpha)}(x)^2} \, (dx)^\alpha.$$

The Euler–Lagrange equation associated to this problem is

$$\frac{d^\alpha}{dx^\alpha} \frac{y^{(\alpha)}(x)}{\sqrt{1 + y^{(\alpha)}(x)^2}} = 0$$

and the natural boundary conditions are

$$\left. \frac{y^{(\alpha)}(x)}{\sqrt{1 + y^{(\alpha)}(x)^2}} \right|_{x=1} = 0 \quad \text{and} \quad \left. \frac{y^{(\alpha)}(x)}{\sqrt{1 + y^{(\alpha)}(x)^2}} \right|_{x=0} = 0.$$

Since $y^{(\alpha)} = 0$ if y is a constant function, we have that any constant curve is a solution to this problem.

4.3.4 *The Isoperimetric Problem*

The study of isoperimetric problems is an important area inside the calculus of variations. One wants to find the extremizers of a given functional, when restricted to a prescribed integral constraint. Problems of this type have found many applications in differential geometry, discrete and convex geometry, probability, Banach space theory, and multiobjective optimization (see Almeida and Torres, 2009c,d; Malinowska and Torres, 2009 and references therein). We introduce the isoperimetric fractional problem as follows: to maximize or minimize the functional

$$\mathcal{J}(y) = \int_0^1 L(x, y(x), y^{(\alpha)}(x)) \, (dx)^\alpha$$

when restricted to the conditions

$$\mathcal{G}(y) = \int_0^1 f(x, y(x), y^{(\alpha)}(x)) \, (dx)^\alpha = K, \quad K \in \mathbb{R}, \qquad (4.91)$$

and

$$y(0) = y_0 \text{ and } y(1) = y_1. \qquad (4.92)$$

Similarly as before, we assume that $L(\cdot, \cdot, \cdot)$ and $f(\cdot, \cdot, \cdot)$ have continuous partial derivatives with respect to the second and third variables.

Theorem 4.18. *Let y be an extremizer of $\mathcal{J}$ restricted to the set of curves that satisfy conditions (4.91) and (4.92). If y is not an extremal for $\mathcal{G}$, then there exists a constant λ such that the curve y satisfies the equation*

$$\frac{\partial F}{\partial y} - \frac{d^\alpha}{dx^\alpha} \frac{\partial F}{\partial y^{(\alpha)}} = 0$$

for all $x \in [0, 1]$, where $F = L - \lambda f$.

Proof. Can be done similarly to the proof of Theorem 3.49 and can be found in Almeida and Torres, 2011b. □

4.3.5 Holonomic Constraints

In Subsection 4.3.4 the subsidiary conditions that the functions must satisfy are given by integral functionals. We now consider a different type of problem: find functions y_1 and y_2 for which the functional

$$\mathcal{J}(y_1, y_2) = \int_0^1 L\left(x, y_1(x), y_2(x), y_1^{(\alpha)}(x), y_2^{(\alpha)}(x)\right) \, (dx)^\alpha \qquad (4.93)$$

has an extremum, where the admissible functions satisfy the boundary conditions

$$(y_1(0), y_2(0)) = (y_1^0, y_2^0) \text{ and } (y_1(1), y_2(1)) = (y_1^1, y_2^1), \qquad (4.94)$$

and the subsidiary holonomic condition

$$g(x, y_1(x), y_2(x)) = 0. \qquad (4.95)$$

Theorem 4.19. *Given a functional $\mathcal{J}$ as in (4.93), defined on the set of curves that satisfy the boundary conditions (4.94) and lie on the surface (4.95), let (y_1, y_2) be an extremizer for $\mathcal{J}$. If $\partial g/\partial y_2 \neq 0$ for all $x \in [0, 1]$, then there exists a continuous function $\lambda(x)$ such that (y_1, y_2) satisfy the Euler–Lagrange equations*

$$\frac{\partial F}{\partial y_k} - \frac{d^\alpha}{dx^\alpha} \frac{\partial F}{\partial y_k^{(\alpha)}} = 0\,, \qquad (4.96)$$

$k = 1, 2$, for all $x \in [0, 1]$, where $F = L - \lambda g$.

Proof. The proof is similar to the proof of Theorem 3.51 and can be found in Almeida and Torres, 2011b. □

We now state our previous result in its general form.

Theorem 4.20. *Let $\mathcal{J}$ be given by*

$$\mathcal{J}(\boldsymbol{y}) = \int_0^1 L(x, \boldsymbol{y}(x), \boldsymbol{y}^{(\alpha)}(x))\,(dx)^\alpha\,,$$

where $\boldsymbol{y} = (y_1, \ldots, y_n)$ and $\boldsymbol{y}^{(\alpha)} = (y_1^{(\alpha)}, \ldots, y_n^{(\alpha)})$, such that y_k, $k = 1, \ldots, n$, are continuous functions defined on the set of curves that satisfy the boundary conditions $\boldsymbol{y}(0) = \boldsymbol{y_0}$ and $\boldsymbol{y}(1) = \boldsymbol{y_1}$ and satisfy the constraint $g(x, \boldsymbol{y}) = 0$. If $\boldsymbol{y}$ is an extremizer for $\mathcal{J}$, and if $\partial g/\partial y_n \neq 0$ for all $x \in [0, 1]$, then there exists a continuous function $\lambda(x)$ such that $\boldsymbol{y}$ satisfies the Euler–Lagrange equations

$$\frac{\partial F}{\partial y_k} - \frac{d^\alpha}{dx^\alpha} \frac{\partial F}{\partial y_k^{(\alpha)}} = 0\,, \quad k = 1, \ldots, n\,,$$

for all $x \in [0, 1]$, where $F = L - \lambda g$.

4.3.6 Lagrangians Depending on the Free End-Points

Now, let us consider the functional defined by

$$\mathcal{J}(y) = \int_a^b L(x, y(x), y^{(\alpha)}(x), y(a), y(b))\,(dx)^\alpha\,,$$

where $L \in C^1([a, b] \times \mathbb{R}^4; \mathbb{R})$ and $x \to \partial_3 L(t)$ has continuous α-derivative.

The fractional problem of the calculus of variations under consideration has the form

$$\mathcal{J}(y) \longrightarrow \text{extr}$$

$$(y(a) = y_a), \quad (y(b) = y_b) \tag{4.97}$$

$$y(\cdot) \in C^0.$$

Using parentheses around the end-point conditions means that the conditions may or may not be present. As before, we denote by $\partial_i L$, $i = 1, \ldots, 5$, the partial derivative of function L with respect to its ith argument.

Necessary conditions

The next theorem gives the necessary optimality conditions for the problem (4.97).

Theorem 4.21. *Let y be an extremizer to problem* (4.97). *Then y satisfies the fractional Euler–Lagrange equation*

$$\partial_2 L(x, y(x), y^{(\alpha)}(x), y(a), y(b)) = \frac{d^\alpha}{dx^\alpha} \partial_3 L(x, y(x), y^{(\alpha)}(x), y(a), y(b)) \tag{4.98}$$

for all $x \in [a, b]$. Moreover, if $y(a)$ is not specified, then

$$\int_a^b \partial_4 L(x, y(x), y^{(\alpha)}(x), y(a), y(b)) (dx)^\alpha$$

$$= \alpha! \partial_3 L(a, y(a), y^{(\alpha)}(a), y(a), y(b)) \tag{4.99}$$

if $y(b)$ is not specified, then

$$\int_a^b \partial_5 L(x, y(x), y^{(\alpha)}(x), y(a), y(b)) (dx)^\alpha$$

$$= -\alpha! \partial_3 L(b, y(b), y^{(\alpha)}(b), y(a), y(b)). \tag{4.100}$$

Proof. Similar to that of Theorem 3.3 and can be found in Malinowska, 2012a. □

In the case L does not depend on $y(a)$ and $y(b)$, by Theorem 4.21 we obtain Theorem 4.15.

Sufficient conditions

Now we present sufficient conditions for optimality.

Theorem 4.22. *Let $L(\underline{x}, y, z, t, u)$ be a jointly convex (concave) in (y, z, t, u). If y_0 satisfies conditions (4.98)–(4.100), then y_0 is a global minimizer (maximizer) to problem (4.97).*

Proof. Similar to that of Theorem 3.12 and can be found in Malinowska, 2012a. □

Examples

We shall provide examples in order to illustrate our results.

Example 4.9. Consider the following problem:

$$\mathcal{J}(y) = \int_0^1 \left\{ \left[\frac{x^\alpha}{\Gamma(\alpha+1)} (y^{(\alpha)})^2 - 2x^\alpha y^{(\alpha)} \right]^2 \right. $$
$$\left. + (y(0) - 1)^2 + (y(1) - 2)^2 \right\} (dx)^\alpha \longrightarrow \text{extr.}$$

The Euler–Lagrange equation associated to this problem is

$$\frac{d^\alpha}{dx^\alpha} \left(2 \left[\frac{x^\alpha}{\Gamma(\alpha+1)} (y^{(\alpha)})^2 - 2x^\alpha y^{(\alpha)} \right] \cdot \left[\frac{2x^\alpha}{\Gamma(\alpha+1)} y^{(\alpha)} - 2x^\alpha \right] \right) = 0. \tag{4.101}$$

Let $y = x^\alpha + b$, where $b \in \mathbb{R}$. Since $y^{(\alpha)} = \Gamma(\alpha + 1)$, it follows that y is a solution of (4.101). In order to determine b we use the generalized natural boundary conditions (4.99)–(4.100), which can be written for this problem as

$$\int_0^1 (y(0) - 1)(dx)^\alpha = 0,$$

$$\int_0^1 (y(1) - 2)(dx)^\alpha = 0.$$

Hence, $\tilde{y} = x^\alpha + 1$ is a candidate solution. We remark that the $\tilde{y}$ is not differentiable in $[0, 1]$.

Example 4.10. Consider the following problem:

$$\mathcal{J}(y) = \int_0^1 \left[(y^{(\alpha)}(x))^2 + \gamma y^2(0) + \lambda(y(1) - 1)^2 \right] (dx)^\alpha \longrightarrow \min, \tag{4.102}$$

where $\gamma, \lambda \in \mathbb{R}^+$. For this problem, the fractional Euler–Lagrange equation and the generalized natural boundary conditions (see Theorem 4.21) are given, respectively, as

$$2\frac{d^\alpha}{dx^\alpha} y^{(\alpha)}(x) = 0, \tag{4.103}$$

$$\int_0^1 \gamma y(0)(dx)^\alpha = \alpha! y^{(\alpha)}(0), \tag{4.104}$$

$$\int_0^1 \lambda(y(1) - 1)(dx)^\alpha = -\alpha! y^{(\alpha)}(1). \tag{4.105}$$

Solving equations (4.103)–(4.105) we obtain that

$$\bar{y}(x) = \frac{\gamma \lambda \alpha!}{\gamma \lambda + (\alpha!)^2(\lambda + \gamma)} x^\alpha + \frac{(\alpha!)^2 \lambda}{\gamma \lambda + (\alpha!)^2(\lambda + \gamma)}$$

is a candidate for minimizer. Observe that problem (4.102) satisfies assumptions of Theorem 4.22. Therefore $\bar{y}$ is a global minimizer to this problem. We note that when α goes to 1 problem (4.102) tends to

$$\mathcal{K}(y) = \int_0^1 \left[(y'(x))^2 + \gamma y^2(0) + \lambda(y(1) - 1)^2 \right] dx \longrightarrow \min$$

with the solution

$$y(x) = \frac{\gamma \lambda}{\gamma \lambda + \lambda + \gamma} x + \frac{\lambda}{\gamma \lambda + \lambda + \gamma}.$$

4.3.7   Composition Functionals

The general (non-classical) problem of the fractional calculus of variations under our consideration consists of extremizing (i.e., minimizing or maximizing)

$$\mathcal{L}[x] = H\left(\int_a^b f_1(t, x(t), x^{(\alpha_1)}(t))(dt)^{\alpha_1}, \ldots, \int_a^b f_n((t, x(t), x^{(\alpha_n)}(t))(dt)^{\alpha_n} \right)$$

$$(x(a) = x_a) \quad (x(b) = x_b) \tag{4.106}$$

over all $x \in \mathbf{D}$ with

$$\mathbf{D} := \{ x \in C^0 : x^{(\alpha_i)}, i = 1, \ldots, n, \text{ exists and is continuous on } [a, b] \},$$

where $0 < \alpha_i < 1$, $i = 1, \ldots, n$. Using parentheses around the end-point conditions means that these conditions may or may not be present.

We assume that:

(a) the function $H : \mathbb{R}^n \to \mathbb{R}$ has continuous partial derivatives with respect to its arguments and we denote them by H'_i, $i = 1, \ldots, n$;

(b) functions $(t, y, v) \to f_i(t, y, v)$ from $[a, b] \times \mathbb{R}^2$ to $\mathbb{R}$, $i = 1, \ldots, n$, have partial continuous derivatives with respect to y, v for all $t \in [a, b]$ and we denote them by f_{iy}, f_{iv};

(c) f_i, $i = 1, \ldots, n$, and their partial derivatives are continuous in t for all $x \in \mathbf{D}$.

A function $x \in \mathbf{D}$ is said to be an admissible function provided that it satisfies the end-points conditions (if any is given). The following norm in $\mathbf{D}$ is considered:

$$\|x\| = \max_{t \in [a,b]} |x(t)| + \sum_{i=1}^{n} \max_{t \in [a,b]} |x^{(\alpha_i)}(t)|.$$

Definition 4.17. *An admissible function $\tilde{x}$ is said to be a weak local minimizer (resp. weak local maximizer) for (4.106) if there exists $\delta > 0$ such that $\mathcal{L}[\tilde{x}] \leq \mathcal{L}[x]$ (resp. $\mathcal{L}[\tilde{x}] \geq \mathcal{L}[x]$) for all admissible x with $\|x - \tilde{x}\| < \delta$.*

For simplicity of notation we introduce the operator $\langle x \rangle_i$, $i = 1, \ldots, n$, defined by

$$\langle x \rangle_i (t) = (t, x(t), x^{(\alpha_i)}(t)).$$

Then,

$$\mathcal{L}[x] = H \left(\int_a^b f_1 \langle x \rangle_1 (t)(dt)^{\alpha_1}, \ldots, \int_a^b f_n \langle x \rangle_n (t)(dt)^{\alpha_n} \right).$$

The next theorem gives necessary optimality conditions for problem (4.106).

Theorem 4.23. *If $\tilde{x}$ is a weak local solution to problem (4.106), then the Euler–Lagrange equation*

$$\sum_{i=1}^{n} \alpha_i H'_i(\mathcal{F}_1[\tilde{x}], \ldots, \mathcal{F}_n[\tilde{x}])(b - t)^{\alpha_i - 1} \left(f_{iy} \langle \tilde{x} \rangle_i (t) - f_{iv}^{(\alpha_i)} \langle \tilde{x} \rangle_i (t) \right) = 0$$

holds for all $t \in [a, b)$, where $\mathcal{F}_i[\tilde{x}] = \int_a^b f_i \langle \tilde{x} \rangle_i (t)(dt)^{\alpha_i}$, $i = 1, \ldots, n$. Moreover, if $x(a)$ is not specified, then

$$\sum_{i=1}^{n} \alpha_i! H'_i(\mathcal{F}_1[\tilde{x}], \ldots, \mathcal{F}_n[\tilde{x}]) f_{iv} \langle \tilde{x} \rangle_i (a) = 0 \, ; \qquad (4.107)$$

if $x(b)$ is not specified, then

$$\sum_{i=1}^{n} \alpha_i! H_i'(\mathcal{F}_1[\tilde{x}], \ldots, \mathcal{F}_n[\tilde{x}]) f_{iv} \langle \tilde{x} \rangle_i(b) = 0. \tag{4.108}$$

Proof.   Suppose that $\mathcal{L}[x]$ has a weak local extremum at $\tilde{x}$. For an admissible variation $h \in \mathbf{D}$ we define a function $\phi : \mathbb{R} \to \mathbb{R}$ by $\phi(\varepsilon) = \mathcal{L}[\tilde{x} + \varepsilon h]$. We do not require $h(a) = 0$ or $h(b) = 0$ in case $x(a)$ or $x(b)$, respectively, is free (it is possible that both are free). A necessary condition for $\tilde{x}$ to be an extremizer for $\mathcal{L}[x]$ is given by $\phi'(\varepsilon)|_{\varepsilon=0} = 0$. Using the chain rule to obtain the derivative of a composed function, we get

$$\phi'(\varepsilon)|_{\varepsilon=0}$$

$$= \sum_{i=1}^{n} H_i'(\mathcal{F}_1[\tilde{x}], \ldots, \mathcal{F}_n[\tilde{x}]) \int_a^b \left[f_{iy} \langle \tilde{x} \rangle_i(t) h(t) + f_{iv} \langle \tilde{x} \rangle_i(t) h^{(\alpha_i)}(t) \right] (dt)^{\alpha_i}.$$

Integration by parts (see equation (4.85)) of the second term of the integrands, gives

$$\int_a^b f_{iv} \langle \tilde{x} \rangle_i(t) h^{(\alpha_i)}(t) (dt)^{\alpha_i}$$

$$= [\alpha_i! f_{iv} \langle \tilde{x} \rangle_i(t) h(t)]_{t=a}^{t=b} - \int_a^b f_{iv}^{(\alpha_i)} \langle \tilde{x} \rangle_i(t) h(t) (dt)^{\alpha_i}.$$

The necessary condition $\phi'(\varepsilon)|_{\varepsilon=0} = 0$ can be written as

$$0 = \sum_{i=1}^{n} H_i'(\mathcal{F}_1[\tilde{x}], \ldots, \mathcal{F}_n[\tilde{x}]) \int_a^b \left(f_{iy} \langle \tilde{x} \rangle_i(t) - f_{iv}^{(\alpha_i)} \langle \tilde{x} \rangle_i(t) \right) h(t) (dt)^{\alpha_i}$$

$$+ \sum_{i=1}^{n} H_i'(\mathcal{F}_1[\tilde{x}], \ldots, \mathcal{F}_n[\tilde{x}]) [\alpha_i! f_{iv} \langle \tilde{x} \rangle_i(t) h(t)]_{t=a}^{t=b}.$$

Taking into account Lemma 4.3, we have

$$\int_a^b \sum_{i=1}^{n} \alpha_i H_i'(\mathcal{F}_1[\tilde{x}], \ldots, \mathcal{F}_n[\tilde{x}])(b-t)^{\alpha_i-1} \left(f_{iy} \langle \tilde{x} \rangle_i(t) - f_{iv}^{(\alpha_i)} \langle \tilde{x} \rangle_i(t) \right) h(t) dt$$

$$+ \sum_{i=1}^{n} H_i'(\mathcal{F}_1[\tilde{x}], \ldots, \mathcal{F}_n[\tilde{x}]) [\alpha_i! f_{iv} \langle \tilde{x} \rangle_i(t) h(t)]_{t=a}^{t=b} = 0. \tag{4.109}$$

In particular, equation (4.109) holds for all variations which are zero at both ends. For all such h's, the second term in (4.109) is zero and by the Dubois–Reymond Lemma (see, e.g., van Brunt, 2004), we have that

$$\sum_{i=1}^{n} \alpha_i H_i'(\mathcal{F}_1[\tilde{x}], \ldots, \mathcal{F}_n[\tilde{x}])(b-t)^{\alpha_i-1} \left(f_{iy} \langle \tilde{x} \rangle_i(t) - f_{iv}^{(\alpha_i)} \langle \tilde{x} \rangle_i(t) \right) = 0$$

$$\tag{4.110}$$

holds for all $t \in [a, b)$. Equation (4.109) must be satisfied for all admissible values of $h(a)$ and $h(b)$. Consequently, equations (4.109) and (4.110) imply that

$$0 = \sum_{i=1}^{n} H_i'(\mathcal{F}_1[\tilde{x}], \ldots, \mathcal{F}_n[\tilde{x}]) \alpha_i! f_{iv} \langle \tilde{x} \rangle_i(t) h(b)$$

$$- \sum_{i=1}^{n} H_i'(\mathcal{F}_1[\tilde{x}], \ldots, \mathcal{F}_n[\tilde{x}]) \alpha_i! f_{iv} \langle \tilde{x} \rangle_i(t) h(a). \quad (4.111)$$

If x is not preassigned at either end-point, then $h(a)$ and $h(b)$ are both completely arbitrary and we conclude that their coefficients in (4.111) must each vanish. It follows that condition (4.107) holds when $x(a)$ is not given, and condition (4.108) holds when $x(b)$ is not given. □

Note that in the limit, when $\alpha_i \to 1$, $i = 1, \ldots, n$, Theorem 4.23 implies the following result (Theorem 3.1 and Equation (4.1) in Castillo, Luceño and Pedregal, 2008):

Corollary 4.5. *If $\tilde{x}$ is a solution to problem*

$$\mathcal{L}[x] = H\left(\int_a^b f_1(t, x(t), x'(t))dt, \ldots, \int_a^b f_n(t, x(t), x'(t))dt \right) \longrightarrow extr,$$

$$(x(a) = x_a) \quad (x(b) = x_b)$$

then the Euler–Lagrange equation

$$\sum_{i=1}^{n} H_i'(\mathcal{F}_1[\tilde{x}], \ldots, \mathcal{F}_n[\tilde{x}]) \left(f_{iy}(t, \tilde{x}(t), \tilde{x}'(t)) - \frac{d}{dx} f_{iv}(t, \tilde{x}(t), \tilde{x}'(t)) \right) = 0$$

holds for all $t \in [a, b]$, where $\mathcal{F}_i[\tilde{x}] = \int_a^b f_i(t, \tilde{x}(t), \tilde{x}'(t))dt$, $i = 1, \ldots, n$. Moreover, if $x(a)$ is not specified, then

$$\sum_{i=1}^{n} H_i'(\mathcal{F}_1[\tilde{x}], \ldots, \mathcal{F}_n[\tilde{x}]) f_{iv}(a, \tilde{x}(a), \tilde{x}'(a)) = 0;$$

if $x(b)$ is not specified, then

$$\sum_{i=1}^{n} H_i'(\mathcal{F}_1[\tilde{x}], \ldots, \mathcal{F}_n[\tilde{x}]) f_{iv}(b, \tilde{x}(b), \tilde{x}'(b)) = 0.$$

Corollary 4.6. *If $\tilde{x}$ is a solution to problem*

$$\mathcal{L}[x] = \left(\int_a^b f_1 \langle \tilde{x} \rangle_1(t)(dt)^{\alpha_1} \right) \left(\int_a^b f_2 \langle \tilde{x} \rangle_2(t)(dt)^{\alpha_2} \right) \longrightarrow extr,$$

$$(x(a) = x_a) \quad (x(b) = x_b)$$

then the Euler–Lagrange equation

$$\alpha_1 \mathcal{F}_2[\tilde{x}](b-t)^{\alpha_1-1}\left(f_{1y}\langle\tilde{x}\rangle_1(t) - f_{1v}^{(\alpha_1)}\langle\tilde{x}\rangle_1(t)\right)$$
$$+ \alpha_2 \mathcal{F}_1[\tilde{x}](b-t)^{\alpha_2-1}\left(f_{2y}\langle\tilde{x}\rangle_2(t) - f_{2v}^{(\alpha_2)}\langle\tilde{x}\rangle_2(t)\right) = 0$$

holds for all $t \in [a, b)$. Moreover, if $x(a)$ is not specified, then

$$\alpha_1! \mathcal{F}_2[\tilde{x}]f_{1v}\langle\tilde{x}\rangle_1(a) + \alpha_2! \mathcal{F}_1[\tilde{x}]f_{2v}\langle\tilde{x}\rangle_2(a) = 0;$$

if $x(b)$ is not specified, then

$$\alpha_1! \mathcal{F}_2[\tilde{x}]f_{1v}\langle\tilde{x}\rangle_1(b) + \alpha_2! \mathcal{F}_1[\tilde{x}]f_{2v}\langle\tilde{x}\rangle_2(b) = 0.$$

Remark 4.22. In the case $\alpha_i \to 1$, $i = 1, 2$, Corollary 4.6 gives the result of Castillo, Luceño and Pedregal, 2008: the Euler–Lagrange equation associated with the product functional

$$\mathcal{L}[x] = \left(\int_a^b f_1(t, x(t), x'(t))dt\right)\left(\int_a^b f_2(t, x(t), x'(t))dt\right)$$

is

$$\mathcal{F}_2[x]\left(f_{1y}(t, x(t), x'(t)) - \frac{d}{dt}f_{1v}(t, x(t), x'(t))\right)$$
$$+ \mathcal{F}_1[x]\left(f_{2y}(t, x(t), x'(t)) - \frac{d}{dt}f_{2v}(t, x(t), x'(t))\right) = 0$$

and the natural condition at $t = a$, when $x(a)$ is free, becomes

$$\mathcal{F}_2[x]f_{1v}(a, x(a), x'(a)) + \mathcal{F}_1[x]f_{2v}(a, x(a), x'(a)) = 0.$$

Corollary 4.7. *If $\tilde{x}$ is a solution to problem*

$$\mathcal{L}[x] = \frac{\int_a^b f_1\langle\tilde{x}\rangle_1(t)(dt)^{\alpha_1}}{\int_a^b f_2\langle\tilde{x}\rangle_2(t)(dt)^{\alpha_2}} \longrightarrow extr,$$
$$(x(a) = x_a) \quad (x(b) = x_b)$$

then the Euler–Lagrange equation

$$\alpha_1(b-t)^{\alpha_1-1}\left(f_{1y}\langle\tilde{x}\rangle_1(t) - f_{1v}^{(\alpha_1)}\langle\tilde{x}\rangle_1(t)\right)$$
$$- \alpha_2 Q(b-t)^{\alpha_2-1}\left(f_{2y}\langle\tilde{x}\rangle_2(t) - f_{2v}^{(\alpha_2)}\langle\tilde{x}\rangle_2(t)\right) = 0$$

holds for all $t \in [a, b)$, where $Q = \frac{\mathcal{F}_1[\tilde{x}]}{\mathcal{F}_2[\tilde{x}]}$. Moreover, if $x(a)$ is not specified, then $\alpha_1! f_{1v}\langle\tilde{x}\rangle_1(a) - \alpha_2! Q f_{2v}\langle\tilde{x}\rangle_2(a) = 0$; if $x(b)$ is not specified, then $\alpha_1! f_{1v}\langle\tilde{x}\rangle_1(b) - \alpha_2! Q f_{2v}\langle\tilde{x}\rangle_2(b) = 0$.

Remark 4.23. In the case $\alpha_i \to 1$, $i = 1, 2$, Corollary 4.7 gives the following result of Castillo, Luceño and Pedregal, 2008: the Euler–Lagrange equation associated with the quotient functional

$$\mathcal{L}[x] = \frac{\int_a^b f_1(t, x(t), x'(t))dt}{\int_a^b f_2(t, x(t), x'(t))dt}$$

is

$$f_{1y}(t, x(t), x'(t)) - Q f_{2y}(t, x(t), x'(t))$$
$$- \frac{d}{dt} \left[(f_{1v}(t, x(t), x'(t)) - Q f_{2v}(t, x(t), x'(t)) \right] = 0$$

and the natural condition at $t = a$, when $x(a)$ is free, becomes

$$f_{1v}(a, x(a), x'(a)) - Q f_{2v}(a, x(a), x'(a)) = 0.$$

An example

Consider the problem

$$\text{minimize} \quad \mathcal{L}[x] = \left(\int_0^1 (x^{(\frac{1}{2})}(t))^2 (dt)^{\frac{1}{2}} \right) \left(\int_0^1 t^{\frac{1}{2}} x^{(\frac{1}{2})}(t)(dt)^{\frac{1}{2}} \right) \quad (4.112)$$
$$x(0) = 0, \quad x(1) = 1.$$

If $\tilde{x}$ is a local minimizer to (4.112), then by Corollary 4.6

$$\frac{1}{2} Q_2 (1-t)^{-\frac{1}{2}} 2(\tilde{x}^{(\frac{1}{2})}(t))^{(\frac{1}{2})} + \frac{1}{2} Q_1 (1-t)^{-\frac{1}{2}} (t^{\frac{1}{2}})^{(\frac{1}{2})} = 0,$$

where

$$Q_1 = \int_0^1 (\tilde{x}^{(\frac{1}{2})}(t))^2 (dt)^{\frac{1}{2}}, \quad Q_2 = \int_0^1 t^{\frac{1}{2}} \tilde{x}^{(\frac{1}{2})}(t)(dt)^{\frac{1}{2}}.$$

Hence,

$$Q_2 2(\tilde{x}^{(\frac{1}{2})}(t))^{(\frac{1}{2})} + Q_1 \frac{\sqrt{\pi}}{2} = 0. \quad (4.113)$$

If $Q_2 = 0$, then also $Q_1 = 0$. This contradicts the fact that a global minimizer to the problem

$$\text{minimize} \quad \mathcal{F}_1[x] = \int_0^1 (x^{(\frac{1}{2})}(t))^2 (dt)^{\frac{1}{2}}$$
$$x(0) = 0, \quad x(1) = 1$$

is $\bar{x}(t) = t^{\frac{1}{2}}$ and $\mathcal{F}_1[\bar{x}] = (\frac{\sqrt{\pi}}{2})^2$. Hence, $Q_2 \neq 0$ and (4.113) implies that candidate solutions to problem (4.112) are those satisfying the fractional differential equation

$$(\tilde{x}^{(\frac{1}{2})}(t))^{(\frac{1}{2})} = -\frac{Q_1\sqrt{\pi}}{4Q_2} \tag{4.114}$$

subject to the boundary conditions $x(0) = 0$ and $x(1) = 1$. Solving equation (4.114) we obtain

$$x(t) = \frac{1}{\sqrt{\pi}} \int_0^t \left(\frac{Q_1\pi + 4\sqrt{\pi}Q_2}{8Q_2} - \frac{Q_1}{2Q_2}\tau^{\frac{1}{2}} \right)(t-\tau)^{-\frac{1}{2}} d\tau. \tag{4.115}$$

Substituting (4.115) into functionals $\mathcal{F}_1$ and $\mathcal{F}_2$ gives

$$\begin{cases} -\dfrac{1}{192}\dfrac{-32\,Q_1{}^2 - 48\,\pi\,Q_2{}^2 + 3\,Q_1{}^2\pi^2}{Q_2{}^2} = Q_1 \\ \dfrac{1}{96}\dfrac{-32\,Q_1 + 3\,Q_1\pi^2 + 12\,\pi^{3/2}Q_2}{Q_2} = Q_2. \end{cases} \tag{4.116}$$

We obtain the candidate minimizer to problem (4.112) solving the system of equations (4.116):

$$\tilde{x}(t) = \frac{1}{\sqrt{\pi}} \int_0^t \left(\frac{Q_1\pi + 4\sqrt{\pi}Q_2}{8Q_2} - \frac{Q_1}{2Q_2}\tau^{\frac{1}{2}} \right)(t-\tau)^{-\frac{1}{2}} d\tau,$$

where

$$Q_1 = \frac{4}{3}\frac{\pi(\sqrt{\pi}(\frac{1}{4}\pi^{\frac{3}{2}} + \frac{1}{4}\sqrt{\pi^3 - 8\pi}) - 4)}{-32 + 3\pi^2} \quad \text{and} \quad Q_2 = \frac{1}{12}\pi^{\frac{3}{2}} + \frac{1}{12}\sqrt{\pi^3 - 8\pi}.$$

4.3.8   *Fractional Theorems of Green, Gauss, and Stokes*

In this section we introduce some useful fractional integral and fractional differential operators. With them we prove fractional versions of the integral theorems of Green, Gauss, and Stokes. Throughout the text we assume that all integrals and derivatives exist.

Fractional operators

Let us consider a continuous function $f = f(x_1, \ldots, x_n)$ defined on $R = \Pi_{i=1}^n [a_i, b_i] \subset \mathbb{R}^n$. Let us extend Jumarie's fractional derivative and the $(dt)^\alpha$ integral to functions with n variables. For $x_i \in [a_i, b_i]$, $i = 1, \ldots, n$, and $\alpha \in (0, 1)$, we define the fractional integral operator as

$$_{a_i}I_{x_i}^\alpha[i] = \alpha \int_{a_i}^{x_i} (x_i - t)^{\alpha - 1} dt.$$

These operators act on f in the following way:

$$_{a_i}I^\alpha_{x_i}[i]f(x_1,\ldots,x_n) = \alpha \int_{a_i}^{x_i} f(x_1,\ldots,x_{i-1},t,x_{i+1},\ldots,x_n)(x_i - t)^{\alpha-1}\, dt,$$

$i = 1,\ldots,n$. Let $\Xi = \{k_1,\ldots,k_s\}$ be an arbitrary nonempty subset of $\{1,\ldots,n\}$. We define the fractional multiple integral operator over the region $R_\Xi = \Pi^s_{i=1}[a_{k_i},x_{k_i}]$ by

$$I^\alpha_{R_\Xi}[k_1,\ldots,k_s] = {}_{a_{k_1}}I^\alpha_{x_{k_1}}[k_1] \cdots {}_{a_{k_s}}I^\alpha_{x_{k_s}}[k_s]$$

$$= \alpha^s \int_{a_{k_1}}^{x_{k_1}} \cdots \int_{a_{k_s}}^{x_{k_s}} (x_{k_1} - t_{k_1})^{\alpha-1} \cdots (x_{k_s} - t_{k_s})^{\alpha-1}\, dt_{k_s} \ldots dt_{k_1},$$

which acts on f by

$$I^\alpha_{R_\Xi}[k_1,\ldots,k_s]f(x_1,\ldots,x_n)$$

$$= \alpha^s \int_{a_{k_1}}^{x_{k_1}} \cdots \int_{a_{k_s}}^{x_{k_s}} f(\xi_1,\ldots,\xi_n)(x_{k_1}-t_{k_1})^{\alpha-1} \cdots (x_{k_s}-t_{k_s})^{\alpha-1}\, dt_{k_s} \ldots dt_{k_1},$$

where $\xi_j = t_j$ if $j \in \Xi$, and $\xi_j = x_j$ if $j \notin \Xi$, $j = 1,\ldots,n$. The fractional volume integral of f over the whole domain R is given by

$$I^\alpha_R f = \alpha^n \int_{a_1}^{b_1} \cdots \int_{a_n}^{b_n} f(t_1,\ldots,t_n)(b_1 - t_1)^{\alpha-1} \cdots (b_n - t_n)^{\alpha-1}\, dt_n \ldots dt_1.$$

The fractional partial derivative operator with respect to the ith variable x_i, $i = 1,\ldots,n$, of order $\alpha \in (0,1)$ is defined as follows:

$$_{a_i}D^\alpha_{x_i}[i] = \frac{1}{\Gamma(1-\alpha)} \frac{\partial}{\partial x_i} \int_{a_i}^{x_i} (x_i - t)^{-\alpha}\, dt,$$

which act on f by

$$_{a_i}D^\alpha_{x_i}[i]f(x_1,\ldots,x_n)$$

$$= \frac{1}{\Gamma(1-\alpha)} \frac{\partial}{\partial x_i} \int_{a_i}^{x_i} (x_i - t)^{-\alpha} [f(x_1,\ldots,x_{i-1},t,x_{i+1},\ldots,x_n)$$

$$- f(x_1,\ldots,x_{i-1},a_i,x_{i+1},\ldots,x_n)]\, dt,$$

$i = 1,\ldots,n$. We observe that the Jumarie fractional integral and the Jumarie fractional derivative can be obtained putting $n = 1$:

$$_aI^\alpha_x[1]f(x) = \alpha \int_a^x (x-t)^{\alpha-1}f(t)\, dt = \int_a^x f(t)\,(dt)^\alpha$$

and

$$_aD^\alpha_x[1]f(x) = \frac{1}{\Gamma(1-\alpha)} \frac{d}{dx} \int_a^x (x-t)^{-\alpha}(f(t) - f(a))\, dt = f^{(\alpha)}(x).$$

Using these notations, the properties

$$\frac{d^\alpha}{dx^\alpha} \int_0^x f(t)(dt)^\alpha = \alpha! f(x),$$

and

$$\int_0^x f^{(\alpha)}(t)(dt)^\alpha = \alpha!(f(x) - f(0)),$$

which are the Jumarie fractional counterparts of the first and the second fundamental theorems of calculus (Jumarie, 2007, 2009a), can be presented respectively as

$$_aD_x^\alpha[1]_aI_x^\alpha[1]f(x) = \alpha! f(x)$$

and

$$_aI_x^\alpha[1]_aD_x^\alpha[1]f(x) = \alpha!(f(x) - f(a)). \tag{4.117}$$

In the two-dimensional case we define the fractional line integral on ∂R, $R = [a, b] \times [c, d]$, by

$$I_{\partial R}^\alpha f = I_{\partial R}^\alpha[1]f + I_{\partial R}^\alpha[2]f$$

where

$$I_{\partial R}^\alpha[1]f = {}_aI_b^\alpha[1][f(b, c) - f(b, d)]$$
$$= \alpha \int_a^b [f(t, c) - f(t, d)] (b - t)^{\alpha-1} dt$$

and

$$I_{\partial R}^\alpha[2]f = {}_cI_d^\alpha[2][f(b, d) - f(a, d)]$$
$$= \alpha \int_c^d [f(b, t) - f(a, t)] (d - t)^{\alpha-1} dt.$$

Fractional differential vector operations

Let $W_X = [a, x] \times [c, y] \times [e, z]$, $W = [a, b] \times [c, d] \times [e, f]$, and denote (x_1, x_2, x_3) by (x, y, z). We introduce the fractional nabla operator by

$$\nabla_{W_X}^\alpha = i_a D_x^\alpha[1] + j_c D_y^\alpha[2] + k_e D_z^\alpha[3],$$

where the i, j, k define a fixed right-handed orthonormal basis. If $f : \mathbb{R}^3 \to \mathbb{R}$ is a continuous function, then we define its fractional gradient as

$$\text{Grad}_{W_X}^\alpha f = \nabla_{W_X}^\alpha f$$
$$= i_a D_x^\alpha[1]f(x, y, z) + j_c D_y^\alpha[2]f(x, y, z) + k_e D_z^\alpha[3]f(x, y, z).$$

If $F = [F_x, F_y, F_z] : \mathbb{R}^3 \to \mathbb{R}^3$ is a continuous vector field, then we define its fractional divergence and fractional curl by

$$\mathrm{Div}_{W_X}^\alpha F = \nabla_{W_X}^\alpha \circ F$$
$$= {_a}D_x^\alpha[1]F_x(x,y,z) + {_c}D_y^\alpha[2]F_y(x,y,z) + {_e}D_z^\alpha[3]F_z(x,y,z)$$

and

$$\mathrm{Curl}_{W_X}^\alpha F = \nabla_{W_X}^\alpha \times F$$
$$= i\left({_c}D_y^\alpha[2]F_z(x,y,z) - {_e}D_z^\alpha[3]F_y(x,y,z)\right)$$
$$+ j\left({_e}D_z^\alpha[3]F_x(x,y,z) - {_a}D_x^\alpha[1]F_z(x,y,z)\right)$$
$$+ k\left({_a}D_x^\alpha[1]F_y(x,y,z) - {_c}D_y^\alpha[2]F_x(x,y,z)\right).$$

Note that these fractional differential operators are non-local. Therefore, the fractional gradient, divergence, and curl, depend on the region W_X.

For $F : \mathbb{R}^3 \to \mathbb{R}^3$ and $f, g : \mathbb{R}^3 \to \mathbb{R}$ it is easy to check the following relations:

(a) $\mathrm{Div}_{W_X}^\alpha (fF) = f\mathrm{Div}_{W_X}^\alpha F + F \circ \mathrm{Grad}_{W_X}^\alpha f$,
(b) $\mathrm{Curl}_{W_X}^\alpha (\mathrm{Grad}_{W_X}^\alpha f) = [0,0,0]$,
(c) $\mathrm{Div}_{W_X}^\alpha (\mathrm{Curl}_{W_X}^\alpha F) = 0$,
(d) $\mathrm{Grad}_{W_X}^\alpha (fg) = g\mathrm{Grad}_{W_X}^\alpha f + f\mathrm{Grad}_{W_X}^\alpha g$,
(e) $\mathrm{Div}_{W_X}^\alpha (\mathrm{Grad}_{W_X}^\alpha f) = {_a}D_x^\alpha[1]{_a}D_x^\alpha[1]f + {_c}D_y^\alpha[2]{_c}D_y^\alpha[2]f + {_e}D_z^\alpha[3]{_e}D_z^\alpha[3]f$.

Let us recall that in general $(D_{W_X}^\alpha)^2 \neq D_{W_X}^{2\alpha}$ (see Jumarie, 2009a).

A fractional flux of the vector field F across ∂W is a fractional oriented surface integral of the field such that

$$(I_{\partial W}^\alpha, F) = I_{\partial W}^\alpha[2,3]F_x(x,y,z) + I_{\partial W}^\alpha[1,3]F_y(x,y,z) + I_{\partial W}^\alpha[1,2]F_z(x,y,z),$$

where

$$I_{\partial W}^\alpha[1,2]f(x,y,z) = {_a}I_b^\alpha[1]{_c}I_d^\alpha[2][f(b,d,f) - f(b,d,e)],$$

$$I_{\partial W}^\alpha[1,3]f(x,y,z) = {_a}I_b^\alpha[1]{_e}I_f^\alpha[3][f(b,d,f) - f(b,c,f)],$$

and

$$I_{\partial W}^\alpha[2,3]f(x,y,z) = {_c}I_d^\alpha[2]{_e}I_f^\alpha[3][f(b,d,f) - f(a,d,f)].$$

Fractional theorems of Green, Gauss, and Stokes

We now formulate the fractional formulae of Green, Gauss, and Stokes. Analogous results via Caputo fractional derivatives and Riemann–Liouville fractional integrals were obtained by Tarasov, 2010.

Theorem 4.24 (Fractional Green's theorem for a rectangle).
Let f and g be two continuous functions whose domains contain $R = [a,b] \times [c,d] \subset \mathbb{R}^2$. Then,

$$I^\alpha_{\partial R}[1]f + I^\alpha_{\partial R}[2]g = \frac{1}{\alpha!} I^\alpha_R \left[{}_a D^\alpha_b[1]g - {}_c D^\alpha_d[2]f \right].$$

Proof. We have

$$I_{\partial R}{}^\alpha[1]f + I^\alpha_{\partial R}[2]g = {}_a I^\alpha_b[1][f(b,c) - f(b,d)] + {}_c I^\alpha_d[2][g(b,d) - g(a,d)].$$

By equation (4.117),

$$f(b,c) - f(b,d) = -\frac{1}{\alpha!}{}_c I^\alpha_d[2]{}_c D^\alpha_d[2]f(b,d),$$

$$g(b,d) - g(a,d) = \frac{1}{\alpha!}{}_a I^\alpha_b[1]{}_a D^\alpha_b[1]g(b,d).$$

Therefore,

$$I^\alpha_{\partial R}[1]f + I^\alpha_{\partial R}[2]g = -{}_a I^\alpha_b[1]\frac{1}{\alpha!}{}_c I^\alpha_d[2]{}_c D^\alpha_d[2]f(b,d)$$

$$+ {}_c I^\alpha_d[2]\frac{1}{\alpha!}{}_a I^\alpha_b[1]{}_a D^\alpha_b[1]g(b,d) = \frac{1}{\alpha!} I^\alpha_R \left[{}_a D^\alpha_b[1]g - {}_c D^\alpha_d[2]f \right].$$

$\square$

Theorem 4.25 (Fractional Gauss's theorem for a parallelepiped).
Let $F = (F_x, F_y, F_z)$ be a continuous vector field in a domain that contains $W = [a,b] \times [c,d] \times [e,f]$. If the boundary of W is a closed surface ∂W, then

$$(I^\alpha_{\partial W}, F) = \frac{1}{\alpha!} I^\alpha_W \, Div^\alpha_W F. \tag{4.118}$$

Proof. The result follows by direct transformations:

$$(I^\alpha_{\partial W}, F) = I^\alpha_{\partial W}[2,3]F_x + I^\alpha_{\partial W}[1,3]F_y + I^\alpha_{\partial W}[1,2]F_z$$

$$= {}_cI^\alpha_d[2]{}_eI^\alpha_f[3](F_x(b,d,f) - F_x(a,d,f))$$

$$+ {}_aI^\alpha_b[1]{}_eI^\alpha_f[3](F_y(b,d,f) - F_y(b,c,f))$$

$$+ {}_aI^\alpha_b[1]{}_cI^\alpha_d[2](F_z(b,d,f) - F_z(b,d,e))$$

$$= \frac{1}{\alpha!}{}_aI^\alpha_b[1]{}_cI^\alpha_d[2]{}_eI^\alpha_f[3]({}_aD^\alpha_b[1]F_x(b,d,f)$$

$$+ {}_cD^\alpha_d[2]F_y(b,d,f) + {}_eD^\alpha_f[3]F_z(b,d,f))$$

$$= \frac{1}{\alpha!}I^\alpha_W({}_aD^\alpha_b[1]F_x + {}_cD^\alpha_d[2]F_y + {}_eD^\alpha_f[3]F_z)$$

$$= \frac{1}{\alpha!}I^\alpha_W \cdot \mathrm{Div}^\alpha_W F. \qquad \square$$

Let S be an open, oriented, and nonintersecting surface, bounded by a simple and closed curve ∂S. Let $F = [F_x, F_y, F_z]$ be a continuous vector field. Divide up S by curves into N subregions $S_1, S_2, \ldots, S_N$. Assume that for small enough subregions each S_j can be approximated by a plane rectangle A_j bounded by curves $C_1, C_2, \ldots, C_N$. Apply Green's theorem to each individual rectangle A_j. Then, summing over the subregions,

$$\sum_j \frac{1}{\alpha!}I^\alpha_{A_j}(\nabla^\alpha_{A_j} \times F) = \sum_j I^\alpha_{\partial A_j}F.$$

Furthermore, letting $N \to \infty$

$$\sum_j \frac{1}{\alpha!}I^\alpha_{A_j}(\nabla^\alpha_{A_j} \times F) \to \frac{1}{\alpha!}(I^\alpha_S, \mathrm{Curl}^\alpha_S F)$$

while

$$\sum_j I^\alpha_{\partial A_j}F \to I^\alpha_{\partial S}F.$$

We conclude with the fractional Stokes formula

$$\frac{1}{\alpha!}(I^\alpha_S, \mathrm{Curl}^\alpha_S F) = I^\alpha_{\partial S}F.$$

4.3.9 *Fractional Variational Calculus with Multiple Integrals*

Consider a function $w = w(x,y)$ with two variables. Assume that the domain of w contains the rectangle $R = [a,b] \times [c,d]$ and that w is continuous on R. We introduce the variational functional defined by

$$J(w) = I_R^\alpha L \left(x, y, w(x,y), {}_aD_x^\alpha[1]w(x,y), {}_cD_y^\alpha[2]w(x,y) \right)$$

$$:= \alpha^2 \int_a^b \int_c^d L \left(x, y, w, {}_aD_x^\alpha[1]w, {}_cD_y^\alpha[2]w \right) (b-x)^{\alpha-1}(d-y)^{\alpha-1} \, dydx.$$

$$(4.119)$$

We assume that the lagrangian L is at least of class C^1. Observe that, using the notation of the $(dt)^\alpha$ integral as presented in Jumarie, 2009a, (4.119) can be written as

$$J(w) = \int_a^b \int_c^d L \left(x, y, w(x,y), {}_aD_x^\alpha[1]w(x,y), {}_cD_y^\alpha[2]w(x,y) \right) (dy)^\alpha (dx)^\alpha.$$

$$(4.120)$$

Consider the following FCV problem, which we address as problem (P).

Problem (P): *minimize (or maximize) functional J defined by (4.120) with respect to the set of continuous functions $w(x,y)$ such that $w|_{\partial R} = \varphi(x,y)$ for some given function φ.*

The continuous functions $w(x,y)$ that assume the prescribed values $w|_{\partial R} = \varphi(x,y)$ at all points of the boundary curve of R are said to be admissible. In order to prove necessary optimality conditions for problem (P) we use a two-dimensional analogue of fractional integration by parts. Lemma 4.5 provides the necessary fractional rule.

Lemma 4.5. *Let F, G, and h be continuous functions whose domains contain R. If $h \equiv 0$ on ∂R, then*

$$\int_a^b \int_c^d [G(x,y) {}_aD_x^\alpha[1]h(x,y) - F(x,y) {}_cD_y^\alpha[2]h(x,y)]$$

$$\times (b-x)^{\alpha-1}(d-y)^{\alpha-1} \, dydx$$

$$= -\int_a^b \int_c^d [({}_aD_x^\alpha[1]G(x,y) - {}_cD_y^\alpha[2]F(x,y))h(x,y)]$$

$$\times (b-x)^{\alpha-1}(d-y)^{\alpha-1} \, dydx. \quad (4.121)$$

Proof. By choosing $f = F \cdot h$ and $g = G \cdot h$ in Green's formula, we obtain

$$I_{\partial R}^\alpha[1](F\,h) + I_{\partial R}^\alpha[2](G\,h)$$

$$= \frac{1}{\alpha!} I_R^\alpha [{}_aD_b^\alpha[1]G \cdot h + G \cdot {}_aD_b^\alpha[1]h - {}_cD_d^\alpha[2]F \cdot h - F \cdot {}_cD_d^\alpha[2]h],$$

which is equivalent to

$$\frac{1}{\alpha!}I_R^\alpha\left[G\cdot {}_aD_b^\alpha[1]h - F\cdot {}_cD_d^\alpha[2]h\right]$$

$$= I_{\partial R}^\alpha[1](F\,h) + I_{\partial R}^\alpha[2](G\,h) - \frac{1}{\alpha!}I_R^\alpha\left[({}_aD_b^\alpha[1]G - {}_cD_d^\alpha[2]F)h\right] .$$

In addition, since $h \equiv 0$ on ∂R, we deduce that

$$I_R^\alpha\left[G\cdot {}_aD_b^\alpha[1]h - F\cdot {}_cD_d^\alpha[2]h\right] = -I_R^\alpha\left[({}_aD_b^\alpha[1]G - {}_cD_d^\alpha[2]F)h\right] .$$

The lemma is proved. $\square$

Theorem 4.26 (Fractional Euler–Lagrange equation). *Let w be a solution to problem (P). Then w is a solution of the fractional partial differential equation*

$$\partial_3 L - {}_aD_x^\alpha[1]\partial_4 L - {}_cD_y^\alpha[2]\partial_5 L = 0 , \tag{4.122}$$

where by $\partial_i L$, $i = 1,\ldots,5$, we denote the usual partial derivative of $L(\cdot,\cdot,\cdot,\cdot,\cdot)$ with respect to its ith argument.

Proof. Let h be a continuous function on R such that $h \equiv 0$ on ∂R, and consider an admissible variation $w + \epsilon h$, for ϵ taking values on a sufficient small neighborhood of zero. Let $j(\epsilon) = J(w + \epsilon h)$. Then $j'(0) = 0$, i.e.,

$$\alpha^2 \int_a^b \int_c^d \left(\partial_3 L\,h + \partial_4 L\,{}_aD_x^\alpha[1]h + \partial_5 L\,{}_cD_y^\alpha[2]h\right)$$
$$\times (b-x)^{\alpha-1}(d-y)^{\alpha-1}\,dy dx = 0.$$

Using Lemma 4.5, we obtain

$$\alpha^2 \int_a^b \int_c^d \left(\partial_3 L - {}_aD_x^\alpha[1]\partial_4 L - {}_cD_y^\alpha[2]\partial_5 L\right) h(b-x)^{\alpha-1}(d-y)^{\alpha-1}\,dy dx = 0 .$$

Since h is an arbitrary function, by the fundamental lemma of the calculus of variations we deduce equation (4.122). $\square$

Let us consider now the situation where we do not impose admissible functions w to be of fixed values on ∂R.

Problem (P'): *minimize (or maximize) J among the set of all continuous curves w whose domain contains R.*

Theorem 4.27 (Fractional natural boundary conditions). *Let w be a solution to problem (P'). Then w is a solution of the fractional differential equation (4.122) and satisfies the following equations:*

(i) $\partial_4 L(a, y, w(a, y), {}_aD_a^\alpha[1]w(a, y), {}_cD_y^\alpha[2]w(a, y)) = 0$ *for all* $y \in [c, d]$;

(ii) $\partial_4 L(b, y, w(b, y), {}_aD_b^\alpha[1]w(b, y), {}_cD_y^\alpha[2]w(b, y)) = 0$ *for all* $y \in [c, d]$;

(iii) $\partial_5 L(x, c, w(x, c), {}_aD_x^\alpha[1]w(x, c), {}_cD_c^\alpha[2]w(x, c)) = 0$ *for all* $x \in [a, b]$;

(iv) $\partial_5 L(x, d, w(x, d), {}_aD_x^\alpha[1]w(x, d), {}_cD_d^\alpha[2]w(x, d)) = 0$ *for all* $x \in [a, b]$.

Proof.   Proceeding as in the proof of Theorem 4.26 (see also Lemma 4.5), we obtain

$$0 = \alpha^2 \int_a^b \int_c^d \left(\partial_3 L\, h + \partial_4 L\, {}_aD_x^\alpha[1]h + \partial_5 L\, {}_cD_y^\alpha[2]h\right)$$
$$\times (b - x)^{\alpha-1}(d - y)^{\alpha-1}\, dy dx$$
$$= \alpha^2 \int_a^b \int_c^d \left(\partial_3 L - {}_aD_x^\alpha[1]\partial_4 L - {}_cD_y^\alpha[2]\partial_5 L\right) h\, (b - x)^{\alpha-1}(d - y)^{\alpha-1}\, dy dx$$
$$+ \alpha! I_{\partial R}^\alpha[2](\partial_4 L\, h) - \alpha! I_{\partial R}^\alpha[1](\partial_5 L\, h), \quad (4.123)$$

where h is an arbitrary continuous function. In particular the above equation holds for $h \equiv 0$ on ∂R. For such h the second member of (4.123) vanishes and by the fundamental lemma of the calculus of variations we deduce equation (4.122). With this result, equation (4.123) takes the form

$$0 = \int_c^d \partial_4 L(b, y, w(b, y), {}_aD_b^\alpha[1]w(b, y), {}_cD_y^\alpha[2]w(b, y))\, h(b, y)(d - y)^{\alpha-1} dy$$
$$- \int_c^d \partial_4 L(a, y, w(a, y), {}_aD_a^\alpha[1]w(a, y), {}_cD_y^\alpha[2]w(a, y))\, h(a, y)(d - y)^{\alpha-1} dy$$
$$- \int_a^b \partial_5 L(x, c, w(x, c), {}_aD_x^\alpha[1]w(x, c), {}_cD_c^\alpha[2]w(x, c))\, h(x, c)(b - x)^{\alpha-1} dx$$
$$+ \int_a^b \partial_5 L(x, d, w(x, d), {}_aD_x^\alpha[1]w(x, d), {}_cD_d^\alpha[2]w(x, d))\, h(x, d)(b - x)^{\alpha-1} dx.$$
$$(4.124)$$

Since h is an arbitrary function, we can consider the subclass of functions for which $h \equiv 0$ on

$$[a, b] \times \{c\} \cup [a, b] \times \{d\} \cup \{b\} \times [c, d].$$

For such h equation (4.124) reduce to

$$0 = \int_c^d \partial_4 L(a, y, w(a, y), {}_aD_a^\alpha[1]w(a, y), {}_cD_y^\alpha[2]w(a, y))\, h(a, y)(d - y)^{\alpha-1} dy.$$

By the fundamental lemma of calculus of variations, we obtain

$$\partial_4 L(a, y, w(a, y), {}_aD_a^\alpha[1]w(a, y), {}_cD_y^\alpha[2]w(a, y)) = 0 \quad \text{for all } y \in [c, d].$$

The other natural boundary conditions are proved similarly, by appropriate choices of h. □

We can generalize Lemma 4.5 and Theorem 4.26 to the three-dimensional case in the following way.

Lemma 4.6. *Let A, B, C, and η be continuous functions whose domains contain the parallelepiped W. If $\eta \equiv 0$ on ∂W, then*

$$I_W^\alpha \left(A \cdot {}_a D_b^\alpha[1]\eta + B \cdot {}_c D_d^\alpha[2]\eta + C \cdot {}_e D_f^\alpha[3]\eta \right)$$
$$= -I_W^\alpha \left(\left[{}_a D_b^\alpha[1]A + {}_c D_d^\alpha[2]B + {}_e D_f^\alpha[3]C \right] \eta \right). \quad (4.125)$$

Proof. By choosing $F_x = \eta A$, $F_y = \eta B$, and $F_z = \eta C$ in (4.118), we obtain the three-dimensional analogue of integrating by parts:

$$I_W^\alpha \left(A \cdot {}_a D_b^\alpha[1]\eta + B \cdot {}_c D_d^\alpha[2]\eta + C \cdot {}_e D_f^\alpha[3]\eta \right)$$
$$= -I_W^\alpha \left(\left[{}_a D_b^\alpha[1]A + {}_c D_d^\alpha[2]B + {}_e D_f^\alpha[3]C \right] \eta \right) + \alpha! (I_{\partial W}^\alpha, [\eta A, \eta B, \eta C]).$$

In addition, if we assume that $\eta \equiv 0$ on ∂W, we have formula (4.125). □

Theorem 4.28 (Euler–Lagrange equation for triple integrals).
Let $w = w(x, y, z)$ be a continuous function whose domain contains $W = [a, b] \times [c, d] \times [e, f]$. Consider the functional

$$J(w) = I_W^\alpha L\Big(x, y, z, w(x,y,z), {}_a D_x^\alpha[1]w(x,y,z), {}_c D_y^\alpha[2]w(x,y,z),$$
$$\qquad {}_e D_z^\alpha[3]w(x,y,z) \Big)$$

$$= \int_a^b \int_c^d \int_e^f L\left(x, y, z, w, {}_a D_x^\alpha[1]w, {}_c D_y^\alpha[2]w, {}_e D_z^\alpha[3]w \right) (dz)^\alpha (dy)^\alpha (dx)^\alpha$$

defined on the set of continuous curves such that their values on ∂W take prescribed values. Let L be at least of class C^1. If w is a minimizer (or maximizer) of J, then w satisfies the fractional partial differential equation

$$\partial_4 L - {}_a D_b^\alpha[1]\partial_5 L - {}_c D_d^\alpha[2]\partial_6 L - {}_e D_f^\alpha[3]\partial_7 L = 0.$$

Proof. A proof can be done similarly to the proof of Theorem 4.26, where instead of using Lemma 4.5 we apply Lemma 4.6. □

Isoperimetric problems

Let us consider functions $u = u(x, y)$. We assume that the domain of functions u contain the rectangle $R = [a, b] \times [c, d]$ and are continuous on R. Moreover, functions u under our consideration are such that the fractional partial derivatives ${}_a D_x^\alpha[1]u$ and ${}_c D_y^\alpha[2]u$ are continuous on R,

$\alpha \in (0,1)$. We investigate now the following fractional problem of the calculus of variations: to minimize a given functional

$$J[u(\cdot,\cdot)] = \alpha^2 \int\limits_a^b \int\limits_c^d f\left(x,y,u,_aD_x^\alpha[1]u,_cD_y^\alpha[2]u\right)(b-x)^{\alpha-1}(d-y)^{\alpha-1}dydx$$

(4.126)

subject to an isoperimetric constraint

$$\alpha^2 \int\limits_a^b \int\limits_c^d g\left(x,y,u,_aD_x^\alpha[1]u,_cD_y^\alpha[2]u\right)(b-x)^{\alpha-1}(d-y)^{\alpha-1}dydx = K$$

(4.127)

and a boundary condition

$$u(x,y)|_{\partial R} = \psi(x,y). \tag{4.128}$$

We are assuming that ψ is some given function, K is a constant, and f and g are at least of class of C^1. Moreover, we assume that $\partial_4 f$ and $\partial_4 g$ have continuous fractional partial derivatives $_aD_x^\alpha[1]$; and $\partial_5 f$ and $\partial_5 g$ have continuous fractional partial derivatives $_cD_y^\alpha[2]$. Along the work, we denote by $\partial_i f$ and $\partial_i g$ the standard partial derivatives of f and g with respect to their i-th argument, $i = 1, \ldots, 5$.

Definition 4.18. *A continuous function $u = u(x,y)$ that satisfies the given isoperimetric constraint* (4.127) *and boundary condition* (4.128), *is said to be admissible for problem* (4.126)–(4.128).

Definition 4.19 (Local minimizer to (4.126)–(4.128)). *An admissible function $u = u(x,y)$ is said to be a local minimizer to problem* (4.126)–(4.128) *if there exists some $\gamma > 0$ such that for all admissible functions $\hat{u}$ with $\|\hat{u} - u\|_{1,\infty} < \gamma$ one has $J[\hat{u}] - J[u] \geq 0$, where*

$$\|u\|_{1,\infty} := \max_{(x,y)\in R} |u(x,y)| + \max_{(x,y)\in R} |_aD_x^\alpha[1]u(x,y)| + \max_{(x,y)\in R} |_cD_y^\alpha[2]u(x,y)|.$$

Necessary optimality condition

The next theorem gives a necessary optimality condition for u to be a solution of the fractional isoperimetric problem defined by (4.126)–(4.128).

Theorem 4.29 (Euler–Lagrange equation to (4.126)–(4.128)). *If u is a local minimizer to problem* (4.126)–(4.128), *then there exists a nonzero pair of constants (λ_0, λ) such that u satisfies the fractional PDE*

$$\partial_3 H\{u\}(x,y) -_aD_x^\alpha[1]\partial_4 H\{u\}(x,y) -_cD_y^\alpha[2]\partial_5 H\{u\}(x,y) = 0 \quad (4.129)$$

for all $(x, y) \in R$, where

$$H(x, y, u, v, w, \lambda_0, \lambda) := \lambda_0 f(x, y, u, v, w) + \lambda g(x, y, u, v, w)$$

and, for simplicity of notation, we use the operator $\{\cdot\}$ defined by

$$\{u\}(x, y) := \left(x, y, u(x, y), _a D_x^\alpha[1]u(x, y), _c D_y^\alpha[2]u(x, y), \lambda_0, \lambda\right).$$

Proof. Let us define the function

$$\hat{u}_\varepsilon(x, y) = u(x, y) + \varepsilon \eta(x, y), \tag{4.130}$$

where η is such that $\eta \in C^0(R)$,

$$\eta(x, y)|_{\partial R} = 0,$$

and $\varepsilon \in \mathbb{R}$. If ε take values sufficiently close to zero, then (4.130) is included into the first order neighborhood of u, i.e., there exists $\delta > 0$ such that $\hat{u}_\varepsilon \in U_1(u, \delta)$, where

$$U_1(u, \delta) := \left\{\hat{u}(x, y) : \|u - \hat{u}\|_{1,\infty} < \delta\right\}.$$

On the other hand,

$$\hat{u}_0(x, y) = u, \quad \frac{\partial \hat{u}_\varepsilon(x, y)}{\partial \varepsilon} = \eta, \quad \frac{\partial_a D_x^\alpha[1]\hat{u}_\varepsilon(x, y)}{\partial \varepsilon} =_a D_x^\alpha[1]\eta,$$

$$\frac{\partial_c D_y^\alpha[2]\hat{u}_\varepsilon(x, y)}{\partial \varepsilon} =_c D_y^\alpha[2]\eta.$$

Let

$$F(\varepsilon) = \alpha^2 \int_a^b \int_c^d f(x, y, \hat{u}_\varepsilon(x, y), _a D_x^\alpha[1]\hat{u}_\varepsilon(x, y), _c D_y^\alpha[2]\hat{u}_\varepsilon(x, y))$$

$$\times (b - x)^{\alpha - 1}(d - y)^{\alpha - 1} dy dx,$$

and

$$G(\varepsilon) = \alpha^2 \int_a^b \int_c^d g(x, y, \hat{u}_\varepsilon(x, y), _a D_x^\alpha[1]\hat{u}_\varepsilon(x, y), _c D_y^\alpha[2]\hat{u}_\varepsilon(x, y))$$

$$\times (b - x)^{\alpha - 1}(d - y)^{\alpha - 1} dy dx.$$

Define the Lagrange function by

$$L(\varepsilon, \lambda_0, \lambda) = \lambda_0 F(\varepsilon) + \lambda (G(\varepsilon) - K).$$

Then, by the extended Lagrange multiplier rule (see, e.g., van Brunt, 2004), we can choose multipliers λ_0 and λ, not both zero, such that

$$\frac{\partial L(0, \lambda_0, \lambda)}{\partial \varepsilon} = \lambda_0 \frac{\partial F}{\partial \varepsilon}\bigg|_{\varepsilon=0} + \lambda \frac{\partial G}{\partial \varepsilon}\bigg|_{\varepsilon=0} = 0. \tag{4.131}$$

The term $\frac{\partial F}{\partial \varepsilon}\big|_{\varepsilon=0}$ is equal to

$$\alpha^2 \int\limits_a^b \int\limits_c^d \frac{\partial}{\partial \varepsilon}\bigg|_{\varepsilon=0} \left[f(x, y, \hat{u}_{\varepsilon}, {}_aD_x^\alpha[1]\hat{u}_\varepsilon, {}_cD_y^\alpha[2]\hat{u}_\varepsilon)(b-x)^{\alpha-1}(d-y)^{\alpha-1} \right] dydx$$

$$= \alpha^2 \int\limits_a^b \int\limits_c^d \partial_3 f \cdot (b-x)^{\alpha-1}(d-y)^{\alpha-1} dydx$$

$$+ \alpha^2 \int\limits_a^b \int\limits_c^d \left[\partial_4 f \, {}_aD_x^\alpha[1]\eta + \partial_5 f \, {}_cD_y^\alpha[2]\eta \right] (b-x)^{\alpha-1}(d-y)^{\alpha-1} dydx.$$

$$(4.132)$$

By (4.121) the last double integral in (4.132) may be transformed as follows:

$$\alpha^2 \int\limits_a^b \int\limits_c^d \left[\partial_4 f \, {}_aD_x^\alpha[1]\eta + \partial_5 f \, {}_cD_y^\alpha[2]\eta \right] (b-x)^{\alpha-1}(d-y)^{\alpha-1} dydx$$

$$= -\alpha^2 \int\limits_a^b \int\limits_c^d \left[{}_aD_x^\alpha[1]\partial_4 f +_c D_y^\alpha[2]\partial_5 f \right] \eta(b-x)^{\alpha-1}(d-y)^{\alpha-1} dydx.$$

Hence,

$$\frac{\partial F}{\partial \varepsilon}\bigg|_{\varepsilon=0}$$

$$= \alpha^2 \int\limits_a^b \int\limits_c^d \left[\partial_3 f -_a D_x^\alpha[1]\partial_4 f -_c D_y^\alpha[2]\partial_5 f \right] \eta(b-x)^{\alpha-1}(d-y)^{\alpha-1} dydx.$$

$$(4.133)$$

Similarly,

$$\frac{\partial G}{\partial \varepsilon}\bigg|_{\varepsilon=0}$$

$$= \alpha^2 \int\limits_a^b \int\limits_c^d \left[\partial_3 g -_a D_x^\alpha[1]\partial_4 g -_c D_y^\alpha[2]\partial_5 g \right] \eta(b-x)^{\alpha-1}(d-y)^{\alpha-1} dydx.$$

$$(4.134)$$

Substituting (4.133) and (4.134) into (4.131), we get

$$\frac{\partial L(\varepsilon, \lambda_0, \lambda)}{\partial \varepsilon} = \alpha^2 \int\limits_a^b \int\limits_c^d \left[\lambda_0 \left(\partial_3 f -_a D_x^\alpha[1]\partial_4 f -_c D_y^\alpha[2]\partial_5 f \right) \right.$$

$$\left. + \lambda \left(\partial_3 g -_a D_x^\alpha[1]\partial_4 g -_c D_y^\alpha[2]\partial_5 g \right) \right] \eta(b-x)^{\alpha-1}(d-y)^{\alpha-1} dydx = 0.$$

Finally, since $\eta \equiv 0$ on ∂R, the fundamental lemma of the calculus of variations (see, e.g., Marutani, 2003) implies that

$$\partial_3 H\{u\}(x,y) -_a D_x^\alpha[1]\partial_4 H\{u\}(x,y) -_c D_y^\alpha[2]\partial_5 H\{u\}(x,y) = 0.$$ □

Natural boundary conditions

In this section we consider problem (4.126)–(4.127), i.e., we consider the case when the value of function $u = u(x,y)$ is not preassigned on ∂R.

Theorem 4.30 (Natural boundary conditions to (4.126)–(4.127)). *If u is a local minimizer to problem (4.126)–(4.127), then u is a solution of the fractional differential equation (4.129). Moreover, it satisfies the following conditions:*

(i) $\partial_4 H\{u\}(a,y) = 0$ for all $y \in [c,d]$;
(ii) $\partial_4 H\{u\}(b,y) = 0$ for all $y \in [c,d]$;
(iii) $\partial_5 H\{u\}(x,c) = 0$ for all $x \in [a,b]$;
(iv) $\partial_5 H\{u\}(x,d) = 0$ for all $x \in [a,b]$.

Proof. Similar to the one of Theorem 4.27. For details see Odzijewicz and Torres, 2011. □

Sufficient condition

We now prove a sufficient condition for an extremal to be a global minimum under appropriate convexity assumptions.

Theorem 4.31. *Let $H(x,y,u,v,w,\lambda_0,\lambda)$ be a convex function of u, v and w. If $u(x,y)$ satisfies (4.129), then for an arbitrary admissible function $\hat{u}(\cdot,\cdot)$ the following holds:*

$$J[\hat{u}(\cdot,\cdot)] \geq J[u(\cdot,\cdot)],$$

i.e., $u(\cdot,\cdot)$ minimizes (4.126).

Proof. Define the following function:

$$\mu(x,y) := \hat{u}(x,y) - u(x,y).$$

Obviously,

$$\mu(x,y)|_{\partial R} = 0.$$

Since $H\{\hat{u}\}(x,y)$ is convex and $_aD_x^\alpha[1]$, $_cD_y^\alpha[2]$ are linear operators, we obtain that

$$H\{\hat{u}\}(x,y) - H\{u\}(x,y)$$

$$\geq (\hat{u}(x,y) - u(x,y))\partial_3 H\{u\}(x,y)$$

$$+ (_aD_a^\alpha[1]\hat{u}(x,y) -_a D_x^\alpha[1]u(x,y))\,\partial_4 H\{u\}(x,y)$$

$$+ \left(_cD_y^\alpha[2]\hat{u}(x,y) -_c D_y^\alpha[2]u(x,y)\right)\partial_5 H\{u\}(x,y)$$

$$= (\hat{u}(x,y) - u(x,y))\partial_3 H\{u\}(x,y) \qquad (4.135)$$

$$+_a D_x^\alpha[1]\,(\hat{u}(x,y) - u(x,y))\,\partial_4 H\{u\}(x,y)$$

$$+_c D_y^\alpha[2]\,(\hat{u}(x,y) - u(x,y))\,\partial_5 H\{u\}(x,y)$$

$$= \mu(x,y)\partial_3 H\{u\}(x,y) +_a D_x^\alpha[1]\mu(x,y)\partial_4 H\{u\}(x,y)$$

$$+_c D_y^\alpha[2]\mu(x,y)\partial_5 H\{u\}(x,y),$$

where the λ_0 and λ that appear in $\{u\}(x,y)$ are constants whose existence is assured by Theorem 4.29. Therefore,

$$J[\hat{u}(\cdot,\cdot)] - J[u(\cdot,\cdot)]$$

$$= \alpha^2 \int_a^b \int_c^d f(x,y,\hat{u},_a D_x^\alpha[1]\hat{u},_c D_y^\alpha[2]\hat{u})(b-x)^{\alpha-1}(d-y)^{\alpha-1}dydx$$

$$- \alpha^2 \int_a^b \int_c^d f(x,y,u,_a D_x^\alpha[1]u,_c D_y^\alpha[2]u)(b-x)^{\alpha-1}(d-y)^{\alpha-1}dydx$$

$$+ \lambda_0 \left(\alpha^2 \int_a^b \int_c^d g(x,y,\hat{u},_a D_x^\alpha[1]\hat{u},_c D_y^\alpha[2]\hat{u})(b-x)^{\alpha-1}(d-y)^{\alpha-1}dydx - K\right)$$

$$- \lambda_0 \left(\alpha^2 \int_a^b \int_c^d g(x,y,\hat{u},_a D_x^\alpha[1]\hat{u},_c D_y^\alpha[2]u)(b-x)^{\alpha-1}(d-y)^{\alpha-1}dydx - K\right)$$

$$= \alpha^2 \int_a^b \int_c^d (H\{\hat{u}\} - H\{u\})(b-x)^{\alpha-1}(d-y)^{\alpha-1}dydx.$$

Using (4.135) and (4.121), we get

$$\alpha^2 \int_a^b \int_c^d \left(H\{\hat{u}\} - H\{u\}\right)(b-x)^{\alpha-1}(d-y)^{\alpha-1}dydx$$

$$\geq \alpha^2 \int_a^b \int_c^d \mu\partial_3 H\{u\}(b-x)^{\alpha-1}(d-y)^{\alpha-1}dydx$$

$$+ \alpha^2 \int_a^b \int_c^d \left({}_aD_x^\alpha[1]\mu\partial_4 H\{u\} +_c D_y^\alpha[2]\mu\partial_5 H\{u\}\right)(b-x)^{\alpha-1}(d-y)^{\alpha-1}dydx$$

$$= \alpha^2 \int_a^b \int_c^d \mu\partial_3 H\{u\}(b-x)^{\alpha-1}(d-y)^{\alpha-1}dydx$$

$$+ \alpha^2 \int_a^b \int_c^d \left(-{}_aD_x^\alpha[1]\partial_4 H\{u\} -_c D_y^\alpha[2]\partial_5 H\{u\}\right)\mu(b-x)^{\alpha-1}(d-y)^{\alpha-1}dydx$$

$$= \alpha^2 \int_a^b \int_c^d \left(\partial_3 H\{u\} -_a D_x^\alpha[1]\partial_4 H\{u\}\right.$$

$$\left. -_c D_y^\alpha[2]\partial_5 H\{u\}\right)\mu(b-x)^{\alpha-1}(d-y)^{\alpha-1}dydx$$

$$= 0.$$

Thus, $J[\hat{u}(\cdot,\cdot)] \geq J[u(\cdot,\cdot)]$. $\qquad\square$

4.3.10 *Applications and Possible Extensions*

The fractional calculus provides a very useful framework to deal with non-local dynamics: if one wants to include memory effects, i.e., the influence of the past on the behavior of the system at the present time, then one may use fractional derivatives. The proof of fractional Euler–Lagrange equations is a subject of strong current study because of its numerous applications. However, while the single time case is well developed, the multitime fractional variational theory is in its childhood, and much remains to be done. In classical mechanics, functionals that depend on functions of two or more variables arise in a natural way, e.g., in mechanical problems involving systems with infinitely many degrees of freedom (string, membranes, etc.). Let us consider a flexible elastic string stretched under constant tension τ along the x axis with its end-points fixed at $x = 0$ and $x = L$. Let us denote the transverse displacement of the particle at time t, $t_1 \leq t \leq t_2$, whose equilibrium position is characterized by its distance x from the end of the string at $x = 0$ by the function $w = w(x, t)$. Thus $w(x, t)$, with $0 \leq x \leq L$, describes the shape of the string during the course of the vibration. Assume a distribution of mass along the string of density $\sigma = \sigma(x)$. Then the function that describes the actual motion of the string is one which renders

$$J(w) = \frac{1}{2} \int_{t_1}^{t_2} \int_0^L (\sigma w_t^2 - \tau w_x^2) \, dx \, dt$$

an extremum with respect to functions $w(x, t)$ which describe the actual configuration at $t = t_1$ and $t = t_2$ and which vanish, for all t, at $x = 0$ and $x = L$ (see p. 95 of Weinstock, 1974, for more details).

We discuss the description of the motion of the string within the framework of the fractional differential calculus. One may assume that, due to some constraints of physical nature, the dynamics do not depend on the usual partial derivatives but on some fractional derivatives ${}_0D_x^\alpha[1]w$ and ${}_{t_1}D_t^\alpha[2]w$. For example, we can assume that there is some coarse graining phenomenon – see details in Jumarie, 2010a. In this condition, one is entitled to assume again that the actual motion of the system, according to the principle of Hamilton, is such as to render the action function

$$J(w) = \frac{1}{2} I_R^\alpha (\sigma \, ({}_{t_1}D_t^\alpha[2]w)^2 - \tau \, ({}_0D_x^\alpha[1]w)^2),$$

where $R = [0, L] \times [t_1, t_2]$, an extremum. Note that we recover the classical problem of the vibrating string when $\alpha \to 1^-$. Applying Theorem 4.26 we obtain the fractional equation of motion for the vibrating string:

$$ {}_0D_x^\alpha[1] {}_0D_1^\alpha[1]w = \frac{\sigma}{\tau} {}_{t_1}D_t^\alpha[2] {}_{t_1}D_2^\alpha[2]w \, .$$

This equation becomes the classical equation of the vibrating string (cf., e.g., p. 97 of Weinstock, 1974) if $\alpha \to 1^-$.

In the above example, we discussed the application of the fractional differential calculus to the vibrating string. We started with a variational formulation of the physical process in which we modify the Lagrangian density by replacing integer order derivatives with fractional ones. Then the action integral in the sense of Hamilton was minimized and the governing equation of the physical process was obtained in terms of fractional derivatives. Similarly, many other physical fields can be derived from a suitably defined action functional. This gives several possible applications of the fractional calculus of variations with multiple integrals, e.g., in describing non-local properties of physical systems in mechanics (see, e.g., Baleanu *et al.*, 2010; Carpinteri and Mainardi, 1997; Klimek, 2002; Rabei and Ababneh, 2008; Tarasov, 2006) or electrodynamics (see, e.g., Baleanu *et al.*, 2009; Tarasov, 2010). Some examples were given in Sections 2.6, 3.9 and 3.10.

We end with some open problems for further investigation. It has been recognized that fractional calculus is useful in the study of scaling in physical systems (Cresson, 2002; Cresson, Frederico and Torres, 2009). In particular, there is a direct connection between local fractional differentiability properties and the dimensions of Holder exponents of nowhere differentiable functions, which provide a powerful tool to analyze the behavior of irregular signals and functions on a fractal set (Almeida and Torres, 2009c; Kolwankar and Gangal, 1997). Fractional calculus appear naturally, e.g., when working with fractal sets and coarse-graining spaces (Jumarie, 2010a,b), and fractal patterns of deformation and vibration in porous media and heterogeneous materials (Alexander and Orbach, 1988). The importance of vibrating strings to the fractional calculus has been given in Dzhaparidze, van Zanten and Zareba, 2005, where it is shown that a fractional Brownian motion can be identified with a string. The usefulness of our fractional theory of the calculus of variations with multiple integrals in physics, to deal with fractal and coarse-graining spaces, porous media, and Brownian motions, are questions to be studied.

Chapter 5

Towards a Combined Fractional Mechanics and Quantization

In this chapter a fractional Hamiltonian formalism is introduced for combined fractional calculus of variations. The Hamilton–Jacobi partial differential equation is generalized to be applicable for systems containing combined Caputo fractional derivatives. The results provide the necessary tools to carry out the quantization of non-conservative problems through combined fractional canonical equations of the Hamilton type.

In Section 5.1 we recall the combined fractional Euler–Lagrange equations. The new results are then given in the following sections. In Section 5.2 we show that combined fractional Hamiltonian equations of motion can be obtained for non-conservative systems. Constants of motion cease to be valid and a new notion is introduced in Section 5.3. Canonical fractional transformations of the first and second kind are studied in Section 5.4 and Section 5.5, respectively. Subsequently, in Section 5.6 a Hamilton–Jacobi type equation is derived and a fractional quantum wave equation suggested.

The understanding of Hamiltonian dynamical systems, classical or quantum, has been a long-standing theoretical question since Dirac's quantization of the classical electromagnetic field. The combined fractional Hamiltonian approach here introduced seems to be a promising direction of research.

5.1 Preliminaries

Let $\alpha, \beta \in (0, 1]$ and $\gamma \in [0, 1]$. The fractional derivative operator $^{C}D_{\gamma}^{\alpha,\beta}$ was introduced in Section 3.11.1 by $^{C}D_{\gamma}^{\alpha,\beta} := \gamma\, {}_{a}^{C}D_{x}^{\alpha} + (1 - \gamma)\, {}_{x}^{C}D_{b}^{\beta}$. For $\gamma = 0$ and $\gamma = 1$ we obtain the standard Caputo operators: $^{C}D_{0}^{\alpha,\beta} = {}_{x}^{C}D_{b}^{\beta}$

and $^{C}D_{1}^{\alpha,\beta} = {}_{a}^{C}D_{x}^{\alpha}$. Consider a fractional Lagrangian

$$L\left(t, \mathbf{q}, {}^{C}D_{\gamma}^{\alpha,\beta}\mathbf{q}\right) \tag{5.1}$$

depending on time t, the coordinates $\mathbf{q} = [q_1, \ldots, q_N]$ and their fractional velocities $^{C}D_{\gamma}^{\alpha,\beta}\mathbf{q} = \left[{}^{C}D_{\gamma_1}^{\alpha_1,\beta_1}q_1, \ldots, {}^{C}D_{\gamma_N}^{\alpha_N,\beta_N}q_N\right]$. The Euler–Lagrange equations corresponding to (5.1) were given in Theorem 3.34 and form the following system of N fractional differential equations:

$$\frac{\partial L}{\partial q_i} + D_{1-\gamma_i}^{\beta_i,\alpha_i} \frac{\partial L}{\partial^{C}D_{\gamma_i}^{\alpha_i,\beta_i}q_i} = 0, \quad i = 1, \ldots N, \tag{5.2}$$

where $D_{1-\gamma_i}^{\beta_i,\alpha_i} := (1 - \gamma_i)_a D_x^{\beta_i} + \gamma_{ix}D_b^{\alpha_i}$ with $_a D_x^{\beta_i}$ and $_x D_b^{\alpha_i}$ denoting the classical left and right Riemann–Liouville fractional derivatives.

Remark 5.1. With the notation of Section 3.11.1, $\partial_{i+1}L = \frac{\partial L}{\partial q_i}$ and $\partial_{N+1+i}L = \frac{\partial L}{\partial^{C}D_{\gamma_i}^{\alpha_i,\beta_i}q_i}$, $i = 1, \ldots N$. In contrast, we adopt here the notation used in mechanics (Goldstein, 1951; Riewe, 1997).

We shall show that (5.2) can be replaced by a special Hamiltonian system of $2N$ fractional differential equations: the combined canonical fractional equations.

5.2   Hamiltonian Formulation of the Combined Euler–Lagrange Equations

In analogy with classical mechanics, let us introduce the canonical momenta p_i by

$$p_i = \frac{\partial L}{\partial^{C}D_{\gamma_i}^{\alpha_i,\beta_i}q_i}, \quad i = 1, \ldots N. \tag{5.3}$$

Assume that

$$\left| \frac{\partial(p_1, \ldots, p_N)}{\partial\left({}^{C}D_{\gamma_1}^{\alpha_1,\beta_1}q_1, \ldots, {}^{C}D_{\gamma_N}^{\alpha_N,\beta_N}q_N\right)} \right| \neq 0.$$

Then, by the implicit function theorem, we can locally solve equations (5.3) with respect $^{C}D_{\gamma_1}^{\alpha_1,\beta_1}q_1, \ldots, {}^{C}D_{\gamma_N}^{\alpha_N,\beta_N}q_N$. The fractional Hamiltonian is defined by

$$H(t, \mathbf{q}, \mathbf{p}) = \sum_{i=1}^{N} p_i {}^{C}D_{\gamma_i}^{\alpha_i,\beta_i}q_i - L(t, \mathbf{q}, {}^{C}D_{\gamma}^{\alpha,\beta}\mathbf{q}), \tag{5.4}$$

where $^C D_{\gamma_i}^{\alpha_i,\beta_i} q_i$ are regarded as functions of variables t, q_1, ..., q_N, p_1, ..., p_N. Therefore, the differential of H is given by

$$dH = \sum_{i=1}^{N} \frac{\partial H}{\partial q_i} dq_i + \sum_{i=1}^{N} \frac{\partial H}{\partial p_i} dp_i + \frac{\partial H}{\partial t} dt. \tag{5.5}$$

From the defining equality (5.4) we can also write

$$dH = \sum_{i=1}^{N} {}^C D_{\gamma_i}^{\alpha_i,\beta_i} q_i dp_i + \sum_{i=1}^{N} p_i d^C D_{\gamma_i}^{\alpha_i,\beta_i} q_i - \sum_{i=1}^{N} \frac{\partial L}{\partial q_i} dq_i$$

$$- \sum_{i=1}^{N} \frac{\partial L}{\partial^C D_{\gamma_i}^{\alpha_i,\beta_i} q_i} d^C D_{\gamma_i}^{\alpha_i,\beta_i} q_i - \frac{\partial L}{\partial t} dt. \tag{5.6}$$

The terms containing $d^C D_{\gamma_i}^{\alpha_i,\beta_i} q_i$ in (5.6) cancel because of the definition of canonical momenta. Applying relations (5.2) we get

$$dH = \sum_{i=1}^{N} {}^C D_{\gamma_i}^{\alpha_i,\beta_i} q_i dp_i + \sum_{i=1}^{N} D_{1-\gamma_i}^{\beta_i,\alpha_i} p_i dq_i - \frac{\partial L}{\partial t} dt.$$

Comparison with (5.5) furnishes the following set of $2N$ relations:

$$\frac{\partial H}{\partial p_i} = {}^C D_{\gamma_i}^{\alpha_i,\beta_i} q_i, \quad \frac{\partial H}{\partial q_i} = D_{1-\gamma_i}^{\beta_i,\alpha_i} p_i, \quad i = 1, \ldots, N, \tag{5.7}$$

which we can call the *combined fractional canonical equations of Hamilton*. They constitute a set of $2N$ fractional order equations of motion replacing the Euler–Lagrange equations (5.2). Moreover,

$$\frac{\partial H}{\partial t} = -\frac{\partial L}{\partial t}. \tag{5.8}$$

For integer-order derivatives, it can be shown (see, e.g., p. 220 of Goldstein, 1951) that

$$\frac{dH}{dt} = \frac{\partial H}{\partial t} = -\frac{\partial L}{\partial t}.$$

Hence, if L (and in consequence of (5.8), also H) is not an explicit function of t, i.e., in the autonomous case, then H is a constant of motion. This is a well-known consequence of Noether's theorem (Torres, 2002, 2004c). With non-integer-order derivatives this is not the case (Frederico and Torres, 2008a, 2010). Observe that using equations (5.7) we can write

$$\frac{dH}{dt} = \sum_{i=1}^{N} \left({}^C D_{\gamma_i}^{\alpha_i,\beta_i} q_i \frac{dp_i}{dt} + D_{1-\gamma_i}^{\beta_i,\alpha_i} p_i \frac{dq_i}{dt} \right) + \frac{\partial H}{\partial t}.$$

In contrast with the integer-order case, the terms containing fractional and classical derivatives of coordinates and momenta do not cancel. Therefore, in general we obtain non-conservative systems, and classical constants of motion cease to be valid (Frederico and Torres, 2007c; Riewe, 1996). To deal with the problem we introduce the notion of the fractional constant of motion (cf. Sections 2.5 and 3.7).

5.3   Fractional Constants of Motion

To account for the presence of dissipative terms, we propose the following definition of the fractional constant of motion for the combined fractional variational calculus (cf. Definitions 2.11 and 3.12).

Definition 5.1. *We say that a function $C(t, \mathbf{q}, \mathbf{p})$ is a fractional constant of motion of order (α, β, γ) if $D_{1-\gamma}^{\beta, \alpha} [t \mapsto C(t, \mathbf{q}(t), \mathbf{p}(t))]$ is the null function along any pair $(\mathbf{q}, \mathbf{p})$ satisfying the $2N$ combined fractional canonical equations* (5.7).

It follows from (5.7) that if q_i is absent in the fractional Hamiltonian, then $D_{1-\gamma_i}^{\beta_i, \alpha_i} p_i = 0$, i.e., if $\frac{\partial L}{\partial q_i} = 0$, then $p_i = \frac{\partial L}{\partial^C D_{\gamma_i}^{\alpha_i, \beta_i} q_i}$ is a fractional constant of motion of order $(\alpha_i, \beta_i, \gamma_i)$.

5.4   Canonical Fractional Transformations of the First Kind

We now look for transformations under which the combined fractional canonical equations of Hamilton (5.7) preserve their canonical form. Consider the simultaneous transformation of independent coordinates and momenta, q_i and p_i, to a new set Q_i and P_i with transformation equations

$$Q_i = Q_i(t, \mathbf{q}, \mathbf{p}), \quad P_i = P_i(t, \mathbf{q}, \mathbf{p}), \quad i = 1, \ldots, N.$$

The new Q_i and P_i are canonical coordinates provided there exists some function $K(t, \mathbf{Q}, \mathbf{P})$ such that the equations of motion in the new set are in Hamiltonian form:

$$\frac{\partial K}{\partial P_i} = {}^C D_{\gamma_i}^{\alpha_i, \beta_i} Q_i, \quad \frac{\partial K}{\partial Q_i} = D_{1-\gamma_i}^{\beta_i, \alpha_i} P_i, \quad i = 1, \ldots, N. \tag{5.9}$$

Transformations for which equations (5.9) are valid are said to be canonical. Function K plays the role of the Hamiltonian in the new coordinates set.

If Q_i and P_i are to be canonical coordinates, then they must satisfy the modified fractional Hamiltonian principle of form

$$\delta \int_a^b \left(\sum_{i=1}^N P_i {}^C D_{\gamma_i}^{\alpha_i,\beta_i} Q_i - K(t,\mathbf{Q},\mathbf{P}) \right) dt = 0. \tag{5.10}$$

At the same time, the original coordinates satisfy the similar principle

$$\delta \int_a^b \left(\sum_{i=1}^N p_i {}^C D_{\gamma_i}^{\alpha_i,\beta_i} q_i - H(t,\mathbf{q},\mathbf{p}) \right) dt = 0. \tag{5.11}$$

The simultaneous validity of equations (5.10) and (5.11) means that integrands of the two integrals can differ at most by a total derivative of an arbitrary function F, sometimes called a gauge term (Torres, 2003):

$$\frac{d}{dt}F + \sum_{i=1}^N P_i {}^C D_{\gamma_i}^{\alpha_i,\beta_i} Q_i - K(t,\mathbf{Q},\mathbf{P}) = \sum_{i=1}^N p_i {}^C D_{\gamma_i}^{\alpha_i,\beta_i} q_i - H(t,\mathbf{q},\mathbf{p}).$$

The arbitrary function F works as a generating function of the transformation, and is very useful in developing direct methods to solve problems of the calculus of variations (Almeida and Torres, 2010; Malinowska and Torres, 2010d) and optimal control (Silva and Torres, 2006; Torres and Leitmann, 2008). For the fractional variational calculus this method was presented in Section 2.9. Function F is only specified up to an additive constant. In order to produce transformations between the two sets of canonical variables, F must be a function of both old and new variables. For mechanics involving fractional derivatives (cf. Riewe, 1996) we need to introduce variables $\bar{q}_i$ and $\bar{Q}_i$, $i = 1, \ldots, N$, satisfying

$$\frac{d\bar{q}_i}{dt} = {}^C D_{\gamma_i}^{\alpha_i,\beta_i} q_i, \qquad \frac{d\bar{Q}_i}{dt} = {}^C D_{\gamma_i}^{\alpha_i,\beta_i} Q_i.$$

For $\gamma_i = 1$, $\alpha_i = 1$, $\beta_i = 0$, $i = 1, \ldots, N$, these new coordinates are the same as the usual canonical coordinates. However, when dealing with fractional derivatives, the coordinates $\bar{q}_i$ and $\bar{Q}_i$ will not be canonical, so all canonical expressions must be written in terms of ${}^C D_{\gamma_i}^{\alpha_i,\beta_i} q_i$ and ${}^C D_{\gamma_i}^{\alpha_i,\beta_i} Q_i$. For a generating function $F_1\left(t,\bar{\mathbf{q}},\bar{\mathbf{Q}}\right)$, the integrands of (5.10) and (5.11) are connected by the relation

$$\frac{d}{dt}F_1(t,\bar{\mathbf{q}},\bar{\mathbf{Q}}) = \sum_{i=1}^N p_i {}^C D_{\gamma_i}^{\alpha_i,\beta_i} q_i - H(t,\mathbf{q},\mathbf{p})$$

$$- \sum_{i=1}^N P_i {}^C D_{\gamma_i}^{\alpha_i,\beta_i} Q_i + K(t,\mathbf{Q},\mathbf{P}). \tag{5.12}$$

Because

$$dF_1 = \sum_{i=1}^{N} \frac{\partial F_1}{\partial \bar{q}_i} d\bar{q}_i + \sum_{i=1}^{N} P_i \frac{\partial F_1}{\partial \bar{Q}_i} d\bar{p}_i + \frac{\partial F_1}{\partial t} dt,$$

we obtain

$$\frac{\partial F_1}{\partial t} = K - H, \quad p_i = \frac{\partial F_1}{\partial \bar{q}_i}, \quad -P_i = \frac{\partial F_1}{\partial \bar{Q}_i}, \tag{5.13}$$

$i = 1, \ldots, N$. We call (5.13) the canonical fractional transformations of the first kind.

5.5 Canonical Fractional Transformations of the Second Kind

We shall now introduce canonical transformations of the second kind with a generating function $F_2(t, \bar{\mathbf{q}}, \mathbf{P})$. We begin by noting that the transformation from $\bar{q}_i$ and $\bar{Q}_i$ to $\bar{q}_i$ and P_i can be accomplished by a Legendre transformation. Indeed, by equations (5.13), $-P_i = \frac{\partial F_1}{\partial \bar{Q}_i}$. This suggests that the generating function F_2 can be defined by

$$F_2(t, \bar{\mathbf{q}}, \mathbf{P}) = F_1(t, \bar{\mathbf{q}}, \bar{\mathbf{Q}}) + \sum_{i=1}^{N} P_i \bar{Q}_i.$$

From equation (5.12) we obtain

$$\sum_{i=1}^{N} p_i \frac{d\bar{q}_i}{dt} - H(t, \mathbf{q}, \mathbf{p}) - \sum_{i=1}^{N} P_i \frac{d\bar{Q}_i}{dt} + K(t, \mathbf{Q}, \mathbf{P})$$

$$= \frac{d}{dt}\left(F_2(t, \bar{\mathbf{q}}, \mathbf{P}) - \sum_{i=1}^{N} P_i \bar{Q}_i \right)$$

$$= \frac{d}{dt} F_2(t, \bar{\mathbf{q}}, \mathbf{P}) - \sum_{i=1}^{N} P_i \frac{d\bar{Q}_i}{dt} - \sum_{i=1}^{N} \bar{Q}_i \frac{dP_i}{dt}.$$

Hence,

$$\left(\sum_{i=1}^{N} p_i \frac{d\bar{q}_i}{dt} - H \right) dt + \left(\sum_{i=1}^{N} \bar{Q}_i \frac{dP_i}{dt} + K \right) dt = dF_2(t, \bar{\mathbf{q}}, \mathbf{P}).$$

Repeating the procedure followed for F_1 in Section 5.4, we obtain the transformation equations

$$\frac{\partial F_2}{\partial t} = K - H, \quad p_i = \frac{\partial F_2}{\partial \bar{q}_i}, \quad \bar{Q}_i = \frac{\partial F_2}{\partial P_i}, \tag{5.14}$$

$i = 1, \ldots, N$.

5.6 Fractional Hamilton–Jacobi Equation

Under the assumption that K is identically zero, for integer order derivatives we know, from equations (5.9), that the new coordinates are constant. The same situation occurs, under appropriate assumptions, in the fractional setting when in the presence of Riemann–Liouville or Caputo derivatives (Golmankhaneh, 2008; Rabei and Ababneh, 2008). In our general combined setting this is not the case, in particular for the new momentum P. To solve the problem we proceed as follows. Assume that the transformed Hamiltonian, K, is identically zero. Then the equations of motion are

$$\frac{\partial K}{\partial P_i} = {}^C D_{\gamma_i}^{\alpha_i, \beta_i} Q_i = 0, \quad \frac{\partial K}{\partial Q_i} = D_{1-\gamma_i}^{\beta_i, \alpha_i} P_i = 0, \quad i = 1, \ldots, N.$$

Function K is related with the old Hamiltonian H and the generating function F by

$$K = H + \frac{\partial F}{\partial t}.$$

Hence, K will be zero if, and only if, F satisfies the equation

$$H(t, \mathbf{q}, \mathbf{p}) + \frac{\partial F}{\partial t} = 0.$$

It is convenient to take F as a function of $\bar{q}_i$ and P_i. Then, from equations (5.14), we can write

$$H\left(t, \mathbf{q}, \frac{\partial F_2}{\partial \bar{\mathbf{q}}}\right) + \frac{\partial F_2}{\partial t} = 0. \tag{5.15}$$

Equation (5.15) is a fractional version of the Hamilton–Jacobi equation. Therefore, the quantum wave equation with the combined ${}^C D_\gamma^{\alpha, \beta}$ fractional derivative is suggested to be

$$H\left(t, \mathbf{q}, -i\hbar \frac{\partial}{\partial \bar{\mathbf{q}}}\right) \psi = i\hbar \frac{\partial \psi}{\partial t},$$

where ψ is a wave equation.

5.7 Conclusion

In this chapter we gave a Hamiltonian formulation to the general fractional Euler–Lagrange equations obtained in Malinowska and Torres, 2011. Motivated by recent results in the literature of physics, dealing with fractional mechanics and non-conservative systems (Abreu and Godinho, 2011; Golmankhaneh, 2008; Rabei and Ababneh, 2008), we illustrated some possible applications of our combined variational calculus in mechanics and quantization. We trust that the obtained results are important for the quantization of fractional variational problems.

Bibliography

E. M. C. Abreu and C. F. L. Godinho, Fractional Dirac bracket and quantization for constrained systems, Phys. Rev. E **84** (2011), 026608.

O. P. Agrawal, A general formulation and solution scheme for fractional optimal control problems, Nonlinear Dynam. **38** (2004), no. 1–4, 323–337.

O. P. Agrawal, Generalized Euler–Lagrange equations and transversality conditions for FVPs in terms of the Caputo derivative, J. Vib. Control **13** (2007), no. 9–10, 1217–1237.

S. Alexander and R. Orbach, Observation of fractons in silica aerogels, Europhys. Lett. **6** (1988), 245–250.

R. Almeida, R. A. C. Ferreira and D. F. M. Torres, Isoperimetric problems of the calculus of variations with fractional derivatives, Acta Math. Sci. Ser. B Engl. Ed. **32** (2012), no. 2, 619–630.

R. Almeida, A. B. Malinowska and D. F. M. Torres, A fractional calculus of variations for multiple integrals with application to vibrating string, J. Math. Phys. **51** (2010), no. 3, 033503.

R. Almeida, A. B. Malinowska and D. F. M. Torres, Fractional Euler-Lagrange differential equations via Caputo derivatives, in: *Fractional Dynamics and Control*, Eds: D. Baleanu, J.A. Tenreiro Machado and A. C. J. Luo, Springer New York (2012), Part 2, 109–118.

R. Almeida, S. Pooseh and D. F. M. Torres, Fractional variational problems depending on indefinite integrals, Nonlinear Anal. **75** (2012), no. 3, 1009–1025.

R. Almeida and D. F. M. Torres, Calculus of variations with fractional derivatives and fractional integrals, *Conference proceedings of the Brazilian Conference on Computational and Applied Mathematics*, September 8–11 (2009a), Cuiabá, Brazil. Anais do XXXII CNMAC, 2009, Vol. 2, pp. 1222–1227.

R. Almeida and D. F. M. Torres, Calculus of variations with fractional derivatives and fractional integrals, Appl. Math. Lett. **22** (2009b), no. 12, 1816–1820.

R. Almeida and D. F. M. Torres, Hölderian variational problems subject to integral constraints, J. Math. Anal. Appl. **359** (2009c), no. 2, 674–681.

R. Almeida and D. F. M. Torres, Isoperimetric problems on time scales with nabla derivatives, J. Vib. Control **15** (2009d), no. 6, 951–958.

R. Almeida and D. F. M. Torres, Leitmann's direct method for fractional optimization problems, Appl. Math. Comput. **217** (2010), no. 3, 956–962.

R. Almeida and D. F. M. Torres, Necessary and sufficient conditions for the fractional calculus of variations with Caputo derivatives, Commun. Nonlinear Sci. Numer. Simulat. **16** (2011a), no. 3, 1490–1500.

R. Almeida and D. F. M. Torres, Fractional variational calculus for nondifferentiable functions, Comput. Math. Appl. **61** (2011b), no. 10, 3097–3104.

T. M. Atanacković, S. Konjik and S. Pilipović, Variational problems with fractional derivatives: Euler–Lagrange equations, J. Phys. A **41** (2008), no. 9, 095201.

T. M. Atanacković, S. Konjik, S. Pilipović *et al.*, Variational problems with fractional derivatives: invariance conditions and Nöther's theorem, Nonlinear Anal. **71** (2009), no. 5–6, 1504–1517.

T. M. Atanackovic and B. Stankovic, On a numerical scheme for solving differential equations of fractional order, Mech. Res. Comm. **35** (2008), no. 7, 429–438.

D. Baleanu, K. Diethelm, E. Scalas *et al.*, *Fractional calculus: models and numerical methods*, World Scientific Publishing, Singapore (2012).

D. Baleanu, A. K. Golmankhaneh, R. Nigmatullin *et al.*, Fractional Newtonian mechanics, Cent. Eur. J. Phys. **8** (2010), no. 1, 120–125.

D. Baleanu, A. K. Golmankhaneh, A. K. Golmankhaneh *et al.*, Fractional electromagnetic equations using fractional forms, Int. J. Theor. Phys. **48** (2009), 3114–3123.

N. R. O. Bastos, R. A. C. Ferreira and D. F. M. Torres, Necessary optimality conditions for fractional difference problems of the calculus of variations, Discrete Contin. Dyn. Syst. **29** (2011a), no. 2, 417–437.

N. R. O. Bastos, R. A. C. Ferreira and D. F. M. Torres, Discrete-time fractional variational problems, Signal Process. **91** (2011b), no. 3, 513–524.

N. R. O. Bastos, D. Mozyrska and D. F. M. Torres, Fractional derivatives and integrals on time scales via the inverse generalized Laplace transform, Int. J. Math. Comput. **11** (2011), no. J11, 1–9.

N. R. O. Bastos and D. F. M. Torres, Combined delta-nabla sum operator in discrete fractional calculus, Commun. Frac. Calc. **1** (2010), no. 1, 41–47.

K. Brading, Which Symmetry Noether, Weyl and conservation of electric charge, Studies in History and Philosophy of Science, Part B, Studies in History and Philosophy of Modern Physics **33** (2002), 3–22.

R. F. Camargo, A. O. Chiacchio, R. Charnet *et al.*, Solution of the fractional Langevin equation and the Mittag-Leffler functions, J. Math. Phys. **50** (2009), no. 6, 063507.

J. F. Carinena, J. A. Lazaro-Cami and E. Martinez, On second Noether's theorem and gauge symmetries in Mechanics, International Journal of Geometric Methods in Modern Physics **3** (2005), no. 3, 1–14.

A. Carpinteri and F. Mainardi, *Fractals and fractional calculus in continuum mechanics*, Springer, Vienna (1997).

E. Castillo, A. Luceño and P. Pedregal, Composition functionals in calculus of variations. Application to products and quotients, Math. Models Methods Appl. Sci. **18** (2008), no. 1, 47–75.

B.C. Chachuat, *Nonlinear and dynamic optimization: From Theory to Practice*, École Polytechnique Fédérale de Lausanne, IC-32: Winter Semester 2006/2007.

J. Cresson, Scale relativity theory for one-dimensional non-differentiable manifolds, Chaos Solitons Fractals **14** (2002), no. 4, 553–562.

J. Cresson, Fractional embedding of differential operators and Lagrangian systems, J. Math. Phys. **48** (2007), no. 3, 033504.

J. Cresson, Inverse problem of fractional calculus of variations for partial differential equations, Commun. Nonlinear Sci. Numer. Simul. **15** (2010), no. 4, 987–996.

J. Cresson, G. S. F. Frederico and D. F. M. Torres, Constants of motion for non-differentiable quantum variational problems, Topol. Methods Nonlinear Anal. **33** (2009), no. 2, 217–231.

P. A. F. Cruz, D. F. M. Torres and A. S. I. Zinober, A non-classical class of variational problems, Int. J. Math. Model. Numer. Optim. **1** (2010), no. 3, 227–236.

K. Diethelm, *The analysis of fractional differential equations*, Lecture Notes in Mathematics, 2004, Springer, Berlin (2010).

K. Dzhaparidze, H. van Zanten and P. Zareba, Representations of fractional Brownian motion using vibrating strings, Stochastic Process. Appl. **115** (2005), no. 12, 1928–1953.

R. A. El-Nabulsi, I. A. Dzenite and D. F. M. Torres, Fractional action functional in classical and quantum field theory, *Scientific Proceedings of Riga Technical University, Series—Computer Science, Boundary Field Problems, and Computer Simulation*, 48th international thematic issue (2006), pp. 189–197.

R. A. El-Nabulsi and D. F. M. Torres, Necessary optimality conditions for fractional action-like integrals of variational calculus with Riemann-Liouville derivatives of order (α, β), Math. Methods Appl. Sci. **30** (2007), no. 15, 1931–1939.

R. A. El-Nabulsi and D. F. M. Torres, Fractional actionlike variational problems, J. Math. Phys. **49** (2008), no. 5, 053521.

R. A. C. Ferreira and D. F. M. Torres, Fractional h-difference equations arising from the calculus of variations, Appl. Anal. Discrete Math. **5** (2011), no. 1, 110–121.

D. Filatova, M. Grzywaczewski and N. Osmolovskii, Optimal control problem with an integral equation as the control object, Nonlinear Anal. **72** (2010), no. 3–4, 1235–1246.

G. S. F. Frederico and D. F. M. Torres, Noether's theorem for fractional optimal control problems, *Proceedings of the 2nd IFAC Workshop on Fractional Differentiation and its Applications*, 19–21 July 2006a, Porto, pp. 142–147.

G. S. F. Frederico and D. F. M. Torres, Constants of motion for fractional actionlike variational problems, Int. J. Appl. Math. **19** (2006b), no. 1, 97–104.

G. S. F. Frederico and D. F. M. Torres, A formulation of Noether's theorem for fractional problems of the calculus of variations, J. Math. Anal. Appl. **334** (2007a), no. 2, 834–846.

G. S. F. Frederico and D. F. M. Torres, Non-conservative Noether's theorem for fractional action-like variational problems with intrinsic and observer times, Int. J. Ecol. Econ. Stat. **9** (2007b), no. F07, 74–82.

G. S. F. Frederico and D. F. M. Torres, Nonconservative Noether's theorem in optimal control, Int. J. Tomogr. Stat. **5** (2007c), no. W07, 109–114.

G. S. F. Frederico and D. F. M. Torres, Fractional conservation laws in optimal control theory, Nonlinear Dynam. **53** (2008a), no. 3, 215–222.

G. S. F. Frederico and D. F. M. Torres, Fractional optimal control in the sense of Caputo and the fractional Noether's theorem, Int. Math. Forum **3** (2008b), no. 9–12, 479–493.

G. S. F. Frederico and D. F. M. Torres, Necessary optimality conditions for fractional action-like problems with intrinsic and observer times, WSEAS Trans. Math. **7** (2008c), no. 1, 6–11.

G. S. F. Frederico and D. F. M. Torres, Fractional Noether's theorem in the Riesz-Caputo sense, Appl. Math. Comput. **217** (2010), no. 3, 1023–1033.

I. M. Gelfand and S. V. Fomin, *Calculus of variations*, Dover Publications, Mineola, NY (2000).

M. Giaquinta and S. Hildebrandt, *Calculus of variations. I*, Springer, Berlin (1996).

H. Goldstein, *Classical Mechanics*, Addison-Wesley Press, Inc., Cambridge, MA (1951).

A. K. Golmankhaneh, Fractional Poisson bracket, Turk. J. Phys. **32** (2008), 241–250.

P. D. F. Gouveia and D. F. M. Torres, Automatic computation of conservation laws in the calculus of variations and optimal control, Comput. Methods Appl. Math. **5** (2005), no. 4, 387–409

P. D. F. Gouveia and D. F. M. Torres, Computing ODE symmetries as abnormal variational symmetries, Nonlinear Anal. **71** (2009), no. 12, e138–e146.

R. Herrmann, *Fractional calculus: an introduction for physicists*, World Scientific Publishing, Singapore (2011).

P. E. Hydon and E. L. Mansfield, Extensions of Noether's Second Theorem: from continuous to discrete systems, Proc. R. Soc. A **467** (2011), no. 2135, 3206–3221.

Z. D. Jelicic and N. Petrovacki, Optimality conditions and a solution scheme for fractional optimal control problems, Struct. Multidiscip. Optim. **38** (2009), no. 6, 571–581.

G. Jumarie, On the representation of fractional Brownian motion as an integral with respect to $(dt)^a$, Appl. Math. Lett. **18** (2005), no. 7, 739–748.

G. Jumarie, Fractional Hamilton-Jacobi equation for the optimal control of nonrandom fractional dynamics with fractional cost function, J. Appl. Math. Comput. **23** (2007), no. 1–2, 215–228.

G. Jumarie, Stock exchange fractional dynamics defined as fractional exponential growth driven by (usual) Gaussian white noise. Application to fractional Black–Scholes equations, Insurance Math. Econom. **42** (2008a), no. 1, 271–287.

G. Jumarie, Modeling fractional stochastic systems as non-random fractional dynamics driven by Brownian motions, Appl. Math. Model. **32** (2008b), no. 5, 836–859.

G. Jumarie, Table of some basic fractional calculus formulae derived from a modified Riemann–Liouville derivative for non-differentiable functions, Appl. Math. Lett. **22** (2009a), no. 3, 378–385.

G. Jumarie, From Lagrangian mechanics fractal in space to space fractal Schrödinger's equation via fractional Taylor's series, Chaos Solitons Fractals **41** (2009b), no. 4, 1590–1604.

G. Jumarie, An approach via fractional analysis to non-linearity induced by coarse-graining in space, Nonlinear Anal. Real World Appl. **11** (2010a), no. 1, 535–546.

G. Jumarie, Analysis of the equilibrium positions of nonlinear dynamical systems in the presence of coarse-graining disturbance in space, J. Appl. Math. Comput. **32** (2010b), no. 2, 329–351.

A. A. Kilbas, H. M. Srivastava and J. J. Trujillo, *Theory and applications of fractional differential equations*, Elsevier, Amsterdam (2006).

V. Kiryakova, *Generalized fractional calculus and applications*, Pitman Research Notes in Mathematics Series, 301, Longman Sci. Tech., Harlow (1994).

J. Klafter, S. C. Lim, R. Metzler, *Fractional dynamics: recent advances*, World Scientific Publishing, Singapore (2011).

M. Klimek, Fractional sequential mechanics—models with symmetric fractional derivative, Czechoslovak J. Phys. **51** (2001), no. 12, 1348–1354.

M. Klimek, Lagrangean and Hamiltonian fractional sequential mechanics, Czechoslovak J. Phys. **52** (2002), no. 11, 1247–1253.

M. Klimek, *On solutions of linear fractional differential equations of a variational type*, The Publishing Office of Czenstochowa University of Technology, Czestochowa (2009).

K. M. Kolwankar, Decomposition of Lebesgue–Cantor devil's staircase, Fractals **12** (2004), no. 4, 375–380.

K. M. Kolwankar and A. D. Gangal, Holder exponents of irregular signals and local fractional derivatives, Pramana J. Phys. **48** (1997), 49–68.

Y. Kosmann-Schwarzbach, *The Noether theorems, Invariance and conservation laws in the twentieth century*, New York, Springer (2010).

G. Leitmann, *The calculus of variations and optimal control*, Mathematical Concepts and Methods in Science and Engineering, 24, Plenum, New York (1981).

J. D. Logan, On variational problems which admit an infinite continuous group, Yokohama Mathematical Journal **22** (1974), 31–42.

J. D. Logan, *Invariant variational principles*, Academic Press [Harcourt Brace Jovanovich Publishers], New York (1977).

F. Mainardi, *Fractional calculus and waves in linear viscoelasticity*, Imperial College Press, London (2010).

A. B. Malinowska, Fractional variational calculus for non-differentiable functions, in: *Fractional Dynamics and Control*, Eds: D. Baleanu, J. A. Tenreiro Machado and A. Luo, Chapter 8, Springer, New York (2012a), Part 2, 97–108.

A. B. Malinowska, On fractional variational problems which admit local transformations, Journal of Vibration and Control (2012b), DOI: 10.1177/1077546312442697

A. B. Malinowska, A formulation of the fractional Noether-type theorem for multidimensional Lagrangians, Appl. Math. Lett. **25** (2012c), no. 11, 1941–1946.

A. B. Malinowska, M. R. Sidi Ammi and D. F. M. Torres, Composition functionals in fractional calculus of variations, Commun. Frac. Calc. **1** (2010), no. 1, 32–40.

A. B. Malinowska and D. F. M. Torres, Necessary and sufficient conditions for local Pareto optimality on time scales, J. Math. Sci. (N. Y.) **161** (2009), no. 6, 803–810.

A. B. Malinowska and D. F. M. Torres, Natural boundary conditions in the calculus of variations, Math. Methods Appl. Sci. **33** (2010a), no. 14, 1712–1722.

A. B. Malinowska and D. F. M. Torres, Generalized natural boundary conditions for fractional variational problems in terms of the Caputo derivative, Comput. Math. Appl. **59** (2010b), no. 9, 3110–3116.

A. B. Malinowska and D. F. M. Torres, Fractional variational calculus in terms of a combined Caputo derivative, *Proceedings of FDA'10, The 4th IFAC Workshop on Fractional Differentiation and its Applications*, Badajoz, Spain, October 18-20 (2010c) (Eds: I. Podlubny, B. M. Vinagre Jara, YQ. Chen *et al.*), Article no. FDA10-084.

A. B. Malinowska and D. F. M. Torres, Leitmann's direct method of optimization for absolute extrema of certain problems of the calculus of variations on time scales, Appl. Math. Comput. **217** (2010d), no. 3, 1158–1162.

A. B. Malinowska and D. F. M. Torres, Fractional calculus of variations for a combined Caputo derivative, Fract. Calc. Appl. Anal. **14** (2011), no. 4, 523–537.

A. B. Malinowska and D. F. M. Torres, Multiobjective fractional variational calculus in terms of a combined Caputo derivative, Appl. Math. Comput. **218** (2012a), no. 9, 5099–5111.

A. B. Malinowska and D. F. M. Torres, Towards a combined fractional mechanics and quantization, Fract. Calc. Appl. Anal. **15** (2012b), no. 3, 407–417.

T. Marutani, Canonical forms of Euler's equation and natural boundary condition—2 dimensional case, The Kwansei Gakuin Economic Review **34** (2003), 21–28.

F. Merrikh-Bayat, Fractional-order differential order equation solver, Matlab Central, File ID: #13866 (2007). Available at: http://www.mathworks.com/matlabcentral/fileexchange/13866. Access date: July 4, 2012.

K. S. Miller and B. Ross, *An introduction to the fractional calculus and fractional differential equations*, Wiley-Interscience, Wiley, New York (1993).

S. Momani, General solutions for the space-and time-fractional diffusion-wave equation, Journal of Physical Sciences **10** (2006), 30–43.

D. Mozyrska and D. F. M. Torres, Modified energy control in the memory domain of fractional continuous-time linear control systems, *Proceedings of Symposium on Fractional Signals and Systems* (FSS09), (Eds: M. Ortigueira *et al.*), Lisbon, Portugal, November 4–6, 2009.

D. Mozyrska and D. F. M. Torres, Minimal modified energy control for fractional linear control systems with the Caputo derivative, Carpathian J. Math. **26** (2010), no. 2, 210–221.

D. Mozyrska and D. F. M. Torres, Modified optimal energy and initial memory of fractional continuous-time linear systems, Signal Process. **91** (2011), no. 3, 379–385.

E. Noether, Invariant variation problems, Transport Theory Statist. Phys. **1** (1971), no. 3, 186–207. English translation of: E. Noether, Invariante variationsprobleme, Gött. Nachr. (1918), 235–257.

T. Odzijewicz, A. B. Malinowska and D. F. M. Torres, Fractional variational calculus of variable order, *Advances in Harmonic Analysis and Operator Theory*, The Stefan Samko Anniversary Volume (Eds: A. Almeida, L. Castro, F.-O. Speck), Operator Theory: Advances and Applications, Birkhäuser Verlag, in press. arXiv:1110.4141 (2012a)

T. Odzijewicz, A. B. Malinowska and D. F. M. Torres, Fractional variational calculus with classical and combined Caputo derivatives, Nonlinear Anal. **75** (2012b), 1507–1515.

T. Odzijewicz, A. B. Malinowska and D. F. M. Torres, Fractional calculus of variations in terms of a generalized fractional integral with applications to Physics, Abstr. Appl. Anal. **2012** (2012c), Article ID 871912.

T. Odzijewicz and D. F. M. Torres, Calculus of variations with fractional and classical derivatives, *Proceedings of FDA'10, The 4th IFAC Workshop on Fractional Differentiation and its Applications*, Badajoz, Spain, October 18–20, 2010 (Eds: I. Podlubny, B. M. Vinagre Jara, YQ. Chen *et al.*), Article no. FDA10-076.

T. Odzijewicz and D. F. M. Torres, Fractional calculus of variations for double integrals, Balkan J. Geom. Appl. **16** (2011), no. 2, 102–113.

T. Odzijewicz and D. F. M. Torres, Calculus of variations with classical and fractional derivatives, Math. Balkanica **26** (2012), no. 1-2, 191–202.

M. D. Ortigueira, *Fractional calculus for scientists and engineers*, Lecture Notes in Electrical Engineering, 84, Springer, Dordrecht (2011).

I. Podlubny, *Fractional differential equations*, Academic Press, San Diego, CA (1999).

L. S. Pontryagin, V. G. Boltyanskii, R. V. Gamkrelidze *et al.*, *The mathematical theory of optimal processes*, Translated from the Russian by K. N. Trirogoff; (Ed: L. W. Neustadt), Interscience Publishers, John Wiley & Sons, Inc. New York (1962).

S. Pooseh, R. Almeida and D. F. M. Torres, Approximation of fractional integrals by means of derivatives, Comput. Math. Appl. (2012a), DOI: 10.1016/j.camwa.2012.01.068

S. Pooseh, R. Almeida and D. F. M. Torres, Expansion formulas in terms of integer-order derivatives for the Hadamard fractional integral and derivative, Numer. Funct. Anal. Optim. **33** (2012b), no. 3, 301–319.

S. Pooseh, H. S. Rodrigues and D. F. M. Torres, Fractional derivatives in Dengue epidemics, AIP Conf. Proc. **1389** (2011), no. 1, 739–742.

E. M. Rabei and B. S. Ababneh, Hamilton–Jacobi fractional mechanics, J. Math. Anal. Appl. **344** (2008), no. 2, 799–805.

F. Riewe, Nonconservative Lagrangian and Hamiltonian mechanics, Phys. Rev. E (3) **53** (1996), no. 2, 1890–1899.

F. Riewe, Mechanics with fractional derivatives, Phys. Rev. E (3) **55** (1997), no. 3, part B, 3581–3592.

S. G. Samko, A. A. Kilbas and O. I. Marichev, *Fractional integrals and derivatives*, Translated from the 1987 Russian original, Gordon and Breach, Yverdon (1993).

M. R. Sidi Ammi and D. F. M. Torres, Existence and uniqueness of a positive solution to generalized nonlocal thermistor problems with fractional-order derivatives, Differ. Equ. Appl. **4** (2012), no. 2, 267–276.

C. J. Silva and D. F. M. Torres, Absolute extrema of invariant optimal control problems, Commun. Appl. Anal. **10** (2006), no. 4, 503–515.

V. E. Tarasov, Fractional variations for dynamical systems: Hamilton and Lagrange approaches, J. Phys. A **39** (2006), no. 26, 8409–8425.

V. E. Tarasov, Fractional vector calculus and fractional Maxwell's equations, Ann. Physics **323** (2008), no. 11, 2756–2778.

V. E. Tarasov, *Fractional Dynamics, Applications of Fractional Calculus to Dynamics of Particles, Fields and Media*, Springer-Verlag, Berlin Heidelberg (2010).

J. A. Tenreiro Machado, V. Kiryakova and F. Mainardi, Recent history of fractional calculus, Commun. Nonlinear Sci. Numer. Simul. **16** (2011), no. 3, 1140–1153.

D. F. M. Torres, On the Noether theorem for optimal control, Eur. J. Control **8** (2002), no. 1, 56–63.

D. F. M. Torres, Gauge symmetries and Noether currents in optimal control, Appl. Math. E-Notes **3** (2003), 49–57.

D. F. M. Torres, Carathéodory equivalence Noether theorems, and Tonelli full-regularity in the calculus of variations and optimal control, J. Math. Sci. (N. Y.) **120** (2004a), no. 1, 1032–1050.

D. F. M. Torres, Quasi-invariant optimal control problems, Port. Math. (N.S.) **61** (2004b), no. 1, 97–114.

D. F. M. Torres, Proper extensions of Noether's symmetry theorem for nonsmooth extremals of the calculus of variations, Commun. Pure Appl. Anal. **3** (2004c), no. 3, 491–500.

D. F. M. Torres and G. Leitmann, Contrasting two transformation-based methods for obtaining absolute extrema, J. Optim. Theory Appl. **137** (2008), no. 1, 53–59.

C. Tricaud and Y. Chen, An approximate method for numerically solving fractional order optimal control problems of general form, Comput. Math. Appl. **59** (2010), no. 5, 1644–1655.

J. L. Troutman, *Variational calculus and optimal control*, Second edition, Springer, New York (1996).

V. Uchaikin and R. Sibatov, *Fractional kinetics in solids: anomalous charge transport in semiconductors, dielectrics and nanosystems*, World Scientific Publishing, Singapore (2012).

B. van Brunt, *The calculus of variations*. Universitext. Springer-Verlag, New York (2004).

R. Weinstock, *Calculus of variations. With applications to physics and engineering*, Reprint of the 1952 edition, Dover, New York (1974).

S. Yakubovich, Eigenfunctions and Fundamental Solutions of the Fractional Two-Parameter Laplacian, Int. J. Math. Math. Sci., **2010** (2010), Article ID 541934.

Index